Food Microbiology and Laboratory Practice

Chris Bell
Consultant Food Microbiologist, UK

Paul Neaves
Consultant Food Microbiologist, UK

Anthony P. Williams
Consultant Food Microbiologist, UK

Blackwell
Science

Editorial offices:
Blackwell Science Ltd, 9600 Garsington Road, Oxford OX4 2DQ, UK
   Tel: +44 (0) 1865 776868
Blackwell Publishing Professional, 2121 State Avenue, Ames, Iowa 50014-8300, USA
   Tel: +1 515 292 0140
Blackwell Science Asia Pty, 550 Swanston Street, Carlton, Victoria 3053, Australia
   Tel: +61 (0)3 8359 1011

First published 2005
2   2007

Library of Congress Cataloging-in-Publication Data
is available

ISBN 978-0-632-06381-9

A catalogue record for this title is available from the British Library

Set in 10 / 12pt Times
by DP Photosetting, Aylesbury, Bucks
Printed and bound in Singapore
by Utopia Press Pte Ltd

The publisher's policy is to use permanent paper from mills that operate a sustainable forestry policy, and which has been manufactured from pulp processed using acid-free and elementary chlorine-free practices. Furthermore, the publisher ensures that the text paper and cover board used have met acceptable environmental accreditation standards.

For further information on Blackwell Publishing, visit our website:
www.blackwellpublishing.com

Contents

Foreword

It gives me great pleasure to write the foreword of this excellent training aid for those working in food industry laboratories and those meeting the training needs of such laboratory personnel. The authors collectively have a vast amount of experience both in the laboratory practice of food microbiology and in the teaching of the subject and are also much-valued consultants in the fields of food hygiene and food safety. Their aim in producing this book has been to provide basic information and sources for further information in order to help in the development of high standards for food microbiology technicians for the next generation and the future generations to come. They have succeeded admirably.

Many texts tend to concentrate on a specific area of laboratory practice, for example food microbiology methods or laboratory design and operation. Here the authors have discussed in a single volume all the practical aspects relevant to the operation of a food microbiology laboratory as would be required for the testing of factory samples of food and their ingredients. In twelve, well illustrated, chapters they describe basic food microbiology including behaviour of organisms, food spoilage, food-borne illness, food preservation and food industry applications, HACCP, laboratory design, equipment, operation and accreditation and methods both conventional and 'alternative'. Each topic is described in terms that are clear and easy to understand and is supported by many figures and tables and also suggestions for further reading. The structure of the book is logical and will take those with a minimal knowledge of the subject forward to a fuller understanding of not only of what to do and how to do it but also that ever important 'why it is done' and finally what the results mean. As a food microbiologist of some 35 years experience I learned many new facts from this book as well as having my memory jogged about the information I had all but forgotten.

My overall impression of this book is that it covers the what, how and why of the basic microbiology of food and its examination in an easy to read and understand manner yet it also stimulates the reader to look further to expand on the knowledge gained. It has a very 'common sense' approach and it will be an invaluable tool to laboratory technicians in the food industry and to those teaching the subject.

Dr Diane Roberts BSc, PhD, CBiol, FIBiol, FIFST.
Former Deputy Director
PHLS Food Safety Microbiology Laboratory
Central Public Health Laboratory
London

Acknowledgements

The authors thank the following companies for their kind help and support in the supply of information for use in this book:

Astec Microflow Ltd
bioMérieux SA
BioTek Instruments Ltd
Celsis Ltd
Don Whitley Scientific Ltd
Hygiena International Ltd
Millipore (UK) Ltd
Norpath Laboratories Ltd
Novatron Ltd
PriorClave Ltd
Pyser-SGI Ltd
Scientific Laboratory Supplies Ltd
Seward Ltd
Sterilin Ltd
Tecra Diagnostics UK
Westward Laboratories Ltd
Woodside Consulting

Preface

The ever-increasing public interest and concern over food safety, as well as commercial pressure to improve food quality and extend product shelf life, has imposed more responsibility and pressure on all those involved in the microbiological examination of foods and related samples. The examinations, whether in public health laboratories, food industry or contract testing service laboratories, serve a number of purposes.

Microbiological examination of foods forms an essential part of product shelf life assessment, food product challenge testing in relation to microbiological safety and new product development. Microbiological examination of foods in which problems have occurred or are suspected to have occurred may help to discover the causes of outbreaks of illness, reasons for, or failure to achieve shelf life or non-conformance to product specifications and the causes of customer complaints. Although in recent years there has rightly been a shift in emphasis away from end-product microbiological testing to process control through the use of HACCP and similar systems, there remains a place for microbiological examination for verifying that the production processes are functioning as planned. This includes supplier performance monitoring, environmental and personal hygiene monitoring and process performance verification to provide data for trend analysis.

Whatever the reason for the examination, there is an absolute need to provide reliable microbiological test results. To do this requires laboratories to develop and maintain high standards in the design and construction of the laboratory facility and the testing environment, and also of equipment, tools and materials, methods, staff practices and documentation.

The customers of any microbiology laboratory, whether the smallest of factory laboratories or the largest of contract testing service laboratories, rely upon the laboratory to provide a 'true' result. As a consequence of microbiological test results that are 'out of specification', particularly where human pathogens are involved, actions may be taken by customers, ranging from increasing the levels of process checking or equipment cleaning to more commercially costly product rejection, cessation of production on a particular manufacturing line or even, factory closure. All actions, though, carry some cost.

It is essential, therefore, that the laboratory staff and management are confident that their facilities, methods, procedures and, importantly, staff competence and practices will, and can be demonstrated to, deliver consistently 'true' information to their customers so that relevant decisions can be taken in relation both to food safety and to wholesomeness.

In order to maintain the high standards required, staff must be suitably trained to understand what they are to do, how they are to do it and why they must do it in a prescribed way. A properly trained microbiology technician provided with the right tools, equipment and environment is a valuable asset and makes a positive contribution to the reliability of test results and ongoing confidence of the laboratory's customers.

Too many incorrect or inaccurate results have been, and still are, attributable to poor laboratory practices, and a lack of knowledge and understanding about microorganisms and related microbiological quality and safety consequences. Such mistakes have in the past damaged the food industry and food microbiology laboratories severely, both commercially, as well as in reputation and lost jobs.

This book is written as an aid for teachers, trainers and trainees in food microbiology laboratories and on practical food microbiology training courses. It aims to provide the basic information and further information sources that will help in the development of high standards for the next and future generations of food microbiology technicians.

Plate section

Illustrations referred to in the text as plates are to be found in one of the two colour plate sections. A number of photomicrographs are taken from real laboratory investigations and reflect the range of shapes and distribution of cells that may be encountered in routine food microbiology work.

1 The Structure and Habit of Microorganisms

1.1 Introduction

Free-living creatures can be divided into five Kingdoms comprising animals, plants, algae and protozoa, fungi and bacteria. Various classifications have been proposed, but the simple grouping as follows is useful:

- Animalia (animals)
- Plantae (plants)
- Protista (algae and protozoa)
- Fungi (yeasts, moulds, mushrooms and toadstools, rusts and smuts, wilts)
- Monera (bacteria and related organisms).

The term 'microorganism' applies to members of three of these Kingdoms, namely the Protista, the Monera and the Fungi, together with other biological forms not included in the free-living Kingdoms, such as viruses and prions. All microorganisms have one essential characteristic in common: all or part of their free-living structure is very small. A precise definition of what constitutes 'small' probably does not exist but most microbiologists would agree that a maximum value of around 100 microns (μm) might be appropriate (1 μm [1 micron] = 1 / 1000 mm), although many micro-organisms are between 1 and 10 μm in size and some (viruses, that are not free-living) can be around one fiftieth (1 / 50) μm or even smaller.

In descending order of size, the major groups of microorganisms comprise the protozoa and algae, the yeasts and the spores of moulds, the bacteria and, finally, the viruses. Single cells of microorganisms cannot be seen with the naked eye, but the protozoa, algae, moulds and yeasts can be seen easily with a relatively low power of magnification, e.g. × 100–400, and indeed when moulds and yeasts are fully grown into a cluster known as a colony, they become clearly visible to the naked eye (mould on bread or cheese is a familiar example). All of these organisms are capable of self-replication through complex mechanisms of sexual and asexual reproduction. They possess a true nucleus enclosed in a nuclear membrane that contains their genetic material within complex chromosomes, i.e. they are eukaryotic.

The bacteria are also capable of self-replication but possess simple chromosomes and no nuclear membrane, i.e. they are prokaryotic; their cell division does not normally involve sexual reproduction, although many are, in fact, capable of some form of sexual activity. Bacteria are smaller than most eukaryotes and can only be seen clearly as individual cells with the aid of the higher powers of a light microscope,

Plates 1–6 are located in the colour plate section at p. 100.

e.g. $\times 1000$ magnification. However, like moulds and yeasts, when bacteria are grown into a cluster known as a colony, they also become clearly visible to the naked eye.

Viruses can only replicate within the cells of a host organism, i.e. they are parasitic. They are parasites of members of the free-living Kingdoms and are even smaller and simpler than bacteria, and it requires the high magnification of an electron microscope ($\times 10\,000$ and higher) in order for them to be seen.

In recent years, prions have been described as the causal agent of bovine spongiform encephalopathy (BSE – 'mad cow disease') and related illnesses. These contain no genetic material in the conventional sense, i.e. DNA (deoxyribonucleic acid) or RNA (ribonucleic acid); they appear to be distorted protein molecules and are so small that, to date, even the most powerful electron microscopes have been unable to allow them to be seen.

Plate 1 (see colour plate sections for all plates) shows some representatives of the Microbial Kingdoms. A comprehensive description of the features of the various microorganisms is outside the scope of this book and the reader is referred to Adams and Moss (2000) for more detailed information.

1.2 Microorganisms associated with foods

Food microbiology is the study of the microorganisms associated with foods. Bacteria, yeasts and moulds are mostly capable of growth in food whilst, although viruses and protozoa cannot grow in food, their physical transmission via food has long been recognised. However, in recent years, our ability to investigate the agents responsible for food-borne illness has developed considerably, and different types of protozoa, the toxins produced by algae and histamine produced in foods through bacterial growth now also need to be understood in relation to food safety by the food microbiologist, as well as prions, even though the latter may not be microorganisms in the true sense of the word.

In practical food microbiology, therefore, it is important to understand the nature and ecology (relationship of organisms to their surroundings) of different microbial groups as well as the basic details of microbial structure and function. Where growth of an organism occurs in a food material, the food microbiologist needs to know whether the outcome could be food spoilage, food-borne illness (from infections or from toxins) or food enhancement, e.g. in the form of a desirable fermentation. It is also important to understand that some organisms can be studied easily in a routine microbiology laboratory, while for others, special facilities such as controlled containment may be needed for safe handling of the organism. Access to databases for help in identifying an organism or for obtaining information relating to outbreaks of illness may also be required.

1.2.1 *Bacteria*

Morphology and cell staining

Bacteria are the single most significant group of microorganisms in food microbiology; they are prokaryotes, with a rather simple structure and relatively little variety in shape ('cell morphology'). They are single-celled organisms, comprising a 'vegetative cell' surrounded by a rigid cell wall that protects the cell from mechanical damage and to some extent, osmotic lysis (cell rupture caused by sudden changes of osmotic pressure). The cell wall is made up of a mixture of proteins, polysaccharides and sometimes lipids. The main contents of the cell are the cytoplasm and nuclear material. In one family of food-borne bacteria (the Bacillaceae), the cell may also contain a developing spore (known as an endospore because it is inside the cell) that, when mature, is highly resistant to heat and chemicals. Outside the cell wall there may also be, in some genera, flagella, which are whiplash-like structures that enable the organism to move actively through liquids (see Figure 1.1).

Figure 1.1   A stylised, rod-shaped bacterial cell, adapted from Hawker, Linton, Folkes *et al.* (1967).

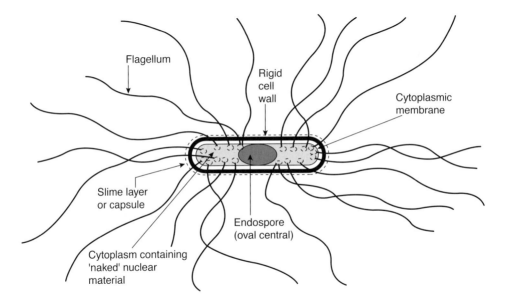

Most bacteria are around 1–5 μm in size and divide by simple binary fission (division). This type of increase, or progression, is known as 'exponential' or 'logarithmic' and mathematically is a $\log_2$ (logarithm to the base 2) increase in numbers, i.e. 1 cell divides into 2, 2 cells divide into 4, 4 cells divide into 8, 8 cells divide into 16, etc. Under optimal growth conditions including nutrients, moisture and temperature, bacteria can multiply very rapidly, some species doubling in numbers every 10–20 minutes (see Chapter 2). This has practical implications in food spoilage as it takes the same time for 1 cell to divide into 2 cells as for 1 million to become 2 million. In other words, when spoilage begins, it usually proceeds very rapidly.

Bacteria are initially characterised according to their shape and their 'Gram reaction'. The Gram stain (named after a Danish microbiologist, Christian Gram) differentiates bacteria into two groups based on differences in their cell wall composition; the Gram reaction identifies a bacterium as either Gram-positive or Gram-negative. The two types of cell wall are quite different. Gram-positive bacteria have a relatively simple but robust cell wall consisting predominantly of peptidoglycan and teichoic acid, while Gram-negative bacteria generally have a weaker, more complex cell wall, both anatomically and chemically, containing phospholipid and lipopoly-saccharide and a higher lipid content than that found in Gram-positive bacteria.

When bacteria are immersed in a solution of the purple aniline dye, crystal violet, the stain binds to the cell and the organisms become visible under the light micro-scope. Both Gram-positive and Gram-negative bacteria are stained in this way; however, in Gram-negative bacteria, the binding reaction is reversible, whilst in Gram-positive organisms, it is not. Thus, by staining a culture preparation with crystal violet, then 'decolourising' it with a solvent that removes the stain from Gram-negative bacteria and re-staining (counter-staining) with a dye of contrasting colour (usually the red stain, safranin), Gram-positive organisms appear blue or purple and are distinguished from Gram-negative organisms that appear pink (Plate 2). The Gram-positive / negative distinction is important in food microbiology because, together with the culture and cell morphology, it provides the starting-point for identifying an unknown bacterial culture.

The Gram staining technique also allows the microbiologist to observe the cell shape and size (morphology) of microorganisms. Bacteria are either rod-shaped ('rods'; like sausages) or spherical ('cocci'; singular, 'coccus'). Different rod-shaped bacteria can have different dimensions and proportions: some may be short and fat, others long and thin and occasionally they may be curved or spiral-shaped. The cells may occur singly, in pairs, in chains or in clumps. Cocci are usually spherical cells, although they sometimes may be slightly ellipsoidal. Like rods, cocci may occur as single cells, in pairs, in chains, in 'packets' of 4 or 8 cells, or in grape-like clusters. Plate 3 shows some of the variety of bacterial cell morphologies that occur. As well as the cell shape, the arrangement of cells under the microscope is often a useful char-acteristic for the identification of bacteria.

Some laboratories that deal with mycobacteria that are pathogenic to animals and can sometimes be transmitted in food to cause illness in humans, may also use the Ziehl-Neelsen or 'acid-fast' stain. This staining technique is similar in principle to the Gram staining method but a more powerful stain is required for these organisms so the dye, basic fuchsin is used. This stains all the bacteria present; however, those with a highly waxy cell wall, e.g. species of *Mycobacterium* that cause tuberculosis in cattle and humans, are able to resist decolourisation using an acidified solvent, i.e. they are acid-fast. Following the decolourisation step, a counter-stain, e.g. methylene blue, is applied so that acid-fast organisms appear red whilst other organisms are blue or green.

Another characteristic possessed by a few bacteria, and only two genera common in foods, is the production of spores. Bacterial endospores are produced inside the vegetative (active / live) cell and are usually approximately 1 μm in diameter (Plate 4). Bacterial endospores evolved as a dormant, resting form of the organism and are highly resistant to harsh environments such as heat, drying, chemicals and irradia-tion. They are found in *Bacillus* species, which can grow and produce spores in the presence of oxygen (aerobe), and *Clostridium* species which only grow and produce spores in the absence of oxygen (anaerobe). A special differential stain, the spore stain, may be used to observe spores produced by the spore-forming genera *Bacillus* and *Clostridium* and to observe the size, shape and location of the spore within the parent cell, helping to determine the identity of the organism. Sometimes this can be observed from a Gram stain but better results are usually obtained by using a spore stain that stains the spores with malachite green (pale green) and counter-stains the vegetative cells with safranin (pink).

When the environmental conditions become unsuitable for the growth of a bacterial cell that is capable of sporing, it produces a single spore that protects the organism's genetic material. The spore is produced inside the normal vegetative cell, which may subsequently disintegrate leaving the spore 'naked'. In this state, the organism is dormant but is able to survive until environmental conditions again become suitable for growth. The spore then germinates into a new vegetative cell that divides to re-populate the environment.

In food microbiology, there are several practical implications and concerns relating to the existence of bacterial spores, including their resistance to many cleaning chemicals and their survival of high heat processes. While most vegetative bacterial cells are killed at temperatures in the range 60–100°C, most bacterial spores survive in boiling water and it takes a high heat process, such as 121°C for 3 minutes, as used in the canning of low-acid (pH 6–7) foods, or 140°C for 2–4 seconds (Ultra Heat Treatment – UHT), as used for the sterilisation of milk, i.e. UHT milk, to kill them. Sometimes, other measures have to be taken to prevent the germination and growth of bacterial spores in foods and these can include the use of preservatives, acidification of food and chilled storage.

Another feature of bacterial spores is that they may be stimulated to germinate by a mild heat treatment known as a 'heat shock'. Thus, a heat-processing activity, such as pasteurisation, could actually facilitate the germination of dormant spores and increase the risk of food spoilage or food-poisoning. In the laboratory, most bacterial spores can be detected and enumerated quite easily by first heating a test sample, usually to around 80°C, which not only stimulates the spores to germinate but also kills any vegetative bacteria present, before allowing spores to germinate and grow on a nutrient medium at the normal incubation temperature for the target organism e.g. 30°C–37°C for *Bacillus cereus*.

Other morphological characteristics

Other characteristics of bacteria can be either of significance in the food processing environment or of use in the food microbiology laboratory. For instance, some species produce extra-cellular polysaccharides known as capsules or capsular material which can be visualised by specific stains known as capsule stains. The capsular material may be simple slime, but of particular significance in some organisms is that the material they produce may act as a glue to attach the bacteria to surfaces. The attached bacteria can build up on surfaces, e.g. the internal surface of pipe-work, especially water pipes, in a layer known as a biofilm that can be very robust and difficult to remove (Figure 1.2). An excessively thick biofilm can actually cause complete pipe blockage and biofilms on food contact surfaces can harbour and protect other organisms that may subsequently contaminate food that comes into contact with the surface.

While some bacteria attach themselves to surfaces, others actively move from place to place and are said to be motile. Their method of movement is through the use of one or more flagella (singular, flagellum), which, as already described, are like

Figure 1.2   A developing biofilm.

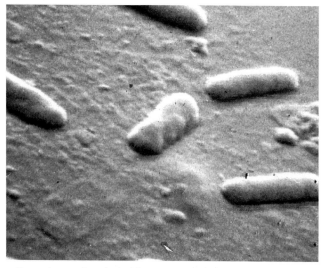

(a) Stage 1: single cells attach to surface (magnification approximately ×10 000).

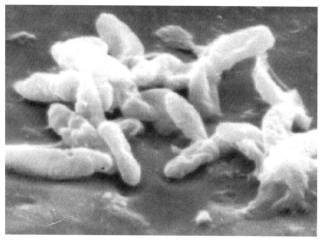

(b) Stage 2: cells multiply and start to form clumps that develop into an irregular layer (biofilm) (magnification approximately ×10 000).

whiplashes protruding from their cells (Figure 1.3). The flagella propel the bacteria through liquids and the movement, which can be extremely rapid, can be seen under a microscope in hanging drops (see Figure 11.1).

   In free suspension in hanging drops, all bacteria can be seen to move to a certain extent, even if they do not possess flagella. This is known as Brownian motion, which causes non-motile organisms to oscillate more or less in one position and is due to bombardment of the cell by the molecules of the liquid in which the bacteria are suspended. However, when bacteria are motile, a 'purposeful' movement is clearly visible and the type of motion observed can help in recognition or identification of the

Figure 1.3 An electron micrograph of *Campylobacter* sp. showing flagella (magnification approximately × 10 000).

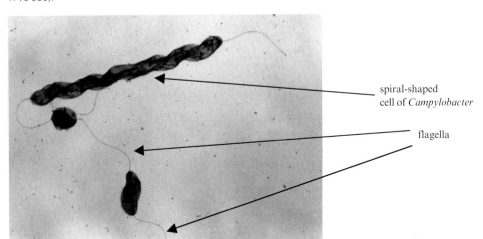

spiral-shaped
cell of *Campylobacter*

flagella

organism. Some bacteria have flagella only at the ends of the cells (almost all motile, food-borne bacteria are rods, the cocci are usually non-motile) and this arrangement propels them in a straight line, often with a darting movement, e.g. *Pseudomonas* spp. and *Campylobacter* spp. Others have flagella all over their surface and this arrangement results in an end over end or 'tumbling' movement; this form of movement is quite common in bacteria and is a particular characteristic used in the identification of *Listeria* spp.

The application of morphological characteristics

A knowledge of a cell's morphology and its Gram reaction can be used to place a bacterial isolate in one of four sub-categories. Of these, **Gram-negative cocci** are the least likely to be encountered in foods as most of these species are more frequently associated with medical microbiology. Conversely, many of the **Gram-positive rods**, **Gram-positive cocci** and **Gram-negative rods** are important in food production, spoilage and safety. The arrangement of cells when viewed by microscopy also provides important information that helps to sub-divide the major groups, for example, whether Gram-positive cocci are arranged (a) singly, or in pairs, (b) in clusters, resembling bunches of grapes, or (c) in long chains etc.

Some bacteria are beneficial in food production, e.g. fermentation processes, whilst others cause deterioration of food quality (spoilage), or are responsible for food-borne disease. Whereas the primary purpose of food production is to produce consistently safe, high quality food, the purpose of food preservation is to maintain within the food, conditions that are sub-optimal for, or hostile to, bacterial growth without diminishing its safety or 'eating quality'. Characterising bacteria that may be found and grow in food by means of Gram reaction, cell morphology and arrangement provides an essential start for use in assessing a food's microbiological safety, quality and shelf life.

1.2.2   *Fungi*

Fungi are eukaryotic organisms, i.e. they have a true nucleus and are larger in size than bacteria; they may be unicellular, e.g. yeasts, or multicellular, e.g. moulds. In many respects, they are a much more diverse group than bacteria but generally grow more slowly than bacteria and under conditions that prevent most bacterial growth. This might suggest that fungi are of greater interest to the food microbiologist than bacteria but, compared with bacteria, fungi are associated mainly with food spoilage rather than with acute food poisoning (in the form of gastroenteritis) so, in practice, they are often considered to be of less concern.

Moulds

In contrast to bacteria, moulds are polymorphic, i.e. they exhibit a variety of shapes and structures; they grow in 'life cycles', starting from and ending with spores; and have asexual (anamorphic) and sometimes also sexual (teleomorphic) forms. Mould spores are quite unlike bacterial endospores and are usually produced in large numbers, e.g. millions, from a single mature colony. As a general rule, a mould spore germinates and produces a thread-like structure known as a hypha (plural, hyphae). The hyphae grow and branch into a complex mass of threads known as a mycelium (Figures 1.4a–d), and this is often visible to the naked eye, e.g. the normal, white, processing mould that grows on a soft, ripened cheese such as Brie. In time, the mycelium develops more complex reproductive structures that start to produce spores, which may be asexual, sexual or a mixture of both, depending on the mould and the environmental conditions. In the sporing stage the colonies often become coloured, e.g. the green spoilage mould on a block of Cheddar cheese or on a slice of bread. Some other examples of mould growth on food are shown in Plate 5. These reproductive structures may be just visible to the naked eye or by low to moderate power (up to × 50) light microscopy and form the basis for the identification and classification of mould genera and species.

   Because moulds have multiple structures that are considerably more varied than those of bacteria, the characteristics of a particular mould at a particular time may depend upon which form the organism happens to be growing in, whether or not it is producing spores and which form of spores is present. This can have important consequences for food processing, as teleomorphs are often more resistant to heat than anamorphs and may survive some pasteurisation processes applied to foods. However it must be emphasised that mould spores are not nearly as heat resistant as bacterial spores and are almost invariably killed by boiling water temperature unless protected in some way, for instance in a fatty food or in a product with a very high concentration of sugar. Moulds require some oxygen for growth but often grow in very low oxygen concentration atmospheres and can be found in or on foods that technically may be in an anaerobic environment but that have microscopic spaces or cracks that provide access to some oxygen.

   The classification of fungi in general, and of moulds in particular, is a complex subject employing / involving various terms according to the individual authors' points of view. However a reasonably good consensus of opinion has been reached in the recent scientific literature. For current purposes, moulds are divided into the micro-fungi and the macro-fungi, the former including Zygomycetes, Ascomycetes and Fungi Imperfecti, whilst the latter includes the Basidiomycetes (mushrooms and toadstools). Basidiomycetes are, of course, of interest to the food microbiologist as many of them can be consumed as food; however, the food microbiologist is rarely concerned with their potential for growth and spoilage of foods and thus they will not be discussed further.

Zygomycetes are among the most simple of the micro-fungi, with a mycelium that has no or only a few septa (transverse walls or 'cross walls'), i.e. they are 'coenocytic', with multiple nuclei in a continuous cytoplasm (gel or viscous fluid making up the substance of the living cell but excluding the nucleus) which is not strictly segregated into individual cells. The reproductive structures, known as 'sporangia' (singular, 'sporangium'), are sacs which contain the spores (correctly called sporangiospores) that are the asexual reproductive units. The sexual forms are rarely encountered in food microbiology.

Ascomycetes are more complex than Zygomycetes; they generally have a septate mycelium (with cross walls that have pores (small holes) in them) and they form naked spores, known as 'conidia' (singular, 'conidium'), as a result of asexual reproduction, as well as sexual spores known as ascospores in structures known as 'asci' (singular, 'ascus'). The majority of food spoilage moulds are Ascomycetes; however most of these have lost the ability to produce sexual spores and are therefore classified in an artificial grouping known as the Fungi Imperfecti or Deuteromycetes.

Many mould spores are very small, usually 2–10 µm in diameter, but there are many different morphological varieties, not only of the spore itself but also of the arrangement of spores and their attachment to the mycelium. Together with the colour and appearance of the mould colony when grown on specific microbiological media, the morphological characteristics of moulds, when viewed by light microscopy are used in their identification. The microscopic diversity of some food-borne moulds is shown in Plate 6.

Yeasts

Yeasts are a group of fungi that often grow optimally when submerged in liquids where conditions are microaerobic, i.e. where concentrations of oxygen are lower than in air, although the so-called 'film yeasts' do grow on the surfaces of liquids. Yeasts are generally unicellular, though some may form 'pseudomycelium' that resembles the micro-fungi (see above). In fact, there is no absolute dividing line between yeasts and moulds because some moulds can grow in a yeast-like form and some yeasts in a mould-like form. Although larger than bacteria, yeasts are still rather small (generally < 10 µm in diameter); they are non-motile and do not possess flagella.

Yeasts generally grow as single often ellipsoidal cells and normally divide by 'budding'. Whereas bacteria divide into two equal cells, most yeasts form a small 'daughter' cell that splits off from the 'mother' cell. Each 'mother' cell may produce one or more 'daughter' cells, so there is no strict mathematical progression that can be

Figure 1.4   Germination of mould spores and growth into a colony.

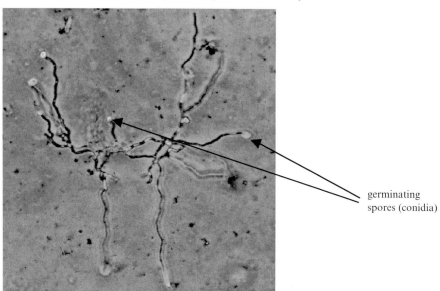

germinating
spores (conidia)

(a) Magnification approximately × 1000.

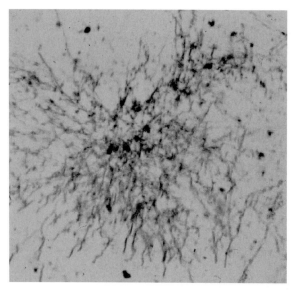

(b) A micro-colony – not yet visible to the naked eye
(magnification approximately × 500).

Figure 1.4 *cont*

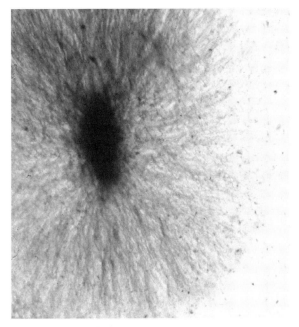

(c) A small colony – just visible to the naked eye
(magnification approximately × 50).

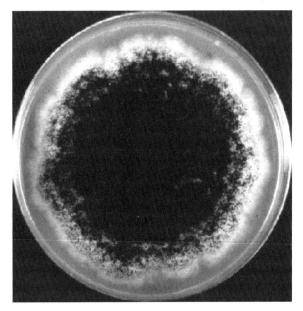

(d) A fully mature and sporulating mould colony
(magnification approximately × 0.8).

applied to all yeast growth, as is the case with bacteria, although the increase in cell numbers during active growth is still more or less exponential.

Many yeast species appear morphologically similar under the light microscope and show very few differences, apart from those that form pseudomycelium and so appear mould-like. Therefore, in general, yeasts cannot be distinguished, one from another, by light microscopy and are more usually differentiated by their different metabolic characteristics and growth habits. Some examples of food-borne yeasts are shown in Figure 1.5.

Figure 1.5   Some food-borne yeasts and yeast-like organisms.

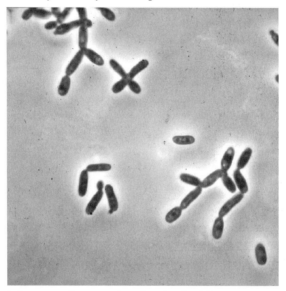

(a) *Pichia membranaefaciens* – a spoilage organism of marinading brines (magnification approximately × 1000).

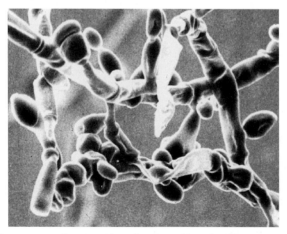

(b) *Moniliella acetoabutans* – a spoilage organism of acetic acid-based pickles (magnification approximately × 5000).

1.2.3 *Viruses*

Viruses are very, very small, being only 20–500 nm in size (1 nm [nanometre] = 1 / 1000 μm or, one millionth of a millimetre (1 / 1000 000 mm)). In fact, they cannot be seen using normal light microscopy and can only be seen by means of an electron microscope which is a highly specialised, and very expensive, form of microscopy only usually employed by research and specialist laboratories.

Unlike bacteria, yeasts and moulds, viruses are not capable of free growth, i.e. they cannot replicate by themselves, and are strictly parasitic on other living organisms. When a virus infects a host, it uses the host cell's genetic processes to generate many copies of the virus genetic material; many replicates of the virus are then produced and the host cell dies liberating the new viral particles to infect new hosts cells. Their parasitism is also rather specific to one or few species of animal, plant or even bacteria (any virus whose host is a bacterium is called a bacteriophage), and so a virus causing human illness is likely to have come originally from a human source.

Although viruses pathogenic to humans are unable to grow in foods they are often able to survive for long periods as infective particles in and on many foods and so food plays an important role as a route for viral infection to humans.

Only low numbers of virus particles are usually necessary to cause an infection (generally fewer than 100 virus particles, but in some cases as few as 1–10 virus particles may cause infection) so there is no point in examining foods directly for their presence by electron microscope because such low numbers cannot be found by this technique. Few viruses of importance in food-borne illness can be cultured and, only then, by the use of cell tissue culture requiring specific expertise. Only new technology methods, e.g. those based on the polymerase chain reaction (see Chapter 12), are likely to become of practical use in detecting these organisms in foods. At present, they are sought, and can be found, in faecal specimens from infected patients, where they can be present in very high numbers. In the investigation of food-associated outbreaks of viral gastroenteritis, public health authorities use epidemiological techniques to ascertain the source of the outbreak and the route of infection.

Although a number of virus types are recognised as being transmitted to humans via foods, most food-borne viral gastroenteritis involves Rotaviruses, mainly in children, or Norovirus (previously known as Norwalk-Like Viruses (NLV) or Small Round Structured Viruses (SRSV)), in all age-groups. Hepatitis A virus can also be food-borne and is a particular cause for concern in ready-to-eat foods that have been handled.

1.2.4 *Protozoa*

Protozoa are small, single-celled animals that often live harmlessly in natural waters and moist environments, but they are also responsible for some of the world's major diseases such as malaria. They have complex life-cycles, some of the stages of which can be resistant to some of the chemical treatments used in water and food processing.

Until the 1980s, the protozoa were mainly considered to be the cause of illness through direct consumption of contaminated water and person-to-person transmission through poor hygienic practices rather than from foods. Many of these illnesses were thought to be of tropical origin, perhaps the best-known example being amoebic dysentery caused by *Entamoeba histolytica*. However, in recent years, a few protozoa have been associated with human food-borne illness, usually from the consumption of raw fruit or salad vegetables that have been irrigated during growth or washed during preparation after harvesting using contaminated natural water. Natural waters can become contaminated with faeces from either human or animal sources, or from the run-off of animal slurries applied to fields as fertiliser. Once present in the water, they are difficult to remove (other than by micro-filtration) or kill, as their cysts can be highly resistant to the agents of water disinfection such as chlorine.

Like viruses, protozoa do not grow in foods and because of the usually very low numbers likely to be present, cannot normally be detected in foods, although, in specialist laboratories, they can be detected in liquid samples. They are concentrated from the liquid by filtration and subjected to staining techniques followed by direct microscopic examination; in public health laboratories, stool samples from infected individuals are examined by direct microscopy.

Protozoa such as *Cryptosporidium parvum* and *Giardia lamblia* are currently the most common protozoa implicated in food-borne and water-borne diarrhoeal illness. Another such organism is *Cyclospora cayetanensis*, which has more recently been associated with food-borne illness in the USA and Canada due to fruit and a few salad vegetables imported from Central and South America. The latter organism is not yet considered to be of major significance in Western Europe but with increasing imports of luxury, out-of-season fresh produce, it may well be a cause for concern in the future.

1.2.5   *Toxic algae*

In aquatic environments, dinoflagellate algae and blue-green photosynthetic bacteria (cyanobacteria, formerly known as blue-green algae) occasionally grow to extremely high numbers; the former often colour sea water and are referred to as a 'red tide'. Some species of dinoflagellate, e.g. *Gonyaulax catanella* and *G. tamarensis*, produce some very powerful toxins (such as saxitoxins and gonyautoxins), which, in humans, can block nerve transmission and result in symptoms ranging from numbness through to paralysis.

Overall, the symptoms of algal and cyanobacterial illness can range from familiar food-poisoning symptoms, for example, nausea, vomiting and diarrhoea, through to the more serious symptoms of paralysis and even amnesia. These organisms rarely cause illness directly because sea-water is not consumed; however the organisms and their toxins can be concentrated by filter-feeding shellfish, indirectly making the shellfish highly toxic to anyone who consumes them. In extreme cases, it is thought that the toxins may be passed further up the food chain when shellfish are consumed by carnivorous fish, such as members of the mackerel and tuna family. There is also a disputed theory that algal toxins may be involved in causing the histamine response in humans that occasionally results from scombrotoxin poisoning following consumption of scombroid fish such as tuna and mackerel.

Algal toxins are very stable and can survive many food processes. Illness is prevented by prohibiting fishing in waters that may contain high levels of algal contamination and governments operate environmental monitoring schemes to detect such hazards as they arise.

1.3 The origin of names

The system of nomenclature used in microbiology today (Kingdom, Order, Family, Genus and Species) was developed by Linnaeus in the eighteenth century. The terms family, genus and species (see Glossary) are used in everyday microbiology, although the term species is sometimes sub-divided to provide a measure of greater definition. Genus and species names are **always** written the same way, i.e. in *italics*, the genus name starting with a capital letter and the species name in lower case, e.g. ***Staphylococcus aureus***. However, when referring to a group of organisms in general, a 'trivial' form can be used, such as 'Staphylococci' or 'Penicillia'. These latter normally start with a capital letter, although there seems to be no universal acceptance of the custom and 'staphylococci' and 'penicillia' are frequently found in the scientific literature. Sometimes the plural is allowed to take an English form, so that the plural of '*Salmonella*' becomes 'salmonellas' rather than 'salmonellae'.

The names of microorganisms are always presented in classical Latin form and often contain words from Latin or Greek that describe something about the organism, e.g. for *Staphylococcus aureus*, 'staphyle' means 'bunch of grapes', 'coccus' means a grain or berry (both relating to the appearance and arrangement of cells when viewed by microscopy) and 'aureus' means 'golden' (the colour of colonies when grown on general purpose media). Also commonly used is the name of a person associated with the discovery or study of the organism, e.g. the genus *Escherichia* after Theodor Escherich.

1.3.1 Bacterial classification

Like any artificial classification system, the one used for bacteria is not only subject to different interpretations but also to change as our knowledge and interpretation of relationships between organisms expands. The result is that several names for one organism may be in use at any one time or several names in the literature may refer to one organism.

Fortunately, it is usually the custom to retain the species name, whilst re-naming the genus. Thus, for example, since about 1980, *Pseudomonas putrefaciens*, a spoilage organism of fish and other proteinaceous foods, has been known as *Alteromonas putrefaciens* and is currently known as *Shewanella putrefaciens* in honour of the British microbiologist James Shewan. Similarly, a species of the genus *Listeria*, *Listeria denitrificans*, has been placed in a new genus *Jonesia* (in honour of another British microbiologist, Dorothy Jones) on the basis of differences in metabolic characteristics that make its retention in the genus *Listeria* incompatible with other members of this genus.

1.4 Microbial grouping in practice

In practice, food-borne microorganisms are often considered in groups relating to their significance, e.g. indicator of hygienic practice or the type of spoilage with which they are associated, as well as the capacity of a test method to discriminate between different species. Thus, it is important to recognise that the practical groupings that occur as the result of the capacity of different organisms to grow on the same medium may be different from the taxonomic discrimination imposed by pure science as described above. Coliforms, for example, are a sub-group of Gram-negative bacteria that is used to indicate the general standard of hygiene in a food processing establishment; they comprise several genera, i.e. *Escherichia, Klebsiella, Enterobacter* and *Citrobacter*, all of which generally grow on 'coliform' media. The limitations, and advantages, of different microbiological methods are described in greater detail in Chapters 10 to 12. Broad groups of organisms are often sought using simple tests to provide an indication of microbial contamination or of potential spoilage of particular food types. These groups include:

- Total colony counts
- Total Enterobacteriaceae
- Gram-negative psychrotrophs
- Lactic acid bacteria
- Yeasts and moulds.

1.4.1 *Total colony counts*

Total colony counts (TCC) are known by a variety of names that are accepted in various parts of the food industry and in different countries. These are described in detail in Chapter 10. The objective of these counts is to provide general information on the numbers of organisms present in a product that grow under the conditions of the test. This can be useful, for example, when assessing the shelf life of a food. Total colony count tests may usefully be applied to most foods, with the exception, perhaps, of fermented products, e.g. cheese, in which microorganisms have been grown intentionally as part of the production process; for such products, a high TCC is normally expected and does not automatically indicate that the product is of poor microbiological quality. Because the objective is to find as many and as wide a range of organisms as possible, a very general nutrient medium is used and incubated at a medium temperature for a moderate length of time, e.g. 30°C for 48 hours.

1.4.2 *Indicator organisms*

Results from tests to detect individual bacterial pathogens, or indeed some specific spoilage organisms, in a food can often take several days to obtain and also the tests can be costly. So, sometimes, it is more effective to use simple tests, that produce results in less time, to seek groups of organisms that may provide some useful indication of the efficacy of the process and manufacturing conditions. The groups of organisms selected, when present, are often present in higher numbers than pathogens and are therefore more likely to be detected than a pathogen. Tests for such 'indicator' organism groups are therefore applied routinely to many foods and food ingredients.

   Members of the Family Enterobacteriaceae are most widely used as indicator organisms, especially to indicate the hygienic status of a process and product. Enterobacteriaceae can be detected and enumerated using a simple colony count technique that takes only 24 hours for a result to be obtained. Conversely, it takes a minimum of 3 days to detect *Salmonella* in a specified quantity of food and the test is much more complex. A test for indicator organisms is therefore of value as a relatively rapid and simple 'screening method' to confirm hygienic processing conditions.

A typical application of tests for indicator organisms is to verify that a heat-treated food product has been processed correctly and has not been re-contaminated after cooking or pasteurisation. This is because members of the Family Enterobacteriaceae are quite heat sensitive and should be absent from a heat-processed product if the correct heat process has been applied and the product has not been re-contaminated. Their presence in a heat-processed product indicates either undercooking or post-process contamination and, thus, that the process or subsequent controls have not been applied as intended.

If enteric indicator organisms are detected in a product in which they should be 'absent', then this is sometimes regarded as an indication that enteric bacterial pathogens may also be present. However, tests for indicator organisms should never be relied upon as an indicator of the presence or absence of a pathogen and should **never** be used as a substitute for a specific pathogen test. Where information concerning pathogens is necessary / required, then a test method targeted at the specific detection of the relevant organism must be carried out.

1.4.3   *Enterobacteriaceae and coliforms*

The Enterobacteriaceae are Gram-negative, non spore-forming, facultatively anaerobic rod-shaped bacteria that are resistant to bile salts and generally catalase positive, oxidase negative, and reduce nitrates to nitrites (see Chapter 11). Although found in various environments, many of those important in food microbiology originate in the intestines of animals.

Members of the Enterobacteriaceae are often present in uncooked foods as a result of their common presence on / in raw food materials, especially of animal and plant origin but, because they are readily destroyed by mild heat treatments, e.g. pasteurisation, they should be absent from heat-processed foods. Their presence in properly cooked foods therefore indicates post heat-process contamination from, for example, the manufacturing environment, equipment, or the food handler.

Coliforms are members of the Enterobacteriaceae that grow on coliform media incubated under specified conditions and produce distinctive colony morphologies and in general, they ferment lactose in the presence of bile salts. The ability to ferment lactose indicates that they are mammal-associated organisms, since lactose is the sugar present in mammalian milk and resistance to bile salts indicates an ability to grow in the animal intestine where bile salts are produced to emulsify fats in the diet to aid digestion. However, coliforms can also colonise the environment, such as watercourses and the drains of food factories, as well as food equipment.

Faecal coliforms are those coliforms most usually associated with a faecal origin. In the laboratory, faecal coliforms are distinguished from the main coliform group by their ability to grow in a selective growth medium incubated at an elevated temperature usually 43–45°C. This distinguishes them from similar organisms that are often found in the agricultural or open environment and, when they are isolated from food, indicates that the food potentially may have been contaminated, either directly or indirectly, by faeces from animal sources.

Escherichia coli Type I is a faecal coliform whose main reservoir is warm-blooded animals, especially humans; its isolation from foods represents a more specific indication of the possibility of faecal contamination from human sources.

1.4.4 *Gram-negative psychrotrophs*

Gram-negative psychrotrophs (mainly *Pseudomonas* spp., but also related organisms such as *Acinetobacter*, *Shewanella* and *Psychrobacter*) are organisms that grow well in cool, moist, low-acid environments such as raw, chilled foods and water, as well as on poorly cleaned food processing equipment. Being strictly aerobic, growth often occurs on the surfaces of foods stored in air, especially on proteinaceous foods. Food spoilage due to Gram-negative psychrotrophs is typified by an unpleasant, putrid odour, sometimes accompanied by slime production and even occasionally, fluorescent green colours. Examples of spoilage caused by these organisms include off odours in uncooked meats, slime on the surfaces of chilled meat and poultry carcases and a bitter taste in milk that has spoiled whilst stored under refrigeration. Gram-negative psychrotrophs also form part of the spoilage flora of salad produce and raw vegetables. Overt spoilage typically becomes apparent when numbers reach 10^7–10^8 cells per gram of food.

1.4.5 *Lactic acid bacteria*

The major lactic acid bacteria (LAB), comprising species of *Lactobacillus*, *Lactococcus* (formerly known as lactic streptococci), *Pediococcus* and *Leuconostoc*, are Gram-positive rods and cocci that are very tolerant of acid conditions. All produce lactic acid as a metabolic end-product and some also produce other metabolites. LAB that only produce lactic acid are called homofermentative whilst those that produce lactic acid and other products such as diacetyl (butter flavour) are called heterofermentative. The LAB are categorised as 'generally recognised as safe' (GRAS), i.e. the consumption of such organisms is not regarded as hazardous to human health. Most organisms in this group are aerotolerant, i.e. they are anaerobes (see Chapter 2) that either survive without growing or grow only slowly in the presence of air.

The LAB are widely used as starter cultures for fermented foods such as:

- Vegetables: olives, dill pickles, sauerkraut
- Dairy products: yoghurt, soft and hard cheeses, kefir, ripened creams and butters
- Meat products: salami, boulogna.

They are also very common food spoilage organisms, especially in reduced oxygen conditions e.g. modified atmosphere packed foods. In fact, the spoilage activity of LAB is very similar to their fermentation activity except that it is considered undesirable in the products in which it occurs.

Examples of LAB spoilage include:

- Vegetables: slime, gas and souring in damaged produce
- Dairy products: sour curdling of inadequately cooled milk
- Meat products: sour spoilage of vacuum packed, cooked or cured meats; greening of ham and bacon (by *Lactobacillus viridescens*).

1.4.6   Yeasts and moulds

Yeasts and moulds are widespread and common food spoilage agents, causing visual deterioration and musty or fruity taints. In general, yeasts and moulds are less fastidious in their growth requirements than bacteria and are able to grow in more harsh, particularly acidic, environments. They grow more slowly than many bacteria and therefore are potentially more problematic to semi-perishable foods that are accorded a longer shelf life than most perishable foods. Having tolerance to high salt and sugar concentrations, low moisture and low pH values means that there are several food types in which yeasts and moulds are the only significant spoilage agents (although some of these may be foods in which pathogenic bacteria survive without growing). Mould and yeast spoilage can be expected in poorly preserved soft drinks, jams and sugar preserves, pickles and sauces and salted foods. It also occurs in bakery goods and cakes, confectionery and dried fruits, cereals, pulses, nuts and nut products, although usually only when a storage or manufacturing problem has occurred e.g. allowing moisture levels to increase. Common causes of mould and yeast spoilage are too much moisture in foods meant to be of low water activity, insufficient levels of salt, sugar, acid or preservative in the food, a heat-processing failure, post-process contamination or contamination by a preservative-resistant species.

Bacteriologists often regard moulds simply as a single group of organisms whereas they are, in fact, diverse. Identifying food-borne mould isolates is often an advantage when investigating a spoilage problem because different mould species grow under different conditions. The spoilage of uncooked pizza provides a good example. On such a composite product, many mould species might be present without causing spoilage, if the pizza is stored properly. However, when spoiled, if the spoilage organisms include a black *Aspergillus*, then temperature abuse must have occurred because black Aspergilli do not grow under proper refrigeration conditions, i.e. < 5°C. Equally, if spoilage is caused only by a blue-grey *Penicillium*, then excessive time in chilled storage would be a likely cause because most blue-grey Penicillia are capable of slow growth in chill conditions. Therefore, identifying a food-borne mould isolate may help to identify the cause of a spoilage incident.

1.5   Further reading

Adams. M.R. & Moss, M.O. (2000) *Food Microbiology*, 2nd edn. The Royal Society of Chemistry, London, UK.
van der Heijden, K., Younes, M., Fishbein, L. *et al.* (eds) (1999) *International Food Safety Handbook: Science, International Regulations and Control*. Marcel-Dekker, Inc., New York, USA.

Hocking, A.D., Arnold, G., Jenson, I., Newton, K. & Sutherland, P. (eds) (1997) *Foodborne Micro-organisms of Public Health Significance*, 5th edn. Australian Institute of Food Science and Technology, NSW Branch, Food Microbiology Group, Sydney, Australia.

Hui, Y.H., Gorham, J.R., Murrell, K.D. *et al.* (eds) (1994) *Foodborne Disease Handbook, Volume 2.* Marcel-Dekker, Inc., New York, USA.

Ingraham, J.L. & Ingraham, C.A. (1999) *Introduction to Microbiology*. Brooks / Cole Publishing Company, Stamford, USA.

Jay, J.M. (2000) *Modern Food Microbiology*, 6th edn. Aspen Publishers, Inc., Maryland, USA.

Pitt, J.I. & Hocking, A.D. (1997) *Fungi and Food Spoilage*, 2nd edn. Blackie Academic & Professional, London, UK.

Samson, R.A., Hoekstra, E.S., Frisvad, J.C. *et al.* (1996) *Introduction to Foodborne Fungi*, 5th edn. Centraalbureau voor Schimmelcultures, Delft and Baarn, The Netherlands.

Singleton, P. & Sainsbury, D. (2001) *Dictionary of Microbiology and Molecular Biology*, 3rd edn. John Wiley & Sons Ltd, Chichester, UK.

Stanier, R.Y., Doudoroff, M. & Adelberg, E. A. (1966) *General Microbiology*. Macmillan, London, UK.

Stanier, R.Y., Ingraham, J.L., Wheelis, M.L. *et al.* (1987) *General Microbiology*, 5th edn. Macmillan Education Ltd, Basingstoke, UK.

2 Factors Affecting the Growth, Survival and Death of Microorganisms

2.1 Introduction

Once a microorganism has gained access to a food, there are a number of possible outcomes depending on whether or not the organism subsequently survives and grows in the food; there may be no consequence, there may be a risk of food-poisoning by intoxication or infection, or there may be a risk of food spoilage.

Some organisms may be unable to grow in the food, or do not cause food spoilage or human illness. In other words they will be of little interest and relevance to food microbiologists and technologists, i.e. of no consequence. Even so, microbiological test methods may allow such organisms to grow and so detect them, giving rise to the need to be able to distinguish them from organisms that are significant in food.

It should be noted, that in some cases, low numbers of viable pathogenic micro-organisms may be naturally present in a food without being a hazard. Their presence would be of particular concern if the normal production processes should have destroyed them.

Among those organisms capable of causing illness, there will be some that cause infections and some that cause intoxication due to the production of toxins (as well as a few that do both). Other organisms that do not cause illness may grow in the food and eventually cause the food to spoil. Food microbiologists, therefore, have not only to distinguish between the various groups of microorganisms but also to understand the characteristics of a food and the factors that affect microbial growth, survival and death in the food and the food-processing environment.

2.2 Some important characteristics of food contaminant microorganisms

2.2.1 *Characteristics that can be studied in the laboratory*

The microorganisms that can be studied in the laboratory have a number of characteristics that make the study possible:

Bacteria

- Grow to produce visible colonies with distinctive colours, in the colony and / or the medium, when grown on appropriate agar media.
- Grow to produce colour changes or turbidity, and sometimes gas, in appropriate liquid media formulations.

Plates 7 and 18 are located in the colour plate section at p. 100.

- Produce by-products of growth that can be assessed by biochemical methods.
- Have a distinctive genetic composition that can be determined by analytical methods.

Yeasts

- Grow to produce visible colonies on laboratory agar media.
- Grow to produce turbidity and gas in appropriate liquid media formulations.
- Produce by-products of growth that can be assessed by biochemical methods.
- Have a distinctive genetic composition that can be determined by analytical methods.

Moulds

- Grow to produce visible characteristic colonies in culture and on foods.
- Possess complex sporing structures that can be recognised microscopically and can be used for identification.

Thus, although yeasts and moulds belong to the Fungal Kingdom, it can be seen that in the laboratory, yeasts and bacteria can be examined in a similar way, but different methods are generally used for mould examination because their more complex structures can be studied under the microscope.

2.2.2 *Characteristics that inhibit study in a (normal / routine) laboratory*

Some groups of organisms cannot be grown in the routine food industry micro-biology laboratory and require special techniques and expertise to study them. The following summarises the chief characteristics of these groups:

Viruses

- Invisible by light microscopy – an electron microscope is needed to observe them.
- Often occur in low numbers in foods, so are difficult to capture for examination.
- Do not grow in foods, therefore no changes due to growth are seen in the food.

Protozoa

- Usually, only occur in low numbers in food so are difficult to capture for examination.
- Do not grow in foods, therefore no changes due to growth are seen in the food.

Toxic algae

- Not directly detectable because they are ingested by filter-feeding shellfish which concentrate them (and the toxin) by filtering them from the water during feeding. When necessary, tests are carried out to detect and quantify the specific toxins in the flesh of shellfish.

Test methods are not yet well-developed or established for detecting viruses and protozoans in food and instead of using methods to detect algae directly, chemical test methods are used for detecting and quantifying algal toxins in shellfish.

2.3 The characteristics of microbial growth

Whether in food or in laboratory growth media the microorganisms that are capable of growth multiply by using energy derived from available nutrients. For those organisms that consist of single cells (bacteria and yeasts), increase in numbers is achieved by cell division. Bacteria and a few types of yeast divide by forming a cross wall which creates two cells in place of the previous single cell and this continues in an exponential progression. Under optimum growth conditions, these divisions occur at more or less constant intervals (often as frequently as every 10–20 minutes); one cell could lead to more than one million cells within 5 hours, assuming a cell division rate of 15 minutes (Table 2.1). It takes the same time for one cell to divide into two as it does for one million to become two million and this has the practical consequence that once spoilage starts in a food, its progression is often very rapid.

In practice, exponential growth is part of a more complex picture known as the microbial growth curve (Figure. 2.1). When a microorganism first contaminates a food there is an initial adjustment period during which its numbers remain constant; this is known as the 'lag phase'. During this period the organism 'senses' the envir-

Table 2.1 Microbial multiplication (one to one million) under optimal conditions.

Cell division number	Number of cells (arithmetic)	Number of cells (logarithm$_2$)	Number of cells (logarithm$_{10}$)*	Time (hours: mins)
	1	0	0.0	
1	2	1	0.3	0:15
2	4	2	0.6	0:30
3	8	3	0.9	0:45
4	16	4	1.2	1:00
5	32	5	1.5	1:15
6	64	6	1.8	1:30
7	128	7	2.1	1:45
8	256	8	2.4	2:00
9	512	9	2.7	2:15
10	1 024	10	3.0	2:30
11	2 048	11	3.3	2:45
12	4 096	12	3.6	3:00
13	8 192	13	3.9	3:15
14	16 384	14	4.2	3:30
15	32 768	15	4.5	3:45
16	65 536	16	4.8	4:00
17	131 072	17	5.1	4:15
18	262 144	18	5.4	4:30
19	524 288	19	5.7	4.45
20	1 048 576	20	6.0	5:00

* to one decimal place

Figure 2.1 The phases of a microbial growth curve.

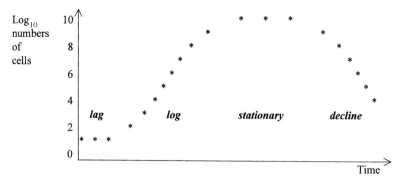

onmental conditions, e.g. moisture, temperature, nutrients, and, if growth is possible, adjusts its metabolism in preparation for growth. In addition, if a solid surface is available, it may attach itself to the surface. During this period, cell division does not occur. When growth is possible, cell division starts and the organism enters the 'exponential growth phase' (sometimes known as the 'logarithmic growth phase'). Exponential growth continues until some change occurs in the environment to inhibit further growth; this might be a decrease in pH value because of acid production by the organism itself, or a shortage of a nutrient or other growth limiting factors such as oxygen depletion. At this point, growth slows and numbers remain constant; this is referred to as the 'stationary phase' of the growth curve. The stationary phase is followed by a decline in numbers of organisms as viability is lost.

2.3.1 The factors that affect microbial growth

In foods, microorganisms may not grow or, when they do, may not grow at their maximum possible rate because one or more of the prevailing factors in their environment is sub-optimal for supporting growth. These are known most simply as 'intrinsic' and 'extrinsic' factors.

Intrinsic factors are the physical and chemical properties of the food itself, i.e. the product characteristics derived from the ingredients, recipe / formulation and the production process of the food that affect the conditions within the food and hence the numbers of any microorganisms present, their survival and rate of growth.

Extrinsic factors, by contrast, are related to the environment that surrounds the food, i.e. the external factors that affect the microorganisms in or on a food. These may include the type of packaging used, the atmosphere around the food (air, vacuum, modified atmosphere) and, probably of greatest importance, the temperature at which the food is stored. A summary of some Intrinsic (I) and Extrinsic (E) factors is given in Table 2.2.

Table 2.2 Examples of physico-chemical contributors to Intrinsic (I) and Extrinsic (E) factors, adapted from Adams and Moss (2000).

Properties of the food	I/E	External factors	I/E	Processing factors	I/E
Nutrients, e.g. proteins, sugars	I	Relative humidity	E	Slicing	I/E
pH value and buffering capacity	I	Temperature	E	Washing	E
Redox potential	I	Gas atmosphere, e.g. carbon dioxide, nitrogen	E	Irradiation	*
Water activity	I			Heat processes	*
Antimicrobial constituents, e.g. salt, nitrite, preservatives	I			Emulsification	*

* these are processes that may be applied to foods and may cause changes to intrinsic factors.

Intrinsic factors – the properties of the food

Nutrients

Most foods are intended to be nutritious to humans and are, therefore, also nutritious for many microorganisms. However, some components of foods may only be utilised by organisms that possess the specific enzymes required to break down the components. For example, starchy foods are exploited by microorganisms, often moulds, that produce amylase to break down the starch, and many of the organisms particularly associated with dairies and dairy products are able to break down and utilise the milk sugar, lactose. Fatty foods are particularly susceptible to the activity of lipolytic microorganisms. There are many types of organism (bacteria, yeasts and moulds) that produce lipolytic enzymes which are fat-splitting enzymes, giving rise to breakdown products that often have strong tastes or smells, so that spoilage due to rancidity of fat is usually very noticeable and offensive.

Likewise, many organisms produce proteolytic enzymes that degrade proteins in foods, the breakdown products of which may smell putrid. In other cases, a severely limited availability of nutrients may inhibit microbial growth, e.g. bottled beverages such as mineral water are usually very low in nitrogen content and will only permit the growth of a very limited range of organisms such as some species of *Pseudomonas*, *Acinetobacter* and some moulds that are characterised by their low nitrogen requirement and the need for only a small amount of organic compounds for growth. Thus, the specific nutrient composition of a food or beverage can significantly affect the type of microorganisms capable of exploiting the nutrient environment.

pH value

Although, in the language of physical chemistry, the pH value is the negative logarithm of the hydrogen ion concentration, in practical terms it has values that are

Figure 2.2   Typical pH limits for growth of microorganisms and pH values of foods.

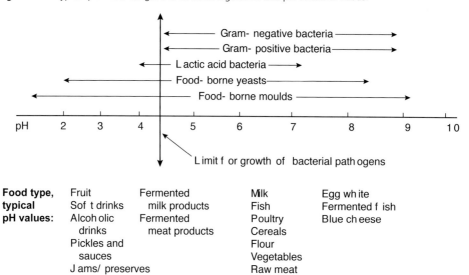

Food type,	Fruit	Fermented	Milk	Egg wh ite
typical	Sof t drinks	milk products	Fish	Fermented f ish
pH values:	Alcoh olic	Fermented	Poultry	Blue ch eese
	drinks	meat products	Cereals	
	Pickles and		Flour	
	sauces		Vegetables	
	J ams/ preserves		Raw meat	

measured on a linear scale from 0 (very acid) to 14 (very alkaline) where 7 is neutral (Figure 2.2). In practice, it is measured using a pH meter (see Plate 18 (see colour plate sections for all plates)) and the use of this equipment is discussed in Chapter 9.

Some alkaline foods occur, in the pH range 7–11, but are rather rare and tend sometimes to have a soapy taste. Examples include:

- Some fermented fish products such as shark fin
- Egg white
- Very ripe blue cheese

Neutral or slightly acid foods, pH 5.5–7.0, are the most common and include many of the world's major agricultural crops and protein sources such as:

- Meat, fish and poultry
- Starchy foods
- Vegetables

Acid foods (pH <2.0–5.5) are also common and are often given a longer shelf life than neutral foods because of the preserving power of the natural or added acids. Acid foods may be sub-divided into moderately acid foods (pH 4.5–5.5) and highly acid foods (pH <2.0–4.5). Moderately acid foods are often capable of supporting the survival and / or growth of some pathogens but extended shelf lives are applied because the acidic conditions inhibit some spoilage organisms. Examples include:

- Many types of cheese other than mould-ripened cheese
- Proprietary sauces and some mild pickles.

Highly acid foods have pH values lower than pH 4.5, which is the lower limit for growth of most pathogens, although sometimes, pathogens do survive at these pH values, as has been reported for *Listeria monocytogenes, Salmonella* and *Escherichia coli* O157. The major examples of highly acid foods are:

- Fruit and fruit products
- Soft drinks
- Jams and preserves
- Fermented foods
- Pickles.

The preserving power of food acids is governed to a large extent by the type and form of acid present. It is the un-dissociated acid which is responsible for the anti-microbial activity of the acid so when organic acids, e.g. acetic, lactic, citric, are used to preserve food, it is important to ensure that the correct concentration of un-dissociated acid is available for microbial inhibition. The proportion of un-dissociated acid present varies with pH (Table 2.3) so this must be taken into account when determining the amount of total acid required at a given pH to yield the required concentration of un-dissociated acid. At pH values near neutral, most organic acids have a limited effect on the growth of pathogens. However, at any given pH value, acetic acid (vinegar) is usually more inhibitory than, for instance, citric acid, particularly in foods of high water activity.

Table 2.3   Percentage of total organic acid that is un-dissociated at different pH values, adapted from International Commission on Microbiological Specifications for Foods (1980).

Organic acid	pH value			
	4	5	6	7
Acetic acid	84.5	34.9	5.1	0.54
Citric acid	18.9	0.41	0.006	<0.001
Lactic acid	39.2	6.05	0.64	0.064

Redox potential (E_h)

The concept of redox potential is less well understood than that of pH and is much less easy to measure. It is, however, known to be a significant factor in controlling the types of microorganism that are capable of growth in a food. It is a measure of the oxidising or reducing power of a food and is a balance of the 'activity' of many food components, although there may be gradients within a food, from highly oxidising conditions on the surface to reducing conditions within the interior. Redox potential is measured in millivolts (mV) and is very dependent on pH value. In aerobic systems such as raw minced meat and some fruit and vegetables, E_h values around $+300\,mV$ are found allowing the growth of microorganisms that are typically aerobes such as

Pseudomonas spp. and moulds. In anaerobic systems such as cooked, vacuum packaged and canned meats, negative E_h values, i.e. less than -150 mV, are found and the typical organisms able to grow include lactic acid bacteria and *Clostridium* spp.

In food microbiology an important species of *Clostridium* is *C. botulinum*. Although this organism is a strict anaerobe that grows optimally at E_h -350 mV, its growth may be initiated in some foods nominally of higher E_h (in the range $+30$ to $+250$ mV). This is because the E_h of food can be significantly reduced locally by the growth of aerobic and facultatively anaerobic spoilage organisms, e.g. yeasts, that deplete the level of oxygen present and produce reducing compounds such as thiols. As the E_h is reduced, growth of *C. botulinum* may be initiated. E_h conditions that permit the growth of *C. botulinum* can develop just millimetres below the surface of fish or meat muscle packed or stored in air and in delicatessen food products in packs only approximately 3 cm deep (Mossel *et al.* 1995; Snyder 1996). This serves to underline the importance of understanding the potential activities of the microbial populations present in different food environments.

Water activity

The water in a food is present in one of two forms. A proportion is present simply in the form of water molecules that if present in sufficient quantity are potentially available to support microbial growth. The rest, however, is unavailable to support microbial growth as it is 'locked up' or 'tied up' by dissolved solutes such as salt or sugar (Figure 2.3). Thus, in many foods, only a proportion of the total water (moisture) content of the food is freely available to support microbial growth, and this is one reason why knowing only the total moisture content of a food is not enough information for the prediction of either the spoilage rate of the food or for understanding the range of organisms that can grow in the food. Information about the salt or sugar content is also required because of their effect on water activity. Water activity provides a means for assessing the amount of moisture available for microbial growth.

Figure 2.3 Simple diagrammatic illustration of how water activity is reduced by the addition of solutes.

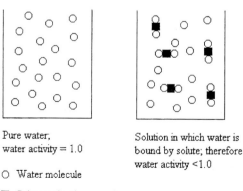

Pure water;
water activity $= 1.0$

Solution in which water is
bound by solute; therefore
water activity <1.0

○ Water molecule

■ Solute molecule, e.g. salt, sugar

However, water activity (a_W) cannot be measured directly; it is a theoretical calculation on a scale of 0 (no water) to 1.0 (pure water) and is equivalent in food to the Equilibrium Relative Humidity (ERH or 'humidity') of air. The latter is measured on a scale of 0–100% and thus:

$$a_W \text{ in food} = \frac{\text{ERH}}{100}$$

Microorganisms are much more versatile than the higher Kingdoms of animals and plants in their ability to withstand low-moisture environments. Whereas animals die, and plants die or permanently wilt, when their internal tissues reach a_W 0.99 and 0.98 respectively, food-borne bacteria, yeasts and moulds are much more resilient, being able to adjust their internal constituents to compensate for external water activities, in some cases to as low as 0.62 (Table 2.4).

Table 2.4 Minimum water activity values allowing cell growth in the Biological Kingdoms.

Organism	Water activity
Animal cells	0.99 (internal value)
Plant cells	0.98 (internal value)
Food-borne bacteria	0.85
Moulds and yeasts	0.62

Table 2.5 shows the approximate water activity of a range of different food types and indicates the minimum values at which growth has been observed, for a range of types of microorganism. As a general rule, the lower the water activity, the more limited is the range of organisms capable of growth and the longer the product shelf life that may be expected, although microbial contaminants might often be expected to survive for long periods without growth.

In some cases, especially for dried products, the moisture content relating to a given water activity is reasonably well known. This is certainly true, for example, of stored cereal grains. It is known that stored wheat is stable for 1–2 years or longer if its water activity is maintained at less than 0.7 and that this corresponds to a moisture content of 13–14%. Where such relationships between moisture content and water activity are known and are more or less constant, then direct moisture determination is used as a monitor for control purposes, as it is a more cost effective and simpler test than measurement of water activity. However when the relationship is not known or for foods of variable composition and, where water activity is important in the control of organisms, it is necessary to make the actual measurement of water activity by an appropriate method.

Various instruments and apparatus are available to determine the water activity of food, ranging from the very simple to the technically sophisticated. Most of the techniques used, in fact, measure the vapour pressure of water within the headspace of a sealed chamber in which the food has been placed, after the vapour pressures in the food and in the headspace have been allowed to equilibrate. As ERH is highly temperature-dependent, very accurate temperature control is essential. One of the best

Table 2.5 Water activity values for some foods compared with the minimum water activity values for microbial growth.

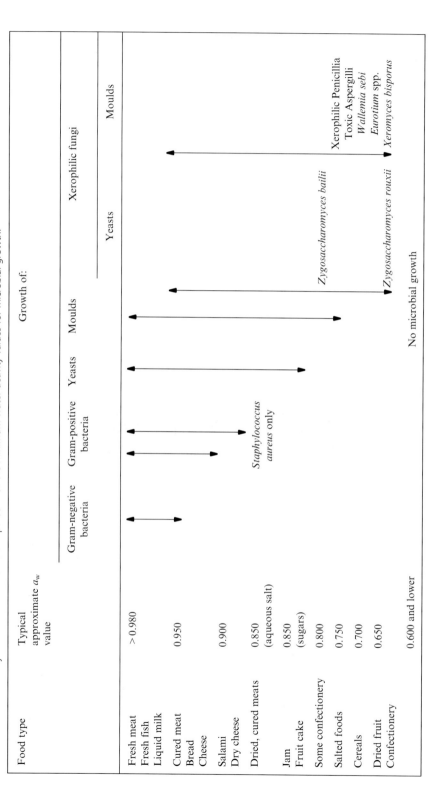

known instruments, the Novasina (Novaton Ltd, UK), employs a sophisticated electrode to measure the electrical impedance within the headspace, the amplitude of this measurement being related to ERH. A diagram of a Novasina sensor cell is shown in Figure 2.4 and the instrument is shown in Plate 7.

Figure 2.4   Diagram of a Novasina sensor cell used to estimate water activity of a food sample.

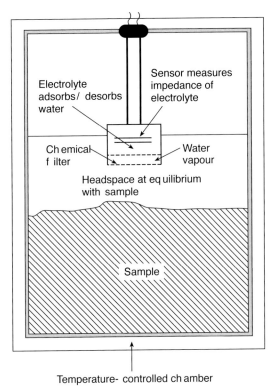

Temperature- controlled ch amber

Notes: The sensor cell contains a very small quantity of an electrolyte liquid. This takes up water from, or loses water to, the air surrounding it as the relative humidity changes, and this alters the electrical properties of the electrolyte. Electrical changes are detected by electrodes, and the signal is monitored by sensitive electronic equipment. Once equilibrium of the relative humidity (ERH) is established, the electrical signal becomes stable and is read on a numerical scale of 0–1 for a_w or 0–100% for ERH.

The measurement of ERH needs very accurate temperature control and the equipment used is constructed especially to achieve this; the international reference temperature for ERH/a_w measurements is 25°C.

Natural antimicrobial constituents

Naturally antimicrobial chemicals are found in many foods. Some or most of these may be part of the original plant or animal's normal defence mechanisms but some may be chemicals with other functions (such as insect-attracting aromas) that just happen also to have antimicrobial properties. Examples include:

- Essential oils in herbs and spices
- Essential oils in citrus fruit

- Components of onion, garlic, mustard and horseradish
- Sorbic and benzoic acids in wild and cultivated fruit.

Most of these materials' antimicrobial effects are not well understood. In any case, it is not usual for the food industry to take account of or rely on such materials' antimicrobial properties in the microbiological safety assessment of a product because, among other things, they can vary in strength from crop to crop and season to season etc. and most are used in such small quantities as to provide no real antimicrobial benefit in relation to food safety. However, much research is still undertaken on this subject as there is considerable consumer interest in the application of natural antimicrobial compounds.

Natural antimicrobial barriers

Almost all foods were, at some stage, alive and many, such as fruit, vegetables and some seeds are still viable and / or performing living functions at the point of consumption. This means that their surfaces still possess the antimicrobial barriers that were necessary to prevent microbial invasion whilst growing. When these become damaged, either in the field, during storage or during processing, microbial invasion usually occurs and spoilage starts.

In the case of internal tissues, the original structural function may continue to prevent or retard the invasion by microorganisms. Structural materials such as cellulose and pectins common in plants, and collagen in animals, are examples of materials that can act as antimicrobial barriers but, where the invading microorganisms (bacteria, yeasts and moulds) produce enzymes that are capable of breaking down such tissues, they become vulnerable to invasion by a succession of organisms, as shown by the rapid destruction of damaged apples and citrus fruit by moulds, sometimes succeeded by yeasts. It is important therefore, that packing and storage systems, especially of fruit and salad vegetables to be sold fresh, are designed and made to protect the food materials from physical damage.

Extrinsic (external) factors

Humidity

The moisture in the atmosphere surrounding the food throughout the manufacturing process and in the final packaging affects its water activity and hence any microbial activity in the food. A high humidity can result in moisture migration from the surrounding atmosphere to the surface layers of the food, increasing the available moisture (water activity) for microbial growth and possibly diluting any, otherwise significant, antimicrobial properties, e.g. preservative compounds that may be present. Where the packaging is in direct contact with the food, as in vacuum-packaged foods, the permeability of the packaging material to water vapour will govern the rate of migration of moisture into or out of the food and, in turn, affect the ability of microorganisms to grow. In some food production processes, e.g. salami production, there is a drying stage in which the product is held in conditions of low humidity; this causes the product to lose moisture to the atmosphere, thus drying the product and reducing its water activity which is an important factor contributing to the safety of these types of product. Control of the moisture conditions around many food types therefore can be important for developing and / or maintaining the product's characteristics, stability and safe shelf life.

Temperature

Different groups of microorganisms grow in different temperature ranges. The names applied to these groups describe the range of temperatures at which they grow, although inevitably, there is some overlap between the groups.

Psychrophiles: These usually grow only at temperatures in the range 0–20°C with an optimum growth temperature of 12–15°C; at present true psychrophiles are of relatively little importance in most foods and few microbiology laboratories carry out routine work on this group of organisms.

Psychrotrophs: These organisms resemble mesophiles with an optimum growth temperature of around 25–30°C but are also capable of growth at temperatures near 0°C. Psychrotrophic microorganisms are of importance in most chilled foods and can include bacteria, yeasts and moulds. The most well known spoilage bacterial psychrotrophs are *Pseudomonas* spp. and for moulds, the species of *Penicillium* and *Cladosporium*; however, some pathogenic organisms are also psychrotrophic, including *Listeria monocytogenes* and non-proteolytic strains of *C. botulinum*.

Mesophiles: Mesophiles grow in the warm range of temperatures, generally from around 10°C to an upper limit of around 45°C, with an optimum growth temperature usually between 25–40°C. This group of organisms includes most of the food-poisoning organisms including enteric pathogens such as *Salmonella*, many lactic acid bacteria, yeasts, and common moulds such as *Aspergillus*.

Thermophiles: Thermophiles only grow in very warm to hot conditions, typically in the range 55–75°C, although some may grow at temperatures as low as 40°C and as high as 90°C. They are mainly spore-forming bacteria although a few mesophilic moulds are capable of growth at extremely high temperatures. Thermophiles are often significant in the canning industry as the extreme heat resistance of their spores allows them to survive the high heat processes applied. These can subsequently germinate and grow, causing product spoilage, especially if the cans are exported to tropical countries and stored in hot conditions.

Gas atmosphere

Many microorganisms have specific oxygen requirements for growth that are used to categorise them into four groups. The groupings have fundamental implications in all areas of food microbiology, both in the food and in the laboratory. Some foods contain plenty of available oxygen, e.g. fresh, minced meat and un-wrapped salad vegetables. Organisms that grow in / on such foods typically require oxygen for their growth. Other foods, such as carbonated soft drinks and vacuum-packaged products, contain little or no oxygen and, to thrive in these environments, any organisms present would have to be able to grow without oxygen. Equally, to grow organisms in the laboratory, it is necessary to supply or remove oxygen according to the specific requirements of the organism. The following four categories of microorganisms are defined by their oxygen requirements for growth:

Aerobes: These require oxygen (an aerobic environment) to grow and are often found on moist, chilled, non-wrapped foods or products packed in air. *Pseudomonas* is a well known bacterial example but most moulds are also strict aerobes.

Micro-aerobes: These require oxygen at levels of 5–10% that are lower than found in the air, which has approximately 21% oxygen, i.e. a microaerobic environment. *Campylobacter* is the most well known food-borne pathogen that requires micro-aerobic conditions for growth.

Facultative anaerobes: Facultative means 'optional life style' and describes the mode of growth that is not the normal mode for the organism, e.g. a facultative anaerobe is usually an aerobe but it can grow anaerobically. Facultative anaerobes are probably the most common type of microorganism associated with foods. Many spoilage bacteria and pathogens fall into this category, including the Enterobacteriaceae, e.g. *Salmonella* and *E. coli*, as well as *Listeria* spp. and the fermentative yeasts.

Anaerobes (or strict anaerobes): This is a group of bacteria with a strict requirement for an oxygen-free environment, as oxygen prevents their growth and may, in fact, kill them. Most of the strict anaerobes in food are spore-formers in the genus *Clostridium*, which includes *Clostridium botulinum* that causes the often fatal illness botulism.

Microenvironments and combined conditions

Microorganisms in foods are inevitably exposed to a combination of intrinsic and extrinsic factors and throughout every solid food product there are many micro-environments caused by a multiplicity of physical and chemical interactions. These include:

- gradients of gases, e.g. oxygen, from the surface to the centre of the food or within sealed packs of respiring food materials, e.g. salad vegetables,
- gradients of chemicals that may only diffuse through the food slowly,
- gradients of E_h and pH value,
- metabolic by-products of microorganisms which are dispersed unevenly throughout the food, creating 'chemical' spheres around individual cells and microcolonies (clusters of organisms growing together),
- gradients of available moisture (a_w),
- localised changes due to surface drying or anaerobic conditions developing at the product / packaging contact interface,

and many other interactions that can occur throughout the food during its shelf life.
  So for the microorganisms in the food, their local 'world' changes constantly,

which in turn affects the types of organisms that can survive and grow at any particular stage in the life of a food. This can cause a microbial progression in the food, i.e. changes in types and numbers over time.

For the sake of maintaining the safety and quality of foods, it is essential that a thorough hazard analysis (see Chapter 5), supported as necessary by microbiological examination is carried out, by relevant and experienced personnel, for each food type and each manufacturing process to ensure that the microbiological profiles (the types and numbers of microorganisms) of the process and the final product throughout its shelf life are clearly understood. This will allow the relevant process checks and monitors to be established and maintained for the identified critical control points.

Implicit factors

Growth rate

The growth rate of a microorganism depends on the physico-chemical conditions prevailing in a particular food; it may, under some circumstances, be very fast. For example, *Clostridium perfringens* can double in numbers every 10 minutes in some foods if storage conditions allow. The spores of this organism can survive normal cooking temperatures and, if a large quantity of, for example, a meat or vegetable mix is cooled slowly after thorough cooking, when competing vegetative bacteria are absent having been destroyed by the cooking, then growth of *Clostridium perfringens* can occur extremely quickly leading to a spoiled product that is also hazardous to health. On the other hand, growth of microorganisms may be very slow, especially in deeply chilled or highly preserved foods and growth might take many months or even some years; for example, in the absence of other organisms that could grow more rapidly, mould can spoil a consignment of dried fruit over a period of many months.

The growth rate of microorganisms is one of the parameters calculated from microbiological growth modelling studies. In these studies, microorganisms are grown, either in microbiological growth media or in food substrates, under a wide variety of conditions of pH value, salt concentration, temperature, etc., and the data generated are used to develop equations to help predict the effects on growth of changes made to food composition and / or storage conditions. Thus, using mathematical models alongside real measurements of pH value, storage temperature and water activity for a particular food can help the food microbiologist to predict which organisms could grow, and therefore whether, and how quickly, spoilage may occur or whether an infectious or toxic dose of a microbial pathogen might develop. Mathematical models therefore can be derived to predict the shelf life and safety of foods. Because it is not practical, or even possible, to take account of all the parameters likely to affect microbial growth in any particular product in such models, they must be used very carefully and, preferably, only by a qualified and experienced food microbiologist who understands the limitations of the information that might be obtained from a mathematical model.

Microbial interactions

In foods, as elsewhere, many microorganisms co-exist without actually affecting each other to any large extent. Sometimes, however, they can have a considerable impact on each other, either to stimulate or inhibit growth, and two terms are used in food microbiology that define these interactions.

Antagonism

Antagonism occurs when a microorganism inhibits the growth of another micro-organism or several other microorganisms. Various examples occur in food micro-biology, often involving the production of lactic and other acids. During the production of fermented pickles, for example, the acid produced by lactic acid bac-teria, coupled with added salt, creates an environment that inhibits the growth of enteric pathogens that might otherwise grow and cause illness. Thus, the lactic acid bacteria are antagonistic to any enteric pathogens that may be present.

Synergism

Synergism is the opposite of antagonism. Sometimes, microorganisms can stimulate each other to grow more vigorously. In the production of yoghurt, *Lactobacillus bulgaricus* and *Streptococcus thermophilus* are grown in mixed culture and co-stimulate each other to produce a more rapid and efficient fermentation than either would achieve individually. This occurs because the *Lactobacillus bulgaricus* breaks down proteins to amino acids, providing nutrients for the *Streptococcus thermophilus* which, in turn, produces formate, pyruvate and carbon dioxide, that stimulate the growth of the lactobacilli.

Processing factors

Many processes and operations are involved in the preparation and manufacture of foods, e.g. washing, squeezing, chopping, dicing, mixing, pumping, slicing, pressing, heating, cooling, freezing, drying, and many more, and each can have a major effect on the types and distribution of microorganisms present on and in a food and upon their capacity to survive and grow. In addition, processing can either add to or reduce the numbers of viable microorganisms in the product. The following describes some of the microbiological effects that can arise from some of the common food production operations.

Cutting operations

Much of food preparation involves forming the food to the correct size, shape or conformation, which frequently involves cutting operations, e.g. slicing, chopping or dicing. In living or metabolically active foods, such as fruit, vegetables and raw meat, the result of cutting is that tissue barriers are broken, releasing nutrients that may help accelerate microbial growth. If equipment is not cleaned properly between production batches, then significant cross-contamination of foods subsequently processed through the same equipment is likely to occur, adding microorganisms that otherwise would not have been present. Thorough and regular equipment cleaning programmes are necessary therefore, to prevent the build-up of food residues and microorganisms on this equipment.

Washing

Washing of foods is intended to reduce contamination from dirt, chemicals and any microorganisms that may be present, but can itself be a source of contamination if dirty water is used. The effectiveness of washing also depends on the geometry of the food so that for example, a tomato is generally easier to clean than lettuce leaves. Disinfectants such as chlorine-based chemicals may be used but they are not completely effective and may only reduce viable microbial numbers without totally eliminating them. At concentrations high enough to be effective, these chemicals may actually bleach and / or taint some foods. It is important that only water of drinking (potable) quality is used for food washing processes and that systems are in place for ensuring that microbial contaminants cannot build up in the washing water.

Packaging

Product packaging is becoming increasingly important in food production. It is necessary to be aware of the protection or indeed absence of protection afforded by any form of packaging. Sometimes no protection is provided, for example in open bags and crates, or the protection may be only partial, as in boxes and over-wrapped trays which are often perforated. The microbiologist must consider not only the possibility of the food becoming contaminated, but also that microbial contamination on the food may pass to other foods in the vicinity. High-risk, e.g. ready-to-eat, foods are especially vulnerable to cross-contamination and great care must be taken in order to protect them from extraneous contamination (re-contamination) after processing.

Increasingly, complete or hermetically-sealed packaging is used. Complete closure has always been necessary for liquid foods and drinks, to prevent spillage, and for canned foods, but is now being used increasingly to provide complete protection to a wide range of foods. Lidded or heat-sealed pots, pouches and dish / tray containers are now widely used to package many chilled and shelf-stable foods. Where hermetic sealing is used there is the added advantage that the enclosed gas atmosphere can be modified, for example, using nitrogen and carbon dioxide which slow down the growth of a number of types of microorganism and can therefore help to extend product shelf life.

Irradiation

Since the early 1990s, following concern that some gas fumigation methods (for example, with ethylene oxide) could leave undesirable residues in foods, the use of ionising radiation has been permitted for use by some countries for some food materials. Irradiation is highly effective for the decontamination of, for example, herbs, spices, fruit, vegetables and shellfish, killing both pathogens and spoilage organisms. However, although the technology is permitted in the UK for various foods, it has not been used because of public and media concerns about safety. Scientifically, however, low level irradiation has been widely researched and found to be safe and effective for a wide variety of food materials.

Heat processing

Many different types of heat process may be applied to food, e.g. boiling, baking, roasting, thermisation, pasteurisation, retorting (canning) or ultra heat treatment, and the technologies can be complex. A good knowledge is needed of the numbers and types of organisms likely to be present from the raw materials and preparation processes in the product to be heated because different organisms have different heat sensitivities and high numbers take longer to kill at a particular temperature than low numbers. However to ensure that consistently safe processes can be achieved, proper process validation is essential. Process validation should establish the conditions necessary to achieve the desired, safe cooking procedure for the specific product under all worst-case conditions (Bell and Kyriakides, 2002). Therefore, it must take into account, amongst other factors, all of the following:

- Largest product size, density and viscosity
- Lowest temperature of product going in to the heat process
- Type of container / packaging, and headspace
- Coldest part of the oven / cooker
- Maximum oven / cooker load
- Shortest process time.

Process validations are normally undertaken as special procedures involving members of the technical and production team and, occasionally, external specialists.

Although very many different combinations of heat processes are used in the food industry, typical examples include:

- Pasteurisation, e.g. for milk, 71.7°C for at least 15 seconds
- Boiling water temperature (near 100°C), for soups or stews
- Retorting, e.g. 121°C for 15 minutes, for canned foods
- Ultra heat treatment, e.g. 140°C for 2 seconds, for UHT-sterilised milk.

Emulsification

Emulsification is a process technology often overlooked as a means of preservation, but is highly effective in controlling microbial growth. It may be used in combination with other factors, such as reduced water activity through the addition of salt. Butter, margarine and related spreads are emulsified products. Emulsification can restrict the amount of nutrients and space available to microorganisms, which prevents their multiplication. In butter, for example, discreet moisture droplets are suspended in a continuous fat phase; the droplets are very small (< 10 μm in diameter) and microorganisms become entrapped and isolated from each other and from the environment by the continuous fat phase. This means that nutrients and space within the droplets are very limited and any microbial cells in the product are unable to multiply.

If salt is added, it is highly concentrated into the water phase i.e. in the droplets, because salt does not dissolve in the fat (butter contains < 16% moisture), and this dramatically reduces the water activity. Thus, if emulsified products are stored in cool ambient or refrigerated conditions, they are usually quite stable as far as microbial growth is concerned.

2.4   Further reading

Adams, M.R. & Moss, M.O. (2000) *Food Microbiology*, 2nd edn. The Royal Society of Chemistry, London, UK.

Bell, C. & Kyriakides, A. (2000) Clostridium botulinum: *A Practical Approach to the Organism and its Control in Foods*. Blackwell Science Ltd, Oxford, UK.

Doyle, M.P. (1989) *Foodborne Bacterial Pathogens*. Marcel-Dekker, Inc., New York, USA.

International Commission on Microbiological Specifications for Foods (ICMSF) (1980) *Microbial Ecology of Foods. Volume I: Factors affecting life and death of Microorganisms*. Academic Press, London, UK.

International Commission on Microbiological Specifications for Foods (ICMSF) (1996) *Microorganisms in Foods 5: Microbiological Specifications of Food Pathogens*. Blackie Academic & Professional, London, UK.

International Commission on Microbiological Specifications for Foods (ICMSF) (1998) *Microorganisms in Foods 6: Microbial Ecology of Food Commodities*. Blackie Academic & Professional, London, UK.

Shapton, D.A. and Shapton, N.F. (eds) (1998) *Principles and practices for the safe processing of Foods*. Woodhead Publishing Ltd, Cambridge, UK.

World Health Organisation (1994) *Safety and Nutritional Adequacy of Irradiated Food*. WHO, Geneva, Switzerland.

World Health Organisation (1999) *High Dose Irradiation: Wholesomeness of Food Irradiated with Doses above 10 kGy*. Joint FAO / IAEA / WHO Study Group, WHO, Geneva, Switzerland.

3 Fundamentals of the Microbial Ecology of Foods I: Food Spoilage and Food-borne Illness

3.1 Introduction

Microorganisms may be associated with foods in a variety of ways, depending on the type and source of the food and the routes whereby microorganisms may gain access to the food material. It is important to have, at least, a broad understanding of how and where food materials are grown or raised and the likely routes of microbial contamination associated with these.

In this chapter, some aspects of the association of microorganisms with foods are discussed, including sources and routes of microbial contamination, food spoilage and food-borne illness and food poisoning due to microorganisms.

3.2 Microbial contamination – sources, routes and control

Microbiological contamination must be expected on food materials that come from natural sources and this contamination may result either in food spoilage, or in food-borne illness if handling or storage conditions are not controlled.

Figure 3.1 summarises the key sources and routes of microbial contamination of foods that can occur. In production environments any movement between low-risk and high-risk areas, including people, equipment, air and poorly designed drainage systems, can increase the possibility of microbial contamination or level of contamination of final products. In addition, the manufacturing environment can become a source of contamination if cleaning procedures are inadequate and food debris or biofilms are allowed to develop. Microorganisms will proliferate on any such residues and can become a significant contamination hazard to any further food products that come into contact with the fouled surfaces. This is a particular problem with food debris containing preservatives.

In any population of microorganisms there are usually a few that are more resistant to a chemical preservative and whereas the majority of the population are killed or prevented from growing, the few are 'selected' and develop a dominant resistant population. Also, some microorganisms can adapt in the presence of preservatives then, build up resistance if they remain in contact with the preservative for extended periods in otherwise good conditions for growth. This can happen when there is a build-up of preservative-containing debris in a food production area and microorganisms capable of surviving become adapted and grow on the debris, acquiring increased resistance. Unless an effective cleaning regime is applied, a cycle of contamination may develop in which new food materials entering the processing environment become contaminated with preservative-resistant organisms and the cycle is repeated until the

Figure 3.1 Key routes and sources of microbial contamination during food production.

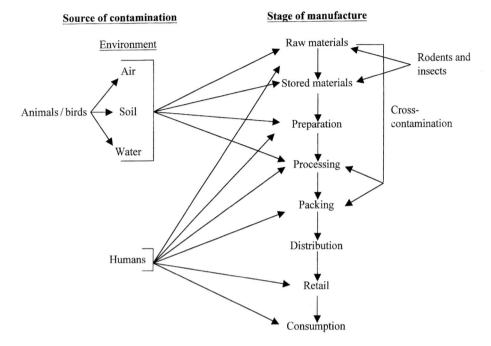

numbers of resistant organisms contaminating the foods being processed can over-come the preservation system, resulting in high levels of product spoilage.

The universal control measure to prevent such problems occurring is the consistent application of good hygienic practice, particularly during cleaning procedures, which should ensure the physical removal of debris prior to applying effective chemical cleaning and disinfection processes. This breaks the cycle of contamination and prevents the build-up of excessive microbial numbers or biofilms within the manu-facturing unit itself (Figure 3.2).

In the case of animal products, some considerations for the control of microbial contamination are:

- for meat-production, maintenance of the animal's health and, the cleanliness of the animal at the time of slaughter or, for milk- and egg-production, maintenance of the health and cleanliness of the animal during its productive life or, for fish and shellfish, the control of water pollution,
- for poultry and egg production maintenance of the microbial quality of the feed and water supplies,
- employment of good hygiene practices in the slaughtering, collecting or harvesting operations as applicable, and for all transportation containers and vehicles,
- the maintenance of hygienic conditions and appropriate cold storage conditions throughout further processing by a food manufacturer, throughout food service and retail, as well as in the hands of the consumer during home storage, handling or cooking, up to the point of consumption.

Figure 3.2 Cyclic contamination.

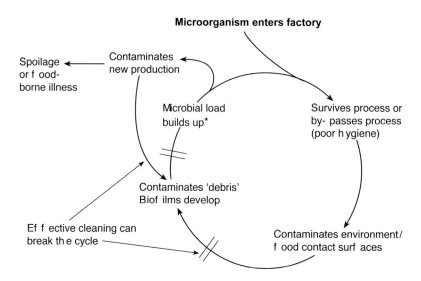

*Where the debris contains preservative, then preservative resistance can develop

For plant products, microbiological control measures include:

● use of potable / treated water supplies where crop irrigation is required,
● correct moisture control before, during and after harvest, particularly for grains,
● judicious and correct use of pesticides, or appropriate alternative controls for organic crop production,
● prevention of damage and maintenance of the necessary storage conditions, e.g. hygiene, temperature and humidity, during transport and storage,
● maintenance of hygienic conditions including the use of potable / treated water supplies and appropriate storage conditions throughout further processing by a food manufacturer.

3.3   The fate of microorganisms in foods

Organisms may die, survive or grow in foods during food production, handling, processing, storage and transport, and there are various potential consequences:

● *Death of organisms:* The numbers of viable spoilage organisms and infectious or toxigenic pathogens present in a food can be reduced, e.g. by cooking processes which destroy the organisms, but microbial toxins already present may not be inactivated by the process applied.
● *Survival:* The numbers of viable organisms do not increase, but those present may

cause illness if consumed by susceptible individuals, or they may contaminate other foods in which they may grow, either directly or indirectly via contact surfaces, if environmental and personal hygiene standards are inadequate.
- *Growth:* The numbers of viable organisms increase, and the likelihood of food spoilage may be increased or the likelihood of infection / intoxication of the consumer may be increased.

The growth of microorganisms both in foods and under laboratory conditions is of interest to the food microbiologist. In fact, many of the factors that affect the ability of microorganisms to grow apply both in the food and in the laboratory.

Of the six categories of microorganism significant in foods (bacteria, yeasts, moulds, viruses, protozoa and algae), only the first three can be grown in the normal food industry microbiology laboratory, and indeed these are the only ones that can normally grow in foods.

3.4   The consequences of microbial growth in foods

When microorganisms grow in foods there may be a number of outcomes; some are undesirable whilst others are highly desirable.

Undesirable outcomes include:

- *Food spoilage:*
  - spoilage is unacceptable to consumers for a variety of reasons but spoiled food is unlikely to cause illness (unless food-poisoning organisms or toxins are already present) because consumers will reject the food without eating it.
- *Food-borne illness:*
  - caused by growth of pathogenic bacteria inside the host's intestines (infection)
  - caused by toxins produced by bacteria (acute intoxication)
  - caused by toxins produced by fungi (mycotoxins) (chronic intoxication).

Desirable outcomes can be seen in the wide variety of foods that are produced by fermentation processes, for example,

- Alcoholic drinks –   beers, wines and ciders
- Dairy products –   cheeses and yoghurt
- Fermented meats –   salamis and boulogna
- Fermented vegetables –   lactic pickles and sauerkraut
- Sauces –   fish and vegetable
- Starches –   breads and sour doughs.

3.4.1   *Food spoilage*

What is food spoilage?

Food can be said to be spoilt when it has changed in a way that makes it unacceptable to a reasonable consumer. The definition has to be rather general because, for various reasons, consumers are not always reasonable. Individual consumers may differ in their tolerances of various flavours and odours. Some people may simply not like a food which is actually in an entirely satisfactory condition whereas others may find the flavours unexpected, for example in dishes containing large quantities of chilli peppers. A genuinely spoilt food has actually deteriorated in some quality such that it does not fulfil the expectations of the reasonable consumer. Spoiled foods are not always unpleasant: fermenting orange juice has a wine-like aroma that is very pleasant to some, but it is spoilt because its properties are inconsistent with those expected. Usually, however, spoilt food is decidedly unpleasant to one of the senses.

The unpleasantness of spoilt food is apparent because the smell, flavour, appearance, texture, or possibly some combination of all four characteristics, has altered to a point where the food has become unpalatable. However, such foods are generally unlikely to cause illness for a number of reasons. Firstly, the microorganisms that cause spoilage generally do not cause food-borne illness. Secondly, spoilt food is likely to be rejected before consumption, because of its unpleasant characteristics. Thirdly, there is strong commercial and legislative pressure not to sell or offer for sale food that is unfit for consumption because, in the words of the UK 1990 Food Safety Act Section 14, it is 'not of the nature or substance or quality demanded by the purchaser'.

What causes food spoilage?

Although the food microbiologist instinctively looks for a microbiological cause of food spoilage, there may be causes other than microbial action. Enzymes can cause changes in the flavour, the texture and the appearance of foods. Enzymes may actually be produced by microorganisms in the food, or they may be natural constituents of the food material itself, e.g. fresh fruit and vegetables, stored grains, pulses and nuts and raw meat, poultry, fish, eggs and milk. Non-enzymatic changes caused by chemicals can also take place in food, affecting its colour or texture.

Spoilage caused by microbial growth or enzymes present in foods

As microorganisms grow and multiply in foods, they utilise some substances and produce others in a process known as metabolism. These activities alter the chemical and sometimes physical composition of the food and can result in spoilage. Also, enzymes derived either from the microorganisms or from the food itself catalyse the breakdown of food constituents to produce other chemicals. The type of spoilage occurring in a food is therefore governed by the food's initial composition and by the microorganisms and enzymes present. The main spoilage changes can be summarised as follows:

- *Carbohydrates*. Carbohydrates of various complexities are broken down in a series of stages by hydrolysing enzymes (carbohydrases), resulting eventually in the production of alcohols, acids, carbon dioxide and water, so the spoilage products tend to be aromatic and often impart a 'fruity' odour and flavour to the food. Very often the spoilage is not unpleasant but simply unexpected. This type of spoilage is typically associated with the growth of yeasts, as well as some bacteria which utilise the simple sugars produced during the early stages of breakdown of complex carbohydrates.

- *Proteins.* The breakdown of proteins is catalysed by proteolytic enzymes (proteases). Proteins may be broken down into a variety of compounds of nitrogen and sulphur, including ammonia and hydrogen sulphide. The resulting spoilage is offensive, often resembling old cabbage or bad eggs and there can also be a putrid odour. This type of spoilage may be associated with a variety of organisms, including organisms that grow anaerobically.

- *Oils and fats.* Oils and fats are triglyceride compounds, the breakdown of which is catalysed by lipolytic enzymes to release free fatty acids. These have a pungent, sometimes 'mousy' odour and contribute to the typical rancid flavour of stale, fatty foods. These enzymes are produced by a wide variety of bacteria, yeasts and moulds.

- *Structural components.* Where foods contain structural materials such as cellulose, pectin or collagen, microorganisms are often able to degrade them into smaller units by enzymatic action. The result is a change in the food's structure, usually resulting in loss of texture, e.g. softening.

Whether or not the enzymes are derived from microorganisms, enzymes can be active at temperatures at which the organisms do not grow. Thus, for example, frozen foods containing high levels of microbial or natural food enzymes, perhaps as a result of freezing poor quality food, can be expected to deteriorate more rapidly than good quality frozen foods. However, some good quality raw foods that are stored frozen for very long periods of time can also deteriorate eventually due to enzyme activity; e.g. lipolytic enzymes in frozen fatty foods can slowly break down the fats to produce rancid flavours.

Other types of spoilage

Physico-chemical changes can occur in foods for reasons other than enzymatic action. The staling of bread, for example, has various causes, one of which is due to structural changes that occur in the gluten component of flour; in other commodities, physico-chemical or electro-chemical changes may be catalysed by trace contaminants such as heavy metals. This can become apparent in high fat products such as mayonnaise or butter when processing equipment is made of copper, or has welds between different types of metal, causing local electrolytic effects that can result in accelerated rancidity.

Although some physical changes can result from enzyme activity, e.g. tenderising of meat or softening during ripening of fruit, other physical changes can be simply structural effects often connected with the loss or movement of water. In fresh produce, this can be seen as wilting, and in frozen foods stored at fluctuating temperatures, as moisture migration leading to the formation of free ice crystals (freezer-burn) that may be associated with toughness, e.g. of meat. In some baked goods, loss of moisture can be a cause of staling, whilst in other products, such as cereals and biscuits, moisture uptake can result in softening (also known as staling).

However, much of food spoilage is due to organoleptic changes brought about by the activity of bacteria, yeasts and moulds, although it is interesting to note that some organisms that cause spoilage in one product may actually be used to produce an organoleptically desirable food product in a different circumstance. For instance, two groups of bacteria, the acetic and lactic acid bacteria, are used for their ability to produce organic acids and other flavour compounds in the manufacture of vinegars, fermented dairy products and pickles. However, when they are unintentionally introduced and allowed to grow in other products, the changes they produce in these are regarded as spoilage, e.g. sour wine and slimy pickles. Table 3.1 gives some other examples of organisms involved in such a dual role, indicating that, micro-biologically, fermentation is only a desirable form of food spoilage!

Table 3.1 Examples of the dual roles of some microorganisms in foods.

Product	Organisms	Result	Acceptability
Milk in a bottle	+ lactic acid bacteria	= spoilt	✗
Milk in a tub	+ lactic acid bacteria	= yoghurt	✓
Cheddar cheese	+ *Penicillium roqueforti*	= spoilt	✗
Soft cheese	+ *Penicillium roqueforti*	= blue cheese	✓
Orange juice in a carton	+ yeast	= spoilt	✗
Orange juice in a fermenter	+ yeast	= orange wine	✓
British sausage	+ lactic acid bacteria	= spoilt	✗
Continental sausage	+ lactic acid bacteria	= salami	✓

Food spoilage patterns

The microbial spoilage of different foods often follows distinct patterns; that is, the mode of spoilage is predictable (Table 3.2). This is why it is essential to know the composition of a food in order to assess its likely spoilage path and hence its potential shelf life under normal storage conditions as well as under temperature abuse conditions. Of course, the manufacturer cannot take into account all possible abuse conditions but guidelines have been produced for evaluating shelf lives of food products, e.g. Campden Food & Drink Research Association (1990), that describe a practical approach to assessing the effects of some storage temperature variations during a product's shelf life. The information gained through such evaluations can greatly assist in determining a realistic shelf life for the product that will neither permit product spoilage nor present a risk of pathogen growth.

Fresh, raw foods

Fresh, raw foods, such as meat, poultry, fish and milk, have fairly neutral pH values and, being highly perishable, have a short natural shelf life. Unless frozen, they are kept in chilled conditions from the point of slaughter or collection until processing or preparation for final consumption. Such storage conditions favour the growth of

Table 3.2   Some types of microbial spoilage of different foods.

Food type (examples)	Spoilage type	Cause
Frozen (vegetables)	Off colour and / or off odour	Enzymatic (microbial or food origin) (Microbially spoiled food can indicate: – poor quality material used for freezing – slow rate of freezing – freezer failure)
Chilled (meat, poultry, fish)	Off-odours	Growth of *Pseudomonas, Shewanella*, etc.
Canned / UHT non-acid foods	Blowing, acid production or liquefaction	Spore-forming bacteria, often thermophiles
Canned acid foods – pH <4.5 (canned or bottled fruit)	Visible growth	Heat-resistant moulds such as *Byssochlamys* and *Talaromyces*
Pasteurised / cooked foods (milk, recipe dishes)	Off-smells and tastes	Growth of post-heating contaminants or spore-formers surviving the process
Low water activity foods – dried, high-salt or high-sugar products (jams, stored cereals)	Visible growth and off-odours	Limited range of yeasts and moulds
Acid foods (pickles, soft drinks)	Turbidity, gas or visible colonies	Lactic acid bacteria, yeasts and moulds
Chemically preserved foods – sorbic and benzoic acid (soft drinks, low fat spreads)	Off-odours and visible growth	Yeasts and moulds, especially *Zygosaccharomyces bailii* and species of *Penicillium*
Vacuum and modified atmosphere packaged foods (cooked, cured meats)	Sour or putrefactive spoilage	Lactic acid bacteria and *Clostridium* spp.

Gram-negative psychrotrophic bacteria, e.g. *Pseudomonas, Psychrobacter, Shewanella* and related species, which can grow and spoil these products causing a variety of obnoxious odours, discoloration or textural changes.

However, there are some bacterial pathogens, including both Gram-positive and Gram-negative bacteria, that can survive and may even grow under the conditions of chilled storage often used for these foods. Low numbers of spoilage bacteria and the vegetative cells of bacterial pathogens should cause no problems in foods that are to be cooked, because these organisms are mostly very heat sensitive. However, food-borne illness can occur through the consumption of undercooked food, or cross-contaminated produce, such as salads and fruit that are intended to be eaten raw, or through the intentional consumption of the raw food, e.g. unpasteurised milk or steak tartare.

Cooking spoiled foods may well kill the spoilage organisms but the products of spoilage are likely to remain and the food will still be unpalatable. It is therefore crucial to ensure that populations of bacteria are kept under control prior to preparation of the food for consumption.

Preserved, raw foods

When perishable foods are preserved using salt, sugar, acid, drying, packing in a modified atmosphere, etc. the bacterial spoilage population tends towards a more Gram-positive flora, e.g. lactic acid bacteria, *Brochothrix*, *Bacillus* and *Clostridium*. The Gram-positive bacteria are generally more resistant to the conditions of pre-servation than the Gram-negative bacteria but they usually grow more slowly than in the same food without any preservation system. Thus, when the growth of Gram-negative bacteria is inhibited by preservative action in foods, the normally micro-biologically limited shelf life is extended, although eventual bacterial spoilage is usually inevitable. In conditions where both Gram-negative and Gram-positive bacteria are inhibited, microbial spoilage tends to be limited to that caused by yeasts and moulds.

Processed foods ('ready-to-eat' or 'ready-to-heat')

Ready-to-eat foods are regarded as high-risk foods and include, among many others, dairy desserts, cooked meats and sandwiches that are prepared in factories that have strict hygiene controls. However, many 'ready-to-heat' (not cook) foods are also manufactured in high care premises and these include some pizzas, recipe dishes (ready meals), savoury and sweet pies and pastries. All of these foods are perishable, although their shelf lives can vary considerably from just a very few days in the case of sandwiches and cream cakes to several weeks for cured, cooked meat products.

The production processes used for such foods change the microbial populations present. For example, cooking kills the vegetative cells of spoilage and pathogenic bacteria which means that any post heat-process contaminants, which could be completely different types of organism from those associated with the original raw materials, are likely to dominate the subsequent microbial population if they are allowed to grow. This may be a particular problem if a pathogen contaminates the processed food in the absence of any spoilage organisms. Because these types of food are often as perishable as fresh, raw foods, reliable temperature control throughout their shelf lives is essential for maintaining microbiological safety, and this includes their time in domestic refrigeration.

Ambient-stable foods

The term ambient-stable refers to those foods that are microbiologically stable at ambient (normal room) temperatures (usually 20–25°C). They are produced with the intention that they will withstand a long shelf life, often in excess of one year, while remaining safe and wholesome to eat. Stability of foods can be achieved by a number of methods, some natural and some artificial. A common method of preservation is drying, either achieved naturally in some crops such as grains and nuts, at the point of harvest or, after harvesting, by direct exposure to the sun, artificial heat sources or dry air, e.g. sultanas and raisins. Dried milk powder, dried sweet or savoury blends, e.g. dessert and soup mixes, and dried cereal products, are often produced in modern, large-scale drying plants.

Ambient stability can also be achieved by chemical preservation methods, using the water-binding properties of materials such as salt or sugar, or the pH / acidity modifying properties of organic acids such as acetic acid, lactic acid or citric acid, as present in many pickles and soft drinks. Many acid-preserved foods are, or contain, liquid and, in these, additional physical protection is achieved by the prevention of extraneous contamination using the sealed / closed containers that are necessary to prevent spillage of the product.

Ultimate deterioration of ambient-stable products is often non-microbial but, where microbiological problems do occur in water activity-controlled products, the cause is frequently a failure of moisture control which allows the water activity of the product to increase. Organisms that otherwise could not multiply if the reduced water activity of the product had been maintained can then grow.

Most bacteria are unable to grow at water activities lower than 0.9 (Table 2.5), especially in acid foods. Such foods, including fruit concentrates, jams and preserves, as well as salted foods, dry, staple foods e.g. rice, wheat, nuts and sweet pickles, if inadequately preserved, are usually spoiled by yeasts and moulds. Soft drinks of high water activity are made ambient stable by combinations of product acidity, a heat-process and the addition of chemical preservatives such as benzoic and / or sorbic acids.

While there is no clear biological distinction between yeasts and moulds (see Chapter 1), in general yeasts grow as single cells whereas moulds grow in a fila-mentous form. This is reflected in the types of foods that they most often spoil. Yeasts are more commonly observed in sweet, somewhat acid, liquid or semi-liquid foods, such as soft drinks, bottled fruit and pickles, in which they cause turbidity, scum formation and, frequently, gassing due to the production of carbon dioxide. Moulds are more commonly found associated with solid surfaces and are typically found on dry foods, such as bread, cakes, cheese, nuts and dry staple foods, when moisture levels are sufficient to permit growth. There is, however, no absolute 'rule'; yeasts are often found on cured meats, for example, and moulds often spoil fruit juices.

Typical spoilage organisms

The following are some of the more common types of food spoilage organisms:

- *Gram-negative bacteria:*
  Pseudomonas and *Psychrobacter* (aerobes)
  Acinetobacter (aerobe)
  Shewanella (aerobe)
  Enterobacteriaceae (facultative anaerobes).
- *Gram-positive bacteria:*
  Lactobacillus (anaerobe / aerotolerant)
  Brochothrix thermosphacta (facultative anaerobe)
  Enterococcus (facultative anaerobe)
  Bacillus (aerobe / facultative anaerobe), forms spores
  Clostridium (anaerobe), forms spores.
- *Yeasts (aerobes or facultative anaerobes):*
  Saccharomyces
  Zygosaccharomyces
  Candida

Pichia
Debaryomyces
Brettanomyces
Rhodotorula.
- *Moulds (all aerobes (or micro-aerotolerant)):*
Penicillium
Aspergillus
Cladosporium
Rhizopus/Mucor
Byssochlamys (heat-resistant).

From information about the composition / formulation, processing, packaging and storage conditions of a food (see Chapter 2), a food microbiologist can usually assess the likely type of food spoilage that could occur (Table 3.2). This in turn enables the microbiologist to determine which organisms or groups of organisms should be sought in microbiological tests, or indeed, whether testing is worthwhile at all.

3.4.2   *Food-borne illness*

Although food-associated illness is largely attributed to microorganisms, other sources may be significant and should also be considered when investigating outbreaks of illness. Some factors may be intrinsic to the food, including potentially toxic plants (potato, red kidney bean), fungi (death cap, magic mushroom) and fish (puffer fish). The histamine reaction that may occur as a result of the consumption of scombroid fish, e.g. mackerel, is a special case in which the naturally occurring amino acid, histidine, in the fish may be converted to histamine during the growth of spoilage bacteria. Chemical contaminants (pesticides, heavy metals) and higher organisms, e.g. parasitic worms, may also cause illness. Pathogenic microorganisms, mainly bacteria, are however, the most significant cause of food-borne illness.

Food-borne intoxications and infections

Microbial food-poisoning is usually the result either of infection, or of an intoxication in which pre-formed toxins may be ingested or toxins may be produced by the organism in the host's intestines.

Microbial toxins

Some microbial toxins are very heat stable and can persist in a food even after the toxin-producing organism has been killed by heat. The onset of illness caused by bacterial toxins in foods is usually rapid when the toxin reaches the stomach (Table 3.3), and the first symptoms are often nausea and vomiting. The most common food-borne toxigenic bacteria of this type are *Staphylococcus aureus* and *Bacillus cereus*.

In cases of food-borne disease caused by intoxication, the cause of illness has to be confirmed by direct examination of the implicated food for the presence of toxin or a clinical sample (as appropriate), e.g. *Clostridium botulinum* toxin in faeces. Often,

Table 3.3 Some organisms associated with food poisoning and some typical sources of the ⌣ ⸏ adapted from Anon (2000).

Microorganism	Usual incubation period	Common symptoms	Usual duration of symptoms	Typical food and food-related sources
Bacillus cereus (emetic toxin)	1–5 hours	nausea vomiting diarrhoea abdominal pain	24 hours	fried and cooked rice pasta pastry products
Bacillus cereus (diarrhoeal toxin)	8–16 hours	diarrhoea nausea vomiting abdominal pain	24 hours	meat soups sauces puddings
Campylobacter spp.	2–5 days	diarrhoea abdominal pain fever blood in stools	2–7 days	raw milk poultry products untreated water
Clostridium perfringens	12–18 hours	diarrhoea abdominal pain	24 hours	cooked meats meat gravies / sauces
Vero-cytotoxin-producing *E. coli*	1–6 days	diarrhoea blood in stools Haemolytic Uraemic Syndrome (HUS)	4–6 days (not HUS)	undercooked meat products raw milk cheese fruit juices salad vegetables faecal–oral transmission*
Salmonella (non-enteric fever / typhoid)	12–72 hours	vomiting diarrhoea fever	<3 weeks	meat products poultry products eggs & egg products raw milk & milk products some high-fat foods
Shigella spp.	1–7 days	diarrhoea blood in stools	<2 weeks	food contaminated by food handler e.g. salad produce faecal–oral transmission water supplies
Staphylococcus aureus	2–4 hours	vomiting abdominal pain fever	<12–48 hours	any food handled directly by a carrier, followed by temperature abuse of the food, e.g. sliced cooked meats, salads, dairy products

Contd.

Table 3.3 *Contd.*

Vibrio spp. (not *V. cholerae* O1 or O139)	12–18 hours	diarrhoea	<7 days	raw fish and seafood from warm water areas
Cyclospora cayetanensis	5–7 days	diarrhoea abdominal pain	variable	fruit salad produce water supplies
Cryptosporidium parvum	2–5 days	diarrhoea	<3 weeks	raw milk water supplies food contaminated by the food handler faecal–oral transmission*
Giardia intestinalis	5–25 days	diarrhoea abdominal pain	variable	water supplies food contaminated by the food handler faecal–oral transmission*
Hepatitis A virus	2–6 weeks	jaundice abdominal pain intermittent nausea diarrhoea	up to several weeks	shellfish raw fruit salads & salad vegetables food contaminated by the food handler faecal–oral transmission*
Norovirus (Norwalk-like viruses (NLV))	1–3 days	vomiting diarrhoea fever	1–3 days	shellfish water supplies raw fruit salad vegetables food contaminated by the food handler faecal–oral transmission* aerosols
Rotavirus	1–2 days	diarrhoea vomiting	4–6 days	faecal–oral transmission*

* person-to-person spread

biochemical methods are used, e.g. enzyme-linked immunosorbent assays (see Chapter 12); fortunately, many proprietary tests are now available for the detection of bacterial toxins, with all the necessary reagents supplied and detailed instructions given.

The toxin formed by *Clostridium botulinum* (the cause of botulism) is somewhat different because it is inactivated after a few minutes in boiling water and it is a neurotoxin. Onset of illness can take several days, depending on the quantity of toxin ingested, and the symptoms include slurred speech, double vision and breathing difficulty. If untreated (treatment is by injection of a neutralising antitoxin and by respiratory support), botulism is frequently fatal, as a result of asphyxia, and, even if treated, recovery may take many months and require life support.

Some moulds also produce toxins, known as mycotoxins, when they grow in foods. These, however, do not usually give the normal symptoms of food poisoning, i.e. gastro-enteritis. Rather, they attack the major internal organs, e.g. the liver or the kidneys, or the blood, especially after a long period of low-level exposure, and cancers may form in some cases. There is no epidemiological evidence at present to indicate that mycotoxins are of significance in the Western diet. In some countries, legislation is in place to control levels of some mycotoxins in foods. For example, one group of mycotoxins, the aflatoxins, often associated with peanuts and maize, is widely controlled in many countries because of its extreme carcinogenicity in the liver and this means that very low maximum permitted levels are enforced, e.g. at µg/kg (parts per billion, ppb) concentrations. Therefore, it is generally considered unwise to consume mouldy food.

Although toxins may be formed by some bacteria and moulds growing in the food, they may also be produced in the intestinal tract once the organism has been ingested, as is the case for *Clostridium perfringens*, or a food material may become toxic by concentrating a toxic agent during feeding, e.g. toxic algae concentrated by filter-feeding shellfish.

Whilst some toxins are destroyed by heat, most are more resistant and persistent than the organism that has produced them so that heat processing to eliminate microorganisms may not destroy their toxins and make the food safe. Hence the way to ensure the absence of toxins in the food supply is, preferably, to prevent the organism from contaminating the food or, if it is inevitably present, to prevent its growth and hence, toxin formation. Even if a toxin can be inactivated by processing, processing methods alone should not be relied upon to eliminate the toxin.

Although the factors that affect the formation of bacterial toxins are likely to vary from one situation to another, it is often the case that sufficient toxin is produced to make a food toxic when the number of toxin-producing bacteria has reached around one million (10^6) in each gram of food. With moulds and toxic algae it is not possible to make such a calculation.

Microbial infections

A number of factors determines whether a food-borne infection can occur. Some organisms, such as viruses and protozoa, cannot, and never do, grow in foods whilst other organisms do not grow in foods at normal food storage temperatures, e.g. *Campylobacter* which cannot grow at temperatures below 32°C. However, for these pathogens, only low numbers of viable organisms (< 100 cells or particles) may be required to cause an infection, so it is essential to ensure that their presence in foods is minimised. Other organisms (bacteria) may be able to grow in foods but are prevented from doing so in a particular commodity by factors such as the food composition or storage conditions so survival often occurs without growth, but the potential to cause infection remains.

For a food-borne bacterial infection to occur, a pathogenic organism has to survive the hydrochloric acid present in the stomach and the addition of bile from the gall bladder and it must pass into the intestines. Sometimes many thousands of organisms may be required to provide an infectious dose (the number of organisms required to cause illness), but on other occasions only a very few, e.g. < 100 cells, may be needed, depending on the health of the individual, the virulence of the organism or the foodstuff in which the organism is ingested. In either case, onset of illness requires growth of the organism in the gut and it may take from several hours to more than a week for symptoms to occur (Table 3.3).

For many years, it was thought that quite high numbers of some pathogens, e.g. *Salmonella*, were needed to cause illness. However, in recent years it has been realised that infectious doses can in fact be very low. In the case of *Salmonella*, many thousands of cells per gram may be needed in some foods, e.g. moist fresh foods, to cause an infection whilst in other foods, e.g. fatty foods such as chocolate, cheese, snack salami or potato snacks, the cells of *Salmonella* may be protected from destruction by the fat / oil in the food during the digestion process and infection can arise after consumption of only a few cells or tens of cells in such foods (Table 3.4). *Campylobacter* commonly has a low infectious dose of a few tens or hundreds of cells and for other infectious agents, such as Noroviruses and *Cryptosporidium*, just a few viral particles or oocysts, respectively, are thought to cause illness.

As well as variations in infectious dose associated with different food types, the virulence of organisms can vary greatly, even within a species in which some strains may be relatively harmless whereas others may cause severe illness. For example, *Escherichia coli* is commonly a harmless member of the normal commensal microflora of the distal (lower) part of the intestinal tract of humans and other warm blooded animals and although most strains of *E. coli* are not pathogenic, the species does contain strains that can cause a number of different types of illness, some severe and even fatal and some of these strains are known to be food-borne, e.g. *E. coli* O157 : H7. Toxin-producing organisms, including bacteria such as *Staphylococcus aureus* and moulds such as *Aspergillus flavus*, can also exhibit strain to strain variation in virulence and it is essential for microbiologists to apply some kind of sub-typing

Table 3.4   Some foods implicated in outbreaks of salmonellosis and the infectious doses of the organism estimated in the food, adapted from D'Aoust (1997).

Food	*Salmonella* serotype	Infectious dose (no. of cells)
Eggnog	S. Meleagridis	10^6–10^7
	S. Anatum	10^5–10^7
Goat's milk cheese	S. Zanzibar	10^5–10^{11}
Chocolate	S. Eastbourne	10^2
	S. Napoli	10^1–10^2
	S. Typhimurium	$\leq 10^1$
Hamburger	S. Newport	10^1–10^2
Cheddar cheese	S. Heidelberg	10^2
	S. Typhimurium	10^0–10^1
Paprika potato crisps	S. Saintpaul	
	S. Javiana	$\left.\right\} \leq 4.5 \times 10^1$
	S. Rubislaw	

scheme in order to identify and characterise very accurately the strains of public health significance (see Chapter 11).

To add to the complexity, there is wide variation in the susceptibility of individuals to attack by infectious pathogens and different sectors of the human population vary in vulnerability to different pathogens. For example, Rotaviruses tend to attack young children, young adults seem more susceptible to *Campylobacter*, and *Salmonella* and Vero cytotoxigenic *E. coli* (VTEC) often cause most severe illness in the very young and the elderly. Even for individuals in the same age group, susceptibility will vary, depending not only on their underlying health but also on their level of stomach acid, which fluctuates widely between and during meals.

The practical consequence of these variations is that there is no single value that can be used reliably as the infectious dose of an organism in food hazard assessment programmes (see Chapter 5) and it is therefore necessary that food specifications and standards require that the organism should not be detected in a given quantity of food (usually 25 grams (g)) rather than to specify a maximum permitted level of detection.

There are two particular circumstances where pathogens may be found in foods in low numbers without giving rise to excessive concern or condemnation of the food. The first is toxin-producing pathogens that may unavoidably be present in foods but would have to grow to high numbers for a 'toxic dose' to be produced; *Staphylococcus aureus* in some cheeses and fermented meats and *Bacillus cereus* in rice and starches are examples. It is however, necessary to prevent their growth and subsequent toxin production in the food and this is achieved by product formulation, production process control and correct product storage temperatures and times.

The second example concerns *Listeria monocytogenes*, which, unlike most other pathogens, is a common environmental organism as well as the cause of listeriosis. This organism is inevitably found in and on many fresh, raw foods because it is common in the natural environment and at present there is no evidence that it causes illness unless ingested in high numbers. The upper limit of numbers that is currently considered acceptable in foods in which it may survive is 100 cfu per gram of a food throughout its shelf life. Obviously, its presence would not be accepted in foods such as cooked meat products because the cooking process destroys the organism and high standards of post-processing hygiene should prevent re-contamination. In such products, specifications relating to this organism are commonly 'not detected in 25 g'.

Generally, infectious pathogens should be undetectable (not detected in 25 g) in foods at the point of consumption; although the presence of low numbers of some toxin-producing and spore-forming bacteria is inevitable, they may be tolerated because specific growth controls are in place. In all cases, however, contamination by these organisms needs to be minimised and product storage conditions or formulation need to be controlled so that pathogens cannot grow to unacceptable numbers.

The prevention of food-poisoning

In the UK, around 100 000 cases of food-related illness, including food-poisoning, are reported annually. However, a detailed report of the study of infectious intestinal disease (IID) in England (Anon, 2000) suggests that the real level of food-borne illness in the community is much higher as only approximately 1 in every 4 cases of salmonellosis is reported, 1 in 8 for campylobacteriosis and for all IID, only 1 in 135 cases is reported. It seems, then, that only 1–2% of cases of all IID are actually reported, although the more severe illnesses would be reported more frequently than the lesser ones. Lost working time due to food-poisoning as part of the IID number must, in fact, represent an enormous economic loss to the country.

Surveys carried out by the Health Protection Agency's Central Public Health Laboratory in Colindale, London, have determined that food-poisoning outbreaks are mainly caused by poor hygienic practice, which allows food-poisoning organisms to contaminate the food, often combined with other poor practices including:

- food preparation too far in advance of consumption,
- keeping food at room temperature for excessive times,
- slow cooling of cooked food, and,
- inadequate re-heating of previously cooked food.

The basic rules for the prevention of food-poisoning are:

- the maintenance of good hygienic practices (personal, equipment and environmental) throughout the food production chain, from the primary producer to the consumer, in order to minimise or prevent contamination of food by pathogens;
- to store prepared and perishable foods at temperatures below 8°C (by law in the UK) and preferably at the lowest temperature that does not damage the food, e.g. 1–4°C;
- for foods requiring cooking, to ensure a minimum internal temperature of at least 70°C for 2 minutes and, if not consumed immediately, to cool the food quickly to a temperature below 8°C, store at temperatures < 8°C and consume as quickly as possible within 48 hours of cooking and;
- if it is necessary to hold the food hot, to hold it at or above 63°C;
- to handle the food under the conditions recommended by the manufacturer or retailer and consume it within the prescribed shelf life;
- to prevent cross-contamination from raw to cooked foods and to clean foods that will be eaten without cooking.

3.5   Further reading

Adams, M.R. & Moss, M.O. (2000) *Food Microbiology*, 2nd edn. The Royal Society of Chemistry, London, UK.

Anon (1993) *Shelf Life of Foods – Guidelines for its Determination and Prediction*. Institute of Food Science & Technology (UK), London, UK.

Anon (1998) *Food and Drink Good Manufacturing Practice – A Guide to its Responsible Management*. Institute of Food Science & Technology (UK), London, UK.

Anon (2000) *Foodborne Disease: a Focus for Health Education*. World Health Organisation, Geneva, Switzerland.

Campden Food & Drink Research Association (1990) *Evaluation of Shelf Life for Chilled Foods. Technical Manual No. 28*. CFDRA, Chipping Campden, UK.

Doyle, M.P. (1989) *Foodborne Bacterial Pathogens*. Marcel-Dekker, Inc., New York, USA.

Hocking, A.D., Arnold, G., Jenson, I., Newton, K. and Sutherland, P. (eds) (1997) *Foodborne Micro-organisms of Public Health Significance*, 5th edn. Australian Institute of Food Science and Technology, NSW Branch, Food Microbiology Group, Sydney, Australia.

International Commission on Microbiological Specifications for Foods (ICMSF) (1996) *Microorganisms in Foods 5: Microbiological Specifications of Food Pathogens.* Blackie Academic & Professional, London, UK.

International Commission on Microbiological Specifications for Foods, (ICMSF) (1998) *Microorganisms in Foods 6: Microbial Ecology of Food Commodities.* Blackie Academic & Professional, London, UK.

Jay, J.M. (2000) *Modern Food Microbiology*, 6th edn. Aspen Publishers, Inc., Maryland, USA.

Labbé, R.G. and García, S. (eds) (2001) *Guide to Foodborne Pathogens.* John Wiley & Sons Ltd, Chichester, UK.

Shapton, D.A. and Shapton, N.F. (eds) (1998) *Principles and practices for the safe processing of Foods.* Woodhead Publishing Ltd, Cambridge, UK.

Varnam, A.H. and Evans, M.G. (1996) *Foodborne Pathogens. An Illustrated Text.* Manson Publishing, London, UK.

4 Fundamentals of the Microbial Ecology of Foods II: Food Preservation and Fermentation

4.1 Introduction

The type of spoilage of a food depends on the constituents of the food, including the presence of preservatives. The type of spoilage of a raw food can be changed by the use of preservatives and different preservation techniques.

Attitudes to food preservation have changed in recent years. Traditionally, the supply of many foods, especially fresh produce, was seasonal and food preservation relied upon a reduction of pH value or water activity, or the addition of chemical preservatives, with relatively little reliance on chilled storage. Many, usually seasonal, foods are now supplied all the year round from sources around the world and thus the life of fresh seasonal produce needs to be extended to facilitate transport. More recently, for health reasons, foods are now required with minimal preservative e.g. salt addition which means that much greater reliance is placed on the chill chain to maintain the safety of the food during its shelf life.

It has become increasingly important to ensure that due attention is given to the assessment of a realistic shelf life of products. For pathogens and a few spoilage organisms, there is a limited range of mathematical models from which databases have been developed to assist the prediction of bacterial growth, survival and death in certain food formulations. However, where spoilage is concerned, much useful practical information can be obtained by comparing changes in the microbiological profile (the types and numbers of organisms) with criteria such as discoloration rate and organoleptic changes under varied storage conditions. Further information on studying the shelf life of foods can be found in *Shelf Life of Foods – Guidelines for its Determination and Prediction* (Institute of Food Science and Technology (UK), 1993) and also in *Technical Manual No. 28* produced by Campden Food & Drink Research Association (1990).

The assessment and application of realistic and safe shelf lives for food products is now an integral part of the hazard analysis process applied to the manufacture and distribution of all chilled and / or ready-to-eat foods, such as chilled ready meals, pizzas, pies, delicatessen foods, sandwiches and chilled desserts. As such, the assessment is subject to audit by customers, e.g. retailers and third party food inspectors, and therefore needs to be thoroughly understood and properly carried out and documented by food manufacturers.

Control of both pathogens and the spoilage microflora during the shelf life of a food remain essential targets, particularly because, at present, there is little effective and reliable control of storage conditions of foods after purchase. Consumers tend to shop less frequently and to purchase items with the latest possible 'Use by' dates. Thus, many products may be kept for a large part of their life in a domestic refrigerator at an unknown and variable temperature, and those that require re-heating may be re-heated in uncontrolled conditions, to an unknown temperature. Microwave ovens are widely used for re-heating foods, some of which heat food unevenly so that some parts of a ready meal, for instance, may appear 'piping hot' whilst other areas are only tepid. Such possibilities are necessarily considered by food microbiologists and technologists during all new product developments and reformulations of existing products which may be made to account for changing consumer requirements.

This chapter discusses some applications of food preservation systems and some beneficial effects to product shelf life resulting from the use of microorganisms in food processes, e.g. fermented foods.

4.2   Controlling shelf life by preservation systems

Sometimes the preservation methods that can be applied to a particular commodity are limited because of the specific nature of the food, e.g. chemical / preservative treatments should not be applied to fresh, raw foods described and sold as such. For example, fresh fruit and vegetables for export are often harvested before they are ripe and transported to the importing country under chilled and / or controlled atmosphere conditions, prior to final ripening in a special warehouse and distribution to stores. Many raw meat, fish and game products can only be transported to the importing country under chilled or frozen conditions; in the latter case, they may be sold or used directly from the frozen state, or defrosted for sale or further processing. In all these cases, care is needed to ensure that the food is always handled hygienically and is well-protected physically to prevent extraneous contamination or damage.

Many foods, especially those that are seasonal, have traditionally been preserved by physical techniques such as drying or chemical processes using humectants (moisture binding agents) e.g. salt, or organic acids. The following examples are typical:

- drying   (cereals, meat, fish, fruit, vegetables)
- salting   (meat, fish, pickles)
- sugaring   (jams, preserved fruit, sweet pickles)
- vinegar   (fruit and vegetable pickles)
- lactic acid   (fermented meats, vegetables and dairy products).

The main objective has been to achieve ambient stability and / or extended shelf life but the resulting flavour can often be strong or the food may be deemed to have adverse health effects, e.g. adding too much salt or sugar to the diet. Modern food preservation techniques use fewer or less of the preservative chemicals but are usually supplemented by other processes and quality systems to maintain their stability and safety, for example,

- purchasing raw materials of a higher microbiological quality
- using packaging materials of high microbiological quality
- inclusion of a mild heat process

- improving containment and segregation of processing operations
- improving equipment design to facilitate cleaning
- improving standards of environmental hygiene, e.g. air management
- improving temperature control throughout processing and distribution
- reduction of shelf life.

Such changes to the production and handling of preserved foods require improved hygiene practices in relation to personnel, equipment and the environment and temperature controls, as well as improvements in facilities for segregating different process operations. In many cases, the product quality and safety is improved by a combination of some or all of these control measures and such approaches are becoming the norm in the preserved foods manufacturing industry.

4.2.1   *Temperature of processing and storage*

Heat processing

Microorganisms vary widely in their heat-resistance but some general rules apply:

- Most organisms, especially microbial pathogens, stop growing at temperatures above approximately 50°C.
- Most vegetative pathogens such as *Salmonella* and *Campylobacter* are killed by pasteurisation processes, e.g. 71.7°C for 15 seconds.
- Boiling water kills all vegetative bacteria and yeasts and almost all moulds. Many bacterial spores survive such treatment and possibly one or two mould species, especially if they are protected by very high sugar concentrations (low water activity).
- Processes such as the retorting processes used in canning (121°C for 15 minutes is a typical process) kill bacterial spores, but some bacterial toxins, e.g. those of *Staphylococcus aureus* and the emetic toxin of *Bacillus cereus*, are able to remain active, even though all the living organisms have been killed.
- Low water activity conditions created by solutes such as salt and especially sugar, may increase the heat resistance of microorganisms so that higher heat processes, i.e. a higher temperature and / or a longer time, may be needed to destroy microorganisms and hence safely preserve these types of product.
- High acid (low pH value) conditions tend to increase the sensitivity of microorganisms to heat. High acidity also inhibits the growth of a range of significant microorganisms, notably the spore-forming bacteria, so that milder heat-treatments may be possible for high acid foods.
- Low-acid canned foods, e.g. vegetables, soups and meats, require heat processes that deliver an internal temperature of at least at 121°C for 3 minutes, commonly referred to as a 'minimum botulinum cook', that is designed to reduce the probability of survival of the most heat-resistant spores of *Clostridium botulinum* by a factor of 10^{12} (1 million million).

Thus, a cooking process such as boiling will normally destroy vegetative spoilage bacteria, yeasts and moulds and only some toxins (bacterial and fungal) and bacterial spores would be expected to remain as potential microbiological hazards. The former should not be significant if systems such as HACCP (see Chapter 5) and Good Manufacturing Practice are properly applied, but the potential for growth of bacterial spores, which are not destroyed but might be activated by the cooking temperature, must be considered in relation to the intrinsic properties of the product, e.g. pH value, any subsequent processing, packaging, storage temperature, shelf life and intended consumer handling.

Because any vegetative organisms present in the raw ingredients should have been killed during the cooking process, any organism that subsequently contaminates the food may be able to grow without competition and high numbers may result if the nutrient and environmental factors are suitable. When such an organism only causes spoilage, the effect is simply to reduce the shelf life (which has commercial but not safety implications), but if the organism is a pathogen that grows without causing noticeable spoilage (a characteristic of many pathogens) then the product may become dangerous to consumer health.

If the food is 'ready-to-eat', e.g. sliced, cooked meats, then there is no consumer step, such as re-heating, to reduce the numbers of pathogens, if present, and any pathogen present could cause illness. If the food is to be re-heated or fully cooked by the consumer, then vegetative pathogens will be reduced (provided the food is re-heated / cooked correctly) although any heat-stable toxins present would be likely to survive the re-heating.

It is important, therefore, that food microbiologists provide detailed and accurate information and guidance to the food industry and the consumer, enabling food processors and handlers to apply relevant monitoring systems and control measures for the consistent production of safe food and to enable the consumer to handle and prepare food safely.

Cold storage

As with heat resistance, microorganisms vary widely in their capacity to grow at low temperatures. Again, a few general points can be made:

- Some organisms, e.g. *Clostridium perfringens*, do not grow at temperatures below 10°C. (In fact, the growth of most, but not all, Gram-positive bacteria is significantly restricted under chilled conditions.)
- Some significant pathogens and spoilage bacteria can grow at temperatures around 0°C; these include *Listeria monocytogenes* and *Pseudomonas* spp.
- Some moulds, e.g. *Cladosporium* spp., grow at temperatures as low as −8°C, but only very slowly.
- 'Deep chill' temperatures, at just above the freezing point of food, i.e. usually −2°C to + 2°C, slow most microbial growth considerably.
- No microbial growth occurs in freezers at temperatures of − 18°C or colder.

This last point is particularly important because, if microbial growth is suspected or demonstrated in a frozen food of initially good microbiological quality, then temperature abuse must have occurred.

During the production or distribution of foods, all processes and storage that involves chilling or freezing must be controlled effectively and monitored routinely. Should such facilities go out of control, the foods will inevitably deteriorate far more quickly than expected and incur serious safety and commercial consequences.

4.3   Microbial fermentations and biotechnology

Some options for controlling microbial growth by preservation using physico-chemical methods are summarised in Table 4.1. In practice, however, with the exception of heat-processing or chilled storage, single preservation systems are rarely applied, as the results are often unpalatable; high-salt foods, very acid products and commodities containing chemical preservatives are, nowadays, often regarded as unpleasant or undesirable by many consumers.

Combinations of preservation factors are therefore often used to achieve the required preservative effect and organoleptic characteristics. The use of combined preservation systems has sometimes been called 'the hurdle effect' but this is a mis-leading and technically incorrect term as it implies a series of unconnected obstacles that microorganisms must overcome one at a time, whereas the preservation systems (intrinsic and extrinsic factors) present for any particular food actually present a single barrier, where all effects combine either, to reduce the rate of microbial growth or, to inhibit it completely.

In some food manufacturing processes, microorganisms are used deliberately to bring about desirable changes to a raw material, e.g. the production of cheese or alcoholic drinks. Their use in foods represents some of our oldest food preservation technologies.

Many foods are made using combined preservation systems and the following are given as two examples:

Smoked sausage

The production of smoked sausage may include:

- addition of salt, to inhibit growth of Gram-negative bacteria,
- fermentation by lactic acid bacteria when lactic acid is produced, which reduces the pH value,
- smoking, often as part of a heat-process, to add flavour and possibly contribute antimicrobial substances to the surface of the product, although this effect is difficult to quantify and may be minimal,
- a heat-process, applied to stop fermentation and kill any surviving vegetative pathogens,
- drying, to reduce the water activity of the sausage,
- inoculation of mould (*Penicillium nalgiovense*) onto the surface to prevent the growth of unwanted moulds and possibly contribute to flavour,
- storage and sale from a chilled delicatessen counter, to minimise the growth of any post-process contaminants, e.g. from a retail slicing operation.

Table 4.1 Summary of commonly-used food preservation methods and their effects.

Method of preservation	Effect
Fermentation e.g. natural production of lactic acid	• Increases in acidity reduces pH value due to production of lactic acid by lactic acid bacteria; principally used in fermented vegetables, dairy products and meat products • Production of bacteriocins such as nisin by lactic acid bacteria in fermented dairy products • Production of alcohol by yeasts such as *Saccharomyces cerevisiae* in the manufacture of wines and beers
Acidification e.g. by addition of acetic, lactic or citric acid	• Increases in acidity reduces pH value of pickles and sauces, including mayonnaise, by addition of vinegar (acetic acid) • Increases in acidity reduces pH value of vegetable products, fruit products and marinated fish by the addition of lactic or citric acids
Addition of solutes, e.g. salt or sugar	• Reduction of water activity (a_w) in salted fish and meat products by addition of sodium chloride • Reduction of water activity (a_w) in jams, sugar preserves and fruit concentrates, sweet bakery goods and confectionery by addition of sucrose
Drying	• Reduction of water activity (a_w) in dehydrated meat, fish, vegetables, fruits and stored cereals, rice, pulses, nuts and their products. Moulds grow if insufficient drying occurs or products become damp or wet.
Heat processing	• Destruction of vegetative bacteria, yeasts and most moulds by processes between pasteurisation (71.7°C/15 seconds) and cooking at the temperature of boiling water (100°C) • Destruction of bacterial spores by a canning process, e.g. 121°C/15 minutes, or UHT (Ultra High Temperature) process, e.g. 140°C/2 seconds • No effect on heat-stable bacterial toxins such as those of *Staphylococcus aureus* or *Bacillus cereus*
Chilling	• All microbial growth (bacteria, yeasts and moulds) is slowed down. Some spoilage and food-poisoning bacteria are unable to grow at refrigeration temperatures but usually do not die • Psychrotrophic spoilage bacteria, yeasts and moulds grow slowly and eventually cause spoilage • Psychrotrophic pathogens (non-proteolytic *Clostridium botulinum*, *Listeria monocytogenes* and *Yersinia enterocolitica*) are able to grow, but only slowly
Freezing	• Most bacterial and yeast growth stops at 0°C to −2°C • A few moulds, e.g. *Cladosporium cladosporioides* grow very slowly at −7°C to −8°C • No microbial growth occurs at temperatures of approximately −10°C but any microorganisms present are likely to survive

Contd.

Table 4.1   *Contd.*

Method of preservation	Effect
Addition of chemical preservatives	• Inhibition of microbial growth occurs, especially at high-acid pH values, but organisms are not necessary killed • Weak acid preservatives (sorbate, benzoate, etc.) inhibit yeasts, moulds and some bacteria but some species are highly resistant, e.g. *Zygosaccharomyces bailii* (yeast) and *Penicillium roqueforti* (mould) • Preservative resistant populations of some organisms may develop during prolonged contact, especially in unhygienic food production areas • Nitrites are responsible for the curing action in the production of ham and bacon. In combination with salt, they inhibit Gram-positive bacteria, e.g. *Clostridium* spp., in cured meat products.
Atmosphere modification, e.g. vacuum, gas flushing	• Atmosphere modification is usually used in combination with chilled storage to extend shelf life • Levels of oxygen higher than those in air are used to retain red colour in red meats; favours growth of aerobic organisms • Vacuum packed food products and modified atmosphere packs containing a high nitrogen content inhibit the growth of aerobes but anaerobes such as lactic acid bacteria and *Clostridium* spp. may grow although spoilage due to these organisms in such packs is slower than if the product was packed in air • High carbon dioxide atmospheres are the most antimicrobial but some lactic acid bacteria, yeasts and clostridia may grow and sour taints may form in some foods
Irradiation	• Effects of increasing irradiation doses are similar to those of increasing heat process • Low doses, e.g. up to 5 kGy (kiloGray), eliminate vegetative bacteria, yeasts and moulds • Higher doses, e.g. up to 10 kGy, can inactivate bacterial spores • One or two highly resistant bacteria are known but are not significant in foods • Among the most resistant food-borne organisms are clostridia such as *Clostridium botulinum*

Fermented pickles

The production of fermented pickles may include:

• addition of salt, to inhibit growth of Gram-negative bacteria,
• fermentation by lactic acid bacteria when lactic acid is produced, which reduces the pH value,
• hermetic closure of the packaging container to prevent re-contamination with spoilage organisms,

- a heat-process, applied to stop fermentation and kill any surviving vegetative pathogens,
- refrigeration of the product, recommended after opening the container to slow the growth of any microorganisms introduced by the retailer, caterer or consumer (refrigeration after opening is often re-enforced by limiting the 'open' shelf life).

There are four main applications of microorganisms in the food industry:

- to produce fermented foods, as indicated above,
- production of chemicals as food ingredients,
- production of food modifying enzymes, and
- direct use as food sources.

The list of existing fermented foods is enormous and all classes of food are represented. Products include beers, wines and vinegars, cheeses and yoghurt, salami and boulogna, pickled cucumbers and sauerkraut, fish and vegetable sauces, breads and sour doughs.

There are many reasons for fermenting a food raw material although the oldest was certainly to preserve seasonal produce and avoid wastage by making use of any surplus. Fermentation can also produce a variety of desirable flavours in foods and improve the palatability, digestive properties and the nutritional quality of some foods. The only class of food that is rarely fermented is nuts. Nuts are already stable produce and command a high price in their own right so fermentation would not add value. Additionally, they contain little fermentable material, such as sugars, being high in fat and protein. A few exceptions are, however, found in regions of the Far East where the press-cake left over from extracting the oil from peanuts is fermented to a soft but solid cake called 'oncom' using the mould *Rhizopus oligosporus* and the press-cake from coconut is fermented to 'bonkrek' using the same mould.

The microorganisms used in food fermentations and other processes are in legal terms 'generally recognised as safe' (GRAS) and come from a number of groups of organisms, including lactic acid bacteria that produce lactic acid from sugars, acetic acid bacteria that produce acetic acid (vinegar) from ethanol, yeasts that produce carbon dioxide which raises doughs for breadmaking, and also produces alcohol from sugars in beer and wine production, and moulds which produce enzymes that break down proteins in the ripening of some cheeses and fermented meats.

A number of food ingredients can be produced by fermentation, of which a major example is citric acid, produced by the mould *Aspergillus niger*. Citric acid is the natural acid found in many fruits, such as *Citrus* spp., as well as in tomatoes; it is used for pH adjustment in many food formulations because its acidity is less astringent than lactic or acetic acids, but it also has an important function for the prevention of spoilage and the maintenance of microbiological safety. The efficient production of citric acid as a food ingredient is therefore of commercial importance. An example of the function of citric acid in food preservation and safety can be seen in the canning of tomatoes. Very ripe tomatoes may not contain sufficient citric acid to guarantee the inhibition of microbial growth and canned tomatoes therefore require the addition of citric acid to reduce the pH value to pH < 4.5 and ensure that surviving spores are unable to germinate, e.g. the spoilage organism *Clostridium butyricum* or the pathogen *Clostridium botulinum*.

Microbial enzymes are used widely to modify many types of food. They are obtained by the large-scale culture of microorganisms that produce highly active forms of the required enzymes in large quantities (Table 4.2). Some of the organisms may have been isolated from natural sources but a number, such as those used to produce chymosin as a substitute for animal rennet in the manufacture of vegetarian cheese, are of necessity genetically modified, in order to produce a mammalian enzyme. A parallel example in the medical field is the production of human insulin for diabetic patients using genetically-modified microorganisms. Microbial proteases (proteolytic enzymes) are produced naturally by the resident microflora in the ripening of cheese and fermented meats; this imparts a specific flavour to these products but proteases are also used to tenderise some meat products. Microbial carbohydrases are used to clarify fruit juices and to soften the centres of filled chocolates whilst microbial isomerases (convert one isomer to another) are used to convert sugars into valuable syrups that are difficult to manufacture naturally.

Over the years many attempts have been made to harness microorganisms, including algae, bacteria, yeasts and micro-fungi, directly as foodstuffs. Part of the challenge was to find organisms that were GRAS organisms which did not produce, in culture, excessive levels of nucleic acids that may be potentially harmful if consumed in large amounts. Of the many attempts, the success story is 'Quorn', a

Table 4.2   Some microbial enzymes used in food production.

Enzyme	Produced by (type of organism)	Use / purpose
Chymosin*	Yeast *Kluyveromyces lactis*	Coagulation of milk protein during cheese production
Chymosin*	Mould *Aspergillus awamori*	Coagulation of milk protein during cheese production
Chymosin*	Bacterium *Escherichia coli*	Coagulation of milk protein during cheese production
Milk coagulating enzymes	Mould *Rhizomucor miehei*	Coagulation of milk protein during cheese production
Milk coagulating enzymes	Mould *Cryphonectria parasitica*	Coagulation of milk protein during the production of Kosher and Halal cheese
Proteases	Lactic acid bacteria	Accelerated ripening of hard cheeses
Lactase	Yeast *Kluyveromyces lactis*	Conversion of lactose present in whey into glucose and galactose; the converted sugars are used as food ingredients
Amylases	Bacteria, e.g. *Bacillus* sp. Moulds, e.g. *Aspergillus niger*, *Rhizopus* sp.	Used in brewing as 'mash enzymes' to break down starch into fermentable 'sugars'

* Produced by an organism that has been genetically-modified to contain the bovine gene for chymosin production mycoprotein produced by the growth of the mould *Fusarium graminearum* on hydrolysed corn starch. It is now marketed as a commercial product and is widely used as a constituent of canned foods and recipe dishes, being particularly popular with vegetarians.

4.4   Further reading

Adams, M.R. & Moss, M.O. (2000) *Food Microbiology*, 2nd edn. The Royal Society of Chemistry, London, UK.

Anon (1993) *Shelf Life of Foods – Guidelines for its Determination and Prediction*. Institute of Food Science and Technology (UK), London, UK.

Anon (1996) *Guide to Food Biotechnology*. Institute of Food Science & Technology (UK), London, UK.

Campbell-Platt, G. (1987) *Fermented Foods of the World. A Dictionary and Guide*. Butterworths, London, UK.

Campden Food & Drink Research Association (1990) *Evaluation of Shelf Life for Chilled Foods. Technical Manual No. 28*. CFDRA, Chipping Campden, UK.

International Commission on Microbiological Specifications for Foods (ICMSF) (1980) *Microbial Ecology of Foods Volume 1: Factors Affecting Life and Death of Microorganisms*. Academic Press, London, UK.

International Commission on Microbiological Specifications for Foods (ICMSF) (1988) *Microorganisms in Foods 4: Application of the hazard analysis critical control point (HACCP) system to ensure microbiological safety and quality*. Blackwell Scientific Publications, Oxford, UK.

International Commission on Microbiological Specifications for Foods (ICMSF) (1996) *Microorganisms in Foods 5: Microbiological Specifications of Food Pathogens*. Blackie Academic & Professional, London, UK.

International Commission on Microbiological Specifications for Foods (ICMSF) (1998) *Microorganisms in Foods 6: Microbial Ecology of Food Commodities*. Blackie Academic & Professional, London, UK.

Wood, B.J.B. (1998) *Microbiology of Fermented Foods Volumes 1 and 2*, 2nd edn. Blackie Academic & Professional, London, UK.

5 Applications of Microbiology in the Food Industry

5.1 Introduction

Every day, tens of thousands of samples of foods and related materials are examined in food industry, contract testing and public health laboratories to determine the presence and numbers of microorganisms that they contain. Microbiological tests for detecting and counting bacteria are by far the most common tests carried out. Food-borne bacterial pathogens, spoilage bacteria and bacteria that indicate the general hygienic status of the sample are all important to food producers and tests devised for their detection and enumeration (counting) form the routine work of most food industry laboratories. Yeasts and moulds play an important role in the spoilage of many foods and specific tests are also carried out to detect and count these organisms. There is, however, little point in carrying out any microbiological examination of a food unless the objectives are clear, i.e. why is the test necessary and how are the results going to be used?

A primary requirement of any microbiological test is to obtain a reliable result that will be of value in informing interested parties about the microbiological status of the sample that, in turn, will reflect the process or environment from which the sample was taken. The test result also indicates the risk of the food becoming a health hazard or spoiled within its allocated shelf life. Such information may well be used as the basis for action to be taken.

The following discusses the main areas of application of microbiological tests in the food industry.

5.2 Hazard Analysis Critical Control Point (HACCP)-based systems and microbiology

In the European Union, North America and elsewhere, legislation requires food business proprietors to produce and supply safe and wholesome foods. For instance, in the UK, Section 4-(1) of *The Food Safety (General Food Hygiene) Regulations*, 1995 (Anon, 1995) (which implements parts of the European Union Directive 93/43/ EEC of 14th June 1993 on the hygiene of foodstuffs (Anon, 1993) states:
'A proprietor of a food business shall ensure that any of the following operations, namely, the preparation, processing, manufacturing, packaging, storing, transportation, distribution, handling and offering for sale or supply, of food are carried out in a hygienic way.' Further, in Section 4-(3):

Plate 13 is located in the colour plate section at p. 100.

'A proprietor of a food business shall identify any step in the activities of the food business which is critical to ensuring food safety and ensure that adequate safety procedures are identified, implemented, maintained and reviewed on the basis of the following principles –

(a) analysis of the potential food hazards in a food business operation;
(b) identification of the points in those operations where food hazards may occur;
(c) deciding which of the points identified are critical to ensuring food safety ('critical points');
(d) identification and implementation of effective control and monitoring procedures at those critical points; and
(e) review of the analysis of food hazards, the critical points and the control and monitoring procedures periodically, and whenever the food business's operations change.'

This responsibility is increasingly discharged by food business proprietors through the operation of a Hazard Analysis Critical Control Point-based system (HACCP) and the microbiological safety of foods they produce depends on the consistently effective operation of the system. Sound microbiological information is important at each step of a HACCP study; the results of the HACCP study are then used to develop appropriate monitoring systems for the Critical Control Points. The monitoring systems employed usually enable a rapid response when process deviations occur, e.g. temperature measurements for cooking processes.

Although HACCP is conventionally understood as 'Hazard Analysis Critical Control Point', it is perhaps more meaningfully interpreted as 'Hazard Analysis and the Control of Critical Points'. In practice, it is a structured and documented approach to the identification of microbial, chemical or physical hazards in the food production chain and then, having determined the relevant hazards, to identify effective critical control points in the process and suitable control measures at these points that are then implemented with formalised in-process monitoring, record keeping and reaction procedures; all together these aim to prevent hazards from occurring.

The definition of a hazard has been broadened to include a biological, chemical or physical agent in, or condition of, food with the potential to cause an adverse health effect (Codex Alimentarius Commission, 1999).

In properly developed HACCP-based systems, each operative's role and responsibility is clearly defined. Reliable operation of such systems gives increased confidence that the level of safety required for all products made is maintained consistently.

For all food production operations, industry standards of Good Manufacturing Practice (now generally referred to as pre-requisite programmes) are required in the areas of:

• Factory production unit construction and biosecurity
• Process design and flow including effective segregation of activities as required
• Environment and equipment cleaning operations
• Personal hygiene and practices
• Elimination of cross-contamination risks between contaminated raw materials and finished goods

- Water and air quality
- Waste disposal.

   Informative guides to good manufacturing practices may be found in the Institute of Food Science and Technology (UK) publication *Good Manufacturing Practice – A Guide to its Responsible Management* (IFST, 1998) and the UK Chilled Food Association's *Guidelines for Good Hygienic Practice in the Manufacture of Chilled Foods.* (Chilled Food Association, 1997).

   From a theoretical point of view, effective implementation of a HACCP-based system of food production control should reduce the need for a laboratory to undertake microbiological testing of in-line materials because HACCP is a preventative system for managing hazards. This would be beneficial, as microbiological test results are almost never available in 'real time' so their use for in-process control is limited. However, the results from the microbiological examination of samples selected as relevant to particular points in a food production process can provide valuable information in support of an effective HACCP-based programme. Such results supplement physico-chemical in-process control information such as temperature measurements, and give added assurance that controls operated have been successful; i.e. they provide independent verification that the system is working as intended.

   The selection of process points relevant for microbiological sampling is determined as part of the primary process in developing a HACCP system, and microbiological tests are usually made to provide independent verification that the HACCP system is under control. Contributions to this process are likely to be required from food technologists, microbiologists, production personnel, engineers and other staff; in other words, it should not be a 'one person' process. All HACCP associated activities must be fully documented, results of all tests and monitoring systems recorded, defects or failures noted and remedial (corrective) actions documented by well-trained and accountable staff.

   Hazard analysis consists of an evaluation of all procedures concerned with the production, distribution and use of all raw materials and finished food products. The aim is to identify potentially hazardous raw materials, part-processed foods and finished products, e.g. ones that may contribute pathogens or allow their survival or growth and / or poisonous substances such as microbial toxins as well as food processes that may allow these hazards to develop. For some production purposes, identification of materials containing large numbers of food spoilage microorganisms or those that can support microbial growth may be required, but such materials would only be included for consideration where safety was concerned.

   By observing each step in the food production process and by considering and discussing practical issues with experienced production personnel, the sources and potential points of contamination can be identified and the potential for microorganisms to survive or multiply during production, processing, distribution, storage and preparation of food for consumption can be determined.

   Following the Hazard Analysis, the Critical Control Points (CCP) are identified; i.e. those parts of the process which, if not correctly controlled, could lead to microbial concerns such as the survival, growth or unwanted addition of microorganisms. Some common steps in a hazard analysis process include:

(1) After making a list of all the raw materials and associated sources of supply, the possible hazards associated with each one are assessed by a variety of means including testing, using the experience, knowledge and reputation of the supplier and the HACCP team members and / or discussion with other experienced colleagues and reference to the scientific literature.

(2) A step-by-step flow chart of the product manufacturing process is then drawn, where possible accompanied by a floor plan of the manufacturing area. It is essential to decide where the HACCP study should start and end for each study as this may affect the critical limits applied to the monitoring systems used at the Critical Control Points; for example, a study often starts with incoming raw materials but may end with despatch of the finished product to another man-ufacturer or a retailer, or consumption by the consumer.

Information that should be included for consideration in the study comprises the manufacturing equipment used, process physico-chemical information, e.g. target temperatures, acidity, humidity, time between process steps, holding locations and conditions for the raw materials, in-process materials and finished product, handling procedures, packaging, expected product shelf life, distribu-tion conditions and the expected conditions of use by the consumer.

A generic food production flow diagram is summarised in Table 5.1; this shows some of those points that need to be considered at each step and may require implementation or control and where microbiological information may be useful. Each step in the production flow should be considered in relation to the microbiological, chemical or physical hazard, if any, presented at that step to the safety of the process or finished product, or that may affect the safe shelf life of the product.

(3) From steps 1 and 2 above, the factors in the process that influence the growth, death, survival or extraneous contamination with microorganisms and the points at which these occur are listed.

(4) The consequences of a failure of any process stage described in step 2 are then considered. This is done in discussion with production, engineering and microbiology personnel as appropriate.

(5) From a consideration of steps 1 to 4, the Critical Control Points of the process are identified. The following are some of the areas that are frequently identified as Critical Control Points in a number of different manufacturing processes, together with an example of an approach to its control:

- *Raw materials:* e.g. it is the responsibility of food manufacturers to check the competence of raw material supply sources and particularly the sources of potentially hazardous materials that may require positive release (holding batches until satisfactory test results are obtained) into production based on the results of microbiological tests. This may not be always possible, due to the short shelf life of some ingredients and the cost of examining a statisti-cally significant number of samples. It is increasingly common (and good

Table 5.1 A generic food production flow and some points requiring consideration at each step.

Production step	Points requiring consideration	Microbiological information required*
Raw materials including packaging	Approved list of raw materials	
	Approved sources of raw materials	
	Supplier quality assurance programme, including audit	*
	Buying specification	
	Transport conditions	
	Microbiological certification (supplier provides microbiology report for each batch of raw material)	*
	Examination at point of receipt (visual, temperature, microbiological, chemical)	*
	Storage life	*
Raw material storage	Temperature	*
	Humidity	
	Pest control	
	Maintenance of storage area	
	Hygiene of storage area	
	Storage life	*
	Stock control	
	Records	
Raw material handling / preparation	Used within life, allowing for the shelf life of the product in which it is used	*
	Defrosting method / control for frozen materials	*
	Food materials washing procedures	*
	Chilled materials kept out of chill for minimum time	*
	Hygiene (personnel and equipment)	*
	Visual checks	
	Analytical checks	*
	Records	
Product recipe	List of materials	
	Weight of each ingredient	
	Check-weigh balances / scales	
	Liquid volume measurement and checks	
	Visual checks	
	Quality control and laboratory checks	*
	Records	
Mixing sequence	Process flow	
	Equipment used	
	Equipment maintenance system	
	Equipment cleaning	*
	Sequence of ingredient addition	
	Temperature of mixing	*
	Times of mixing stages	*
	Storage / holding times of mixes and conditions	*
	Visual checks	
	Analytical checks	
	Records	

Contd.

Table 5.1 *Contd.*

Production step	Points requiring consideration	Microbiological information required*
Processing, e.g. heating, fermentation	Process flow	
	Equipment used	
	Equipment maintenance system	
	Equipment cleaning	*
	Process temperature, e.g. cooking, pasteurisation	*
	Process times, e.g. cooking, drying	*
	Hygiene	*
	Visual checks	
	Analytical checks	*
	Records	
Post-processing e.g. slicing, product assembly	Process flow	
	Storage / holding times and conditions	*
	Equipment used	
	Equipment maintenance system	
	Equipment cleaning	*
	Temperatures	*
	Handling hygiene	*
	Cross-contamination control	*
	Visual checks	
	Analytical checks	*
	Records	
Product packing	Process flow	
	Equipment used	
	Equipment maintenance system	
	Equipment cleaning	*
	Packaging materials and method; specification, integrity and hygiene	*
	Temperatures	*
	Handling hygiene	*
	Cross-contamination control	*
	Product labelling with consumer instructions	*
	Product shelf life, allowing for the life of any perishable raw materials used, e.g. cream in cream cakes	*
	Visual checks	*
	Analytical checks	
	Records	
Product storage	Store conditions	*
	Temperature / humidity	*
	Pest control	
	Maintenance of storage area	
	Hygiene of storage area	
	Timescale	
	Visual checks	
	Records	

Contd.

Table 5.1 *Contd.*

Production step	Points requiring consideration	Microbiological information required*
Distribution	Transport and depot conditions	*
	Temperature / humidity	*
	Pest control	
	Maintenance of vehicles and depots	
	Hygiene of vehicles and depots	
	Timescale	
	Visual checks	
	Records	
Retail outlet	Storage conditions (maintenance and hygiene)	*
	Temperature / humidity	*
	Timescale	*
	In-store handling methods and hygiene, e.g. slicing cooked meats on delicatessen counters	*
Customer / consumer	Instructions for storing product (temperature, shelf life)	*
	Instructions for handling products, e.g. cooking, warming	*
	Warning to vulnerable groups	*

practice) for food manufacturers to source ingredients from suppliers who have achieved 'approved supplier status' through a supplier safety and quality assurance scheme based upon audit, sample examination and reputation, as appropriate. Raw material buying specifications agreed between the supplier and the purchaser can be used to limit identified hazards to acceptable levels and the critical raw material microbiological safety criteria can be checked against the buying specification, e.g. by supplier certification (certificates provided by the supplier containing microbiological information relating to the batch delivered) and on-delivery inspection by the purchaser.

– *Cooking processes:* e.g. minimum time–temperature combinations required to achieve the correct cook may need to be specified.
– *Chilling procedures:* e.g. the maximum time permitted for chilling the product to a particular temperature may need to be specified.
– *Process timescales:* e.g. the maximum time delay permitted between certain process points may need to be specified.
– *Post heat-process conditions and handling procedures:* e.g. methods may need to be specified to prevent re-contamination of cooked or pasteurised products from personnel and equipment.
– *Packaging:* e.g. the type, handling conditions and treatment of packaging materials may need to be specified such as for aseptically filled dairy products.
– *Personal hygiene:* e.g. particular footwear and / or overall changing procedures may be required in high-risk / high-care product areas such as for sandwich making.
– *Equipment hygiene:* e.g. particular methods and frequencies of cleaning may need to be specified.

– *Equipment maintenance:* e.g. a planned maintenance programme for certain pieces of equipment or parts of equipment may need to be specified.
– *Environmental hygiene:* e.g. methods of cleaning the environment that will prevent the spread of microbial contamination to food contact surfaces may need to be specified.
– *Product physico-chemical composition:* e.g. minimum rates of acidity development, decrease in pH value and water activity decrease may need to be specified for production processes such as for raw, fermented dry-cured meats and cheese,
– *Shelf life allocation:* e.g. microbiological challenge tests and / or shelf life studies may be required to demonstrate the microbial stability and safety of the product throughout the proposed shelf life especially when the potential growth of *Clostridium botulinum* is a concern.
– *Customer instructions:* e.g. adequate cooking procedures for raw meats may be required and need to be specified on the packaging.

(6) Following Hazard Analysis and CCP identification, appropriate CCP monitoring systems must be determined. Visual, physical and chemical, but rarely microbiological, tests are used for CCP monitoring purposes; however, as far as practically possible the monitoring methods employed should provide rapid information feedback so that rapid responses to any adverse change can be made if required, for example:

- Visual: such as hand-washing practices, equipment cleaning efficacy
- Physical: such as temperature and time measurement, weight and volume checks
- Chemical: such as measurement of pH values.

If microbiological tests are selected for monitoring CCPs, then it is important to ensure that the numbers and types of samples to be taken, sampling points and frequency, sampling method, test organism(s), test method(s), acceptable limits for test results and the reaction procedure to any out of specification results are all clearly specified.

Clearly it can be seen that microbiological information, which may or may not include practical sample examination, is important at every stage of the HACCP procedure and thereafter for some verification purposes.

5.3   Risk assessment and microbiology

National government departments and international bodies, e.g. Codex Alimentarius Commission, European Union, USA Food and Drug Administration, whose work concerns matters of food safety in relation to human health are increasingly applying the process of risk assessment for determining appropriate policies for the protection of consumers. As the microbiological risk assessment process becomes more fully developed, some food industry microbiologists and technologists are also starting to take account of microbiological risk assessment information and processes to strengthen the HACCP-based systems in operation and to help confirm the appropriateness of the manufacturing process controls selected and the assigned safe shelf life of the finished products.

The most widely accepted guidelines, although a number are still in draft form, in respect of microbiological risk assessment and management are those of the Codex Alimentarius Commission. Risk assessment is defined by this Commission (Codex Alimentarius Commission, 1999) as a scientifically based process consisting of:

(1) Hazard identification
(2) Hazard characterisation
(3) Exposure assessment, and
(4) Risk characterisation.
(See Glossary for definitions.)

Although these specific descriptive terms have only recently come into general use, the general processes involved in risk assessment, facilitating risk management, have really been applied by scientists since the early days of work in the area of food microbiological safety, e.g. the development in the 1920s of adequate heat processes to destroy the spores of *Clostridium botulinum*, which followed the recognition of the link between human cases of botulism and the consumption of inadequately heat-processed bottled and canned foods in which the spores of the organism survived, grew during subsequent storage and produced toxin. In the 1930s, pasteurisation was applied to destroy the agents of a variety of serious human diseases recognised as being milk-borne at that time including those of bovine tuberculosis, diphtheria, scarlet fever, dysentery and acute gastro-enteritis. Later, in the 1980s, the food industry successfully responded to the 'new' pathogen challenge of *Listeria mono-cytogenes*. Once identified as a particular problem, industry and public health microbiologists commenced extensive monitoring of food raw materials, production environments, processes, practices and products for the presence of *Listeria* spp. including *Listeria monocytogenes*. Research provided information and an under-standing of the organism which, combined with the results of work programmes undertaken by the food industry, assisted the development of the control strategies adopted by food manufacturers today. The controls applied resulted in the incidence and levels of *Listeria monocytogenes* being significantly reduced and the continued application of controls by food manufacturers now keeps *Listeria monocytogenes* at a consistently low incidence and level in many processed foods.

Microbiological risk assessment of foods is complicated by a variety of issues (Table 5.2), for which full and complete answers are not, and are probably unlikely ever to be, known. The unknowns indicated in the table also greatly limit the appli-cation of Quantitative Risk Assessment to food safety matters. This, in particular, is because of the current impossibility of obtaining accurate and reliable quantitative data concerning the level (numbers) and distribution of an identified microbial hazard e.g. *Salmonella* spp. in a particular food, including reliable knowledge of the changes in levels of the microbial hazard throughout the assigned shelf life of the food. The latter point is important because the potential exposure of the consumer to the microbial hazard in the food may vary depending on how much of the shelf life of the food has elapsed when the product is consumed.

It is clear from a scrutiny of the definitions of the terms relating to risk assessment

Table 5.2   Unknowns and complications associated with the Microbiological Risk Assessment for foods, adapted from International Life Sciences Institute (1993).

Unknowns
The changes in numbers and types of microorganisms during processing, storage, handling, and preparation for consumption of the food, including effects of cooking
The numbers and distribution of microorganisms within any food at any time
The range of consumer practices
The range of effects of microorganisms on the individual host (consumer)
The range of effects of individual foods on both microorganisms and consumers
Complications
The ability of a single microorganism to cause a wide variety of disease conditions
Person-to-person transmission of infectious microorganisms
Principal consideration is given by both the food industry and public health authorities to the short-term rather than long-term effects of infection, but some food-borne illnesses may have long-term sequelae (morbid condition occurring as a result of the previous illness)

that real and reliable data from the microbiological examination of food materials concerning the types, incidence and numbers of microorganisms (crucially, those pathogenic to humans) present in any food raw material and final food product is essential for obtaining the maximum benefit from microbiological risk assessments.

5.4   Raw food materials / ingredients and microbiology

Food raw materials are now sourced from countries all around the world and most of these materials are now also available throughout the year instead of, as in the past, being seasonally restricted. Food materials not native to many countries are now increasingly available to them, although some foods may still be regarded as 'exotic'. The whole range of food commodities including cereals, sugar, cocoa, poultry and eggs, game birds and game animals, milk, vegetables and salad vegetables, fish and shellfish, oils and fats, red meats, fruit, nuts, herbs and spices are used in increasingly novel combinations to meet commercial and consumer demands.

A number of bacterial pathogens have, from time to time, been associated with food-borne illness, often severe, involving most of these food commodities. Table 5.3 indicates some of these. It has long been considered prudent to assume that all raw food materials may, on occasion, be contaminated with low numbers of bacterial pathogens, albeit that the pathogen types will vary according to the material involved, its source and growing and harvesting or farming conditions. Therefore, sources of supply should be selected that employ and maintain the highest standards of aquaculture, agricultural and / or animal husbandry practices to minimise the incidence and levels of pathogens in the raw materials produced. Raw material supplier quality assurance programmes that include audits of supplier sites (including agricultural

Table 5.3   Some examples of bacterial pathogens and some food types with which they have been associated in outbreaks of food-borne illness.

Bacterial pathogen	Food commodity
Bacillus cereus	Rice
Campylobacter jejuni	Poultry Milk
Clostridium botulinum	Vegetables Fish, e.g. salmon, trout Peanuts Hazelnuts Honey
Clostridium perfringens	Vegetables Red meat
Entero-haemorrhagic *Escherichia coli*	Red meat Fruit (juice) Milk and milk products Salad vegetables
Listeria monocytogenes	Milk (cheese) Meat products (pâté)
Salmonella spp.	Coconut Pepper Poultry meat Chicken eggs Fruit and fruit juices
Vibrio parahaemolyticus	Shellfish

sites) and practices as well as appropriate microbiological examination of samples are valuable elements in the supply selection process.

Once a supplier is selected, a raw material buying specification is drawn up between the vendor and the buyer and this usually includes a full description of the food material to be purchased, appropriate packaging materials, delivery and storage conditions. Shelf life information is also included which is particularly important for perishable raw materials used in products without a further microorganism-controlling process applied. Also, key chemical and microbiological data should be specified including, as appropriate, acceptable and unacceptable levels of specific pathogens, indicator organisms and spoilage organisms and sampling levels (numbers and frequency) to be undertaken by the supplier and microbiological examinations to be carried out.

A raw material supplier may be required to provide a Certificate of Analysis (sometimes referred to as Certificate of Conformance or Certificate of Examination) with each batch of material supplied. This is a formal certificate which usually records the results of chemical analyses or microbiological examinations carried out on samples of the batch of raw material at the point of production or may just indicate that the batch conforms to an agreed specification that is shown on the certificate. The specification against which the results are judged is agreed between the vendor and buyer (See Section 5.7.1). Such certificates are most often required for raw materials regarded as high risk, e.g. liquid egg products, coconut, spices, or even for processed materials such as cooked meats and boiled eggs destined for inclusion in sandwiches, or chocolate for making confectionery products, all of which have, from time to time, been implicated as sources of contamination with *Salmonella* spp.

When raw materials are received onto the manufacturing site of the purchaser, they may be examined visually, temperature checked and sampled for microbiological examination or other purposes as they are being unloaded from the delivery vehicle or once they are placed in the appropriate raw material storage area. The results of any microbiological tests carried out on samples taken at this point can provide a useful monitor of the regular microbiological quality of the incoming materials. The results can be used for trend analysis that can provide an early warning of microbiological problems developing in the supply and, in discussion with the vendor, serious problems may be averted and a high microbiological quality of supply maintained.

5.5   Hygiene monitoring and microbiology

For food industry purposes, a useful definition of hygiene adapted from *The Shorter Oxford English Dictionary on Historical Principles* (1973) is 'practices relating to the maintenance of health'. 'Practices', in the context of a food-processing unit, are the cleaning and disinfection procedures applied to equipment, people and the environment.

Cleaning procedures are developed and operated to achieve a standard of cleaning and disinfection that will ensure that the cleaned surface will not represent a source of microbial contamination to the food materials being processed.

Cleaning procedures are structured to ensure that all food contact items and parts of equipment are accessible for cleaning. This often requires equipment to be dismantled. Cleaning staff have to be properly trained and supplied with the necessary means to carry out their job effectively, e.g. cleaning equipment, adequate hot water supplies, cleaning and disinfection chemicals.

Following any dismantling of equipment that may be required, the cleaning process for each item of equipment usually involves four stages that are carried out manually:

(1)   Rinsing with clean water to remove gross debris
(2)   Washing with hot detergent and physical brushing
(3)   Rinsing with clean water
(4)   Applying a disinfectant / sanitising agent.

In industries where there is a considerable amount of pipework, e.g. milking parlours and dairies, breweries, soft drinks production units, cleaning-in-place (CIP) systems are employed. These automated systems operate the required sequence of cleaning steps via electronically controlled storage tanks, pumps and chemical dosing systems. Some equipment, e.g. in pastry making areas, is not generally wet-cleaned. In these cases, dry-cleaning methods, such as brushing and vacuuming, are used to remove dust and debris.

Monitoring of the effectiveness of cleaning procedures is usually done by visual assessment and by microbiological testing. In the latter case, there are two main approaches:

- Sterile swabs (large sponge swabs or small medical swabs depending on the area / surface to be swabbed; Figure 5.1) are used to capture debris and microorganisms from a specified area or object. The swab is then examined using enumeration or detection tests for specified organisms.
- Samples of final rinsing water or the first quantities of product passing through the cleaned pipes and nozzles, or being transported along conveyor belts etc. are taken for microbiological examination.

Figure 5.1   Examples of different types of swabs used for sampling surfaces and environmental areas for microbiological contamination (courtesy of Bibby Sterilin Ltd.).

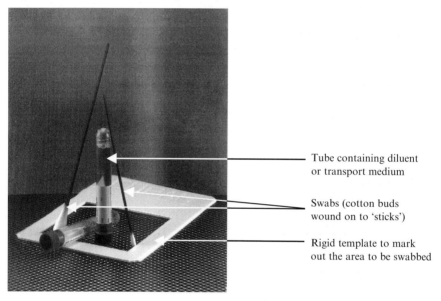

Tube containing diluent
or transport medium

Swabs (cotton buds
wound on to 'sticks')

Rigid template to mark
out the area to be swabbed

Notes:
- The swab type used will depend on the specific application required and the laboratory protocol applied.
- The swab diluent used is (normally) Maximum Recovery Diluent but diluents containing specific neutralising agents may be required (see Table 9.4).

5.5.1   *Swabs*

Swabs normally consist of cotton wool wound around the end of a plastic stick to form a pad or 'bud', or sponge pieces attached to the end of a plastic stick. These are placed into unbreakable tubes with an air-tight seal and sterilised by irradiation (Figure 5.1). Swabs are used to assess the hygienic status of food processing equipment and the production environment. Where swabs are used, the swabbing technique has a significant effect on the result obtained. Because of this personnel required to carry out such work must be properly trained in the correct techniques.

In order to gain the best assessment of the microbiological status of the surface, swabs should be used in a consistently correct manner as follows:

(1) Lightly moisten the swab in a sterile diluent or neutralising solution (contains substances that neutralise any residual disinfectant chemicals present, see Table 9.4) – swabs should not be allowed to drip moisture onto the surface; if necessary squeeze any excess moisture from the swab by pressing and rolling it around the inside surface of the diluent / solution container.

(2) Select an area for swabbing, e.g. $16\,cm^2$ or $100\,cm^2$. A template may be used, see Figure 5.1, or the swab tip can be used to lightly 'draw' around the area to be swabbed to provide a guide for swabbing.

(3) Lay the swab as flat to the surface as possible to ensure maximum contact between the surface and the swab then, maintaining a light pressure (sufficient to cause a slight bend in the stick), rub the swab over the surface while, at the same time, rotating it. In this way, the swab is rubbed over the entire selected surface in two directions (Figure 5.2).

(4) Return the swab to its container for transporting to the laboratory for examination ensuring that the container is labelled with sufficient information for identification and traceability purposes.

It is often necessary to swab areas that are not flat surfaces, e.g. pipes, nozzles, seals, joints between walls and floors, etc. In these cases, it is often not possible to

Figure 5.2   Diagrammatic representation of the technique for swabbing surfaces.

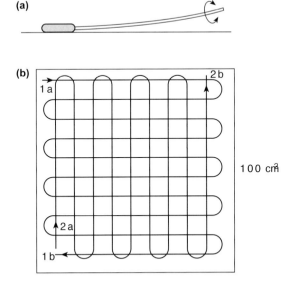

Key points:
- A light pressure should be maintained on the swab stick sufficient to cause a slight bend in the stick and the swab handle should be rotated constantly (see (a) above) during the action of swabbing to ensure that all surfaces of the swab are used.
- The area swabbed should be swabbed in two directions (starting from point 1a and finishing at point 1b, followed immediately by swabbing from point 2a to 2b; see (b) above) using even strokes of the swab as indicated.

prescribe a particular area for swabbing but a procedure should be specified that is relevant to each type of area to be swabbed to establish consistency between the people who do the swabbing and between the different occasions on which the task is carried out. For example, the internal surface of a pipe could be swabbed to a depth / length of 5 cm and the swab rotated around this area 10 times, a floor / wall junction could be swabbed for a length of 100 cm taking the swab over this length 3 times in each direction. Whatever approach is specified for these areas, the swab should be handled in the same way as for the flat surfaces, i.e. light pressure maintained together with rotation to ensure that the entire swab surface is used.

Swabbing is a flexible technique that can be used in the assessment of the hygienic status of any item in the food-processing environment. In addition to food contact surfaces, such as tables, conveyor belts, cutting blades, mixer bowls, depositors, hoppers and a variety of utensils (knives, ladles, scrapers, etc.), all environmental areas can be swabbed including:

- floors, walls and ceilings
- drains
- floor contact items, such as wheels, footwear, cleaning equipment
- overhead pipes, girders and light fittings, and
- chiller unit evaporator fins.

Another technique sometimes used is the 'contact plate' also known as RODAC (Replicate Organism Direct Agar Contact) plates. Contact plates are small diameter Petri dishes (plates) (see Plate 13(b), see colour plate section for all plates) that are carefully filled with an agar medium (selected to detect the target organisms sought) so that the surface of the medium forms a convex shape above the level of the edge of the plate; these are used specifically for the microbiological sampling of flat surfaces by gently pressing the agar surface directly onto the surface to be sampled. After sampling, the lid is replaced, the plate labelled, incubated and assessed for microbial growth.

5.5.2   *Solid or liquid samples*

Physical (solid or liquid) samples may also be taken as part of hygiene monitoring programmes in agricultural as well as food manufacturing units.

Samples of the final rinse water from a cleaning operation can be taken and specific volumes either used directly in or on microbiological growth media, or filtered after which the organisms captured on the filter are detected or enumerated as required. Alternatively, the very first food material of a day's production to pass through a processing system is commonly taken for examination. As this is the first material to come into contact with the equipment surfaces after cleaning, it can provide a practical assessment of the efficacy of the cleaning process.

When assessing the production environment for the presence of microbial hazards, it is possible and often preferable to take physical samples where these can be obtained, e.g. using sterile, plastic disposable pipettes to take liquid samples from drains or puddles on the floor; using sterile spatulas or spoons to take debris from the treads of wellington boots, accumulated debris on floors or accumulated debris on / in production equipment. Debris in / on the equipment and tools used for cleaning, e.g.

the contents of vacuum cleaners, debris trapped in brushes and mops / squeegees, are also useful sources of physical (solid) samples. Such gross samples are particularly useful for assessing food production environments for the presence of pathogens such as *Listeria monocytogenes* or *Salmonella* spp.

5.5.3 *Personnel*

In addition to equipment and environmental swabbing, hand-swabbing is often used as a training aid in programmes dealing with personal hygiene and thereafter to monitor hand-washing practices.

For this purpose, medical swabs are generally used, applying the techniques for swab sample collection already described. Care must be taken to ensure that the entire palm of the hand is swabbed, from the finger-tips to the wrist and especially between the fingers.

Hand-rinsing techniques are occasionally used in which the person being assessed places one or both hands into a bag containing a volume e.g. 100 ml of sterile diluent. The hand is massaged to remove surface organisms then removed from the diluent which is submitted for microbiological examination.

Important note: Personal hygiene and practice is an emotive subject, so assessing personnel in this way requires tact, sensitivity and care in reporting and discussion with the individual, especially when unsatisfactory levels of microbial contamination are detected. The support of senior management, medical advisors and the personnel themselves is essential if such hygiene programmes are to work effectively. The results from these assessments must be interpreted and reported with great care and should always involve senior staff.

5.5.4 *Air*

In many food production environments, it is important that air quality meets particular hygiene standards, e.g. for the production of bakery goods that are at risk from mould spoilage. For these environments, programmes are established for the regular monitoring of air for microbiological quality.

Monitoring air for microbiological contamination is done either using settle plates (also known as exposure plates) or with air sampling machines. Settle plates are Petri dishes containing an agar medium suited to the organisms of interest, that are left exposed in the relevant area for a specified time period. Particles and microbial cells in the air fall (settle) onto the agar surface. After the required exposure time e.g. 1 hour, the lids are replaced and the plates are incubated under appropriate conditions for the target organisms. Following incubation, the number of colonies that have grown is counted. Air sampling machines require a Petri dish (plate) containing a suitable agar medium to be inserted into the machine, which is designed and operated to draw a pre-set volume of air at a specific velocity over the plate. Organisms in the air stream are captured by impact on the agar surface; plates are then incubated and colonies are counted.

Results obtained from these tests are usually used in trend analyses as it is very difficult to set general criteria that would apply to different situations. Should the trend exceed in-house determined limits, action is taken to reduce the microbial load in the air, e.g. change air filters or clean overhead structures.

5.5.5 *Test results and their interpretation*

Each food and beverage production company has its own schedules and methods for assessing the efficacy of equipment, environmental (including air) and personal hygiene programmes. Alongside these, internal systems should be in place for interpreting and reacting to the results from assessments. It is important that the systems to be applied are clearly established and documented because results relating to different surface types, e.g. stainless steel, polypropylene, rubber, and different environments, e.g. high-risk or low-risk, bakery or cannery, etc. may well require different interpretations and reaction procedures.

It is essential to ensure that all methods used e.g. swab, rinse, gross sample are validated for the surfaces / items under examination and that the tests carried out for indicator organisms or specific groups / organisms and interpretation of results are meaningful in relation to the surface / item of equipment / area examined.

5.6 Process monitoring and microbiology

Table 5.1 indicates some key points in a food production process flow and where microbiological information can be of particular value.

As discussed previously, real-time, rapid methods for monitoring production processes, such as temperature, pH value, time and visual assessment, are preferred because the results can be acted on immediately in the event that a process does not meet the required specification. Physical measurements can also be automated and visual or audible alarms can be linked to the measuring equipment to provide a warning when the process parameters are not met.

In addition to the areas indicated in Table 5.1, there are points in processes where microbiological assessments can provide very useful supportive information concerning the efficacy of particular processes. Table 5.4 indicates some examples of these in relation to different types of food production operation. Because microbiological information is generally retrospective, taking at least hours and, more frequently several days to produce results, its main value is to provide confirmation that the systems in place have operated correctly and that microbial growth is properly controlled; this information forms part of the HACCP verification data necessary to demonstrate that the control measures implemented are effective.

There are, nevertheless, a few microbiological test methods that can provide results in minutes. Of such tests, the most widely used are those employing ATP-bioluminescence reactions for monitoring cleaning efficacy. Results from ATP-bioluminescence tests can be available within minutes of the swab being taken, thus allowing immediate action to be implemented, e.g. re-cleaning, if results indicate an 'out of specification' level and they can provide an important tool for food hygiene management. These methods are discussed in more detail in Chapter 12.

Table 5.4   Examples of production process efficacy assessment.

Process stage	Physical / chemical assessment	Microbiological assessment (verification)
Preparation process Mixing, chopping, injecting	Distribution of ingredients by measuring levels of additives, e.g. salt, nitrite	None usually applied
Use of starter cultures	Rate of acidity development	Confirm identity of organisms Activity tests (rate of acid production)
Washing procedures, e.g. vegetables, fruit	Disinfectant e.g. hypochlorite level Water flow rate	Monitor quality of primary water supply Monitor microbial load development / control in wash water
In-process cleaning	Visual Chemical concentration / strength Water temperature and flow rate	Swab tests Monitor quality of final rinse water
Further processing Cooking / pasteurisation	Temperature and time	Test for the presence and level of specified organisms after the process, e.g. Total Enterobacteriaceae after pasteurisation processes – none should survive
Fermentations	Acidity, pH value, water activity	Microbial population (fermentation culture) development Monitor levels of specific indicator organisms and pathogens to ensure control of growth
Cooling / chilling	Temperature	Air quality of the environment Test for the level of growth of residual organisms in recycled water systems Microbiological quality of water used for container cooling
Slicing	Temperature Visual – debris accumulation	Cleanliness of equipment and people Air quality of the environment Microbial load addition / development during the timescale of the slicing operation
Depositing / filling	Temperature Visual – debris accumulation	Cleanliness of equipment Air quality of the environment Microbial load development during the timescale of the operations
Transfer systems	Time Temperature	Cleanliness of equipment and people Air quality of the environment Microbial load development during the timescale of the operations

Contd.

Table 5.4 *Contd.*

Process stage	Physical / chemical assessment	Microbiological assessment (verification)
Packing and finished product		
Handling	Visual checks	Cleanliness of equipment and people
Packaging	Visual checks	Swab or rinse tests on packaging for aseptically filled products
Product	Temperature	Microbial load at the end point of production
Transfer systems e.g. conveyors	Time Temperature	Cleanliness of equipment Microbial load development during the timescale of the operations
Chill / freezer stores	Temperature	Microbial load development during storage (as applicable)

5.7   Finished products and microbiology

Even where food production processes have been developed and operated reliably according to HACCP-based principles, the finished product may still be subjected to a variety of microbiological tests for different purposes.

5.7.1   *Conformance to microbiological criteria*

Microbiological criteria are drawn up in relation to specified foods and consist of:

- a statement of the microorganisms and / or their toxins of concern to the food;
- the methods to be used for their detection and / or enumeration;
- a sampling plan;
- the microbiological limits to be used in assessing the results;
- guidance for interpreting the results; e.g. the number of sample units that should conform to the limits (International Commission on Microbiological Specifications for Foods, 1986).

Microbiological criteria are often used in purchase agreements, e.g. between a raw material supplier and manufacturer or a manufacturer and retailer. They are important documents, as failure to comply with the agreed product description and criteria can lead to rejection of the product by the purchaser and serious commercial losses to both parties.

Microbiological criteria in such agreements may include limits for pathogens, microbial toxins, indicator microorganisms or spoilage microorganisms and, as a minimum, the criterion statement should include notes on the following:

- The frequency of sampling, the point(s) at which samples should be taken, e.g. from the processing or packing line, from the finished product store, etc. and the numbers of samples taken.
- A clear description of the food / sample type to be examined, e.g. individual components or a composite test sample made up from portions of all the com-

ponents of the factory sample. This is especially important for multi-component foods such as recipe dishes that may contain layers of different ingredients, e.g. meat sauce and potato topping.

- A clear description of the relevant microorganism(s) or microbial toxin(s) of concern.
- An unambiguous sampling plan, i.e. a statement of the number and type of factory samples to be taken and quantities to be tested (this is usually derived from the hazard analysis) and realistic microbiological limits above which corrective action is required and will be taken.

Sampling plans in which the enumeration of organisms is required are often formatted according to the three-class plan structure described fully by the International Commission on Microbiological Specifications for Foods (1986). This uses n, m, M and c to describe the number of test samples examined and the acceptable limits, e.g. a typical criterion for coliform organisms enumerated using violet red bile agar, a pour plate technique with agar overlay, incubated at 37°C for 24 hours, in a sandwich containing salad vegetable components might be, $n = 5$, $m = 100$ cfu/g, $M = 1000$ cfu/g, $c = 2$, where

n = the number of test samples examined
m = the maximum level of target organism(s) acceptable under conditions of good manufacturing practice
M = the level of target organism(s) which, if exceeded, is considered unacceptable
c = the maximum number of test sample results that can fall between m and M without the batch being considered unacceptable.

The stringency of a sampling plan depends on n and c, e.g. in the example above, if c is increased to 3 or 4, the plan becomes more lenient; if c is decreased to 1, then the plan becomes more stringent. For foods destined for vulnerable groups of the population, n may be increased to 10 or, in extreme cases, 60 or more.

A clear reporting procedure for test results and a clear action procedure in the event that results show that the product does not comply with the specified criteria are also necessary, and these are normally included in the company technical documentation.

The majority of microbiological testing carried out in food industry or associated laboratories is done to demonstrate conformance to specific microbiological criteria. It is essential, therefore, that the methods and techniques used for this work have been properly validated for the microorganism(s) or toxin(s) of concern in the specific food type being examined (see Chapter 10).

5.7.2 *Product shelf life evaluations*

Food manufacturers must ensure that all of the different products they produce will remain safe and wholesome for the entire life allocated under the specified storage conditions.

It is usual for shelf life studies to be carried out at the product development stage, as the shelf life is a key parameter written into a final product / buying specification. A variety of parameters is used to determine the product quality specification and shelf life, e.g. deterioration in specific organoleptic qualities (flavour, aroma, appearance, texture), microbial growth (bacterial pathogens, bacterial indicator organisms and spoilage organisms), rate of preservative depletion, etc.

Shelf life studies carried out during the development of perishable chilled products usually involve storing finished products both at the specified storage temperature and at a higher, 'abuse' temperature; for example, these could be 5°C and 9°C respectively, however, all companies should have their own regimes for specific product shelf life studies. Products are removed from the storage units periodically throughout the study period and examined for key groups or specific types of microooganisms considered to be relevant as derived from the Hazard Analysis. Shelf life studies are normally carried out three or more times to establish a reliable microbiological profile (the change in numbers of general or specific microbial types over the storage period) that can be used to allocate a commercially acceptable life to the product that ensures its quality and safety.

Table 5.5 shows some examples of different product types within each of the main categories of product shelf life storage condition, i.e. ambient-stable, chilled or frozen conditions.

Products made to be stable under ambient conditions are generally less susceptible to deterioration under adverse storage conditions than food intended for chilled storage. For ambient-stable foods, if all the process control data and final product physico-chemical data show that the process was fully satisfactory then, routine shelf life monitoring and evaluation work is normally limited to visual and organoleptic studies. Examples of such foods may include:

- dried foods,
- canned, bottled or foil-packed, low-acid goods receiving a minimum heat process equivalent to 121°C for 3 minutes ($F_0 = 3$),
- acidic foods, and
- pasteurised and glass-packed foods, e.g. conserves and vegetables in oil, or fermented foods, that have a water activity of 0.85 or less.

However, canned and bottled foods and beverages are also often subjected to 'accelerated incubation' tests, i.e. a percentage of the production batch is incubated at an elevated temperature (25–55°C depending on the product type and intended distribution locations) and, after the pre-determined incubation time, the packs are visually examined for signs of 'blowing' (distension due to gas production caused by microbial growth within the pack), or they may be opened and the contents examined microbiologically, chemically and visually.

During the routine production of chilled foods, in addition to the samples taken for examination against the conformance criteria, a number of product packs are usually taken from each batch produced and kept for at least the full, allocated shelf life under the specified storage conditions. These may be examined microbiologically both during and at the end of the storage period to ensure that the expected normal microbiological profile established at the product development stage is obtained. Most chilled foods are sold and often consumed before the microbiological test results are available. The main use and application of the results however, is for confirmation

Table 5.5 Examples of different product types within each category of product shelf life storage condition.

Storage condition	Products
Ambient-stable	Dried soup mixes Dried baby foods Canned, condensed and evaporated milks Canned meats and fish Canned vegetables Fermented, dry-cured meats (water activity 0.85 or less) Foil packed ready-meals (heat process $F_0 = 3$ or greater) Cereal products Bread, cakes and biscuits Rice and pasta (dry) Savoury snacks, e.g. crisps, nuts Confectionery goods Wine, beer and soft drinks Ultra high temperature (UHT) treated milks
Chilled	Sandwiches Pasteurised milks Dairy desserts and yoghurts Cream cakes Cheese Ready-meals (prepared meals) Raw meats, sausages Raw poultry products Raw and smoked fish products Shellfish Pies and savoury bakery products Some canned, pasteurised, cured meats Pastry Prepared salad vegetables Delicatessen foods, e.g. cooked sliced meats
Frozen	Meat and poultry Vegetables Ice-cream and dairy dessert products Fish products Pastry and pastry products Ready-meals (prepared meals)

that the process parameters were achieved and for trend analysis purposes, whereby any trend away from target levels can be investigated and corrective action taken, where necessary, before any results reach unacceptable levels.

Frozen foods maintained at temperatures colder than $-18°C$ are microbiologically stable because most bacteria do not grow below $0°C$ and only very psychrotrophic organisms, e.g. some fungi, will grow, but only slowly, at a few degrees below the freezing point of water. Therefore, the shelf life of frozen foods is usually determined by physical, e.g. dehydration, and chemical / enzymatic changes, e.g. oxidation or lipolysis of fatty components, and not by microbial growth. However, it is well known that many microorganisms are preserved by freezing and *Salmonella* has been recovered from frozen foods after years of storage. Neither will freezing inactivate bacterial toxins but it does reduce the infectivity of some pathogenic protozoa and nematodes in protein foods such as raw meat and fish (International Commission on Microbiological Specifications for Foods, 1980). To maintain the quality and safety of frozen foods, freezing and thawing processes (where applicable) must always be strictly controlled and microbiological examination of relevant samples can be useful for monitoring these processes. Because of the long shelf life applied to most frozen foods, microbiological test results are sometimes used in a positive release system for quality assurance purposes, e.g. for frozen cream cake products, ice cream and egg products.

The microbiological work carried out during shelf life studies at the product development stage can also help to confirm the relevance of the organisms determined during the Hazard Analysis to the product's safety and durability (microbial spoilage-free life) and thus, those organisms that will be of value for inclusion in the product specification.

5.7.3 *Microbiological challenge testing*

Microbiological challenge tests are undertaken to establish the safety and shelf life of some products by assessing the potential for growth of specific pathogens that may be present from the raw materials and / or toxin production by such organisms. This is important because shelf life is sometimes limited by the potential for pathogen growth, for example,

- *Salmonella* or toxigenic *Escherichia coli* e.g. Vero cytotoxigenic *E. coli* (VTEC) in raw beef used for salami production;
- *Salmonella* or VTEC in raw milk used without pasteurisation for cheese manufacture;
- *Listeria monocytogenes* in salad vegetables used for sandwiches or prepared salad dishes;
- *Bacillus cereus* in rice used for cooked rice dishes;
- *Clostridium botulinum* in raw fish used for cooked, vacuum-packed fish products.

Because of the generally hazardous nature of the organisms used in these tests, the work is usually sub-contracted by the food manufacturer to external laboratories that are not situated on production premises, e.g. private contract testing laboratories or centralised company laboratories, and that have the necessary facilities and trained food microbiologists who are experienced in designing appropriate studies for the product types under examination.

In these tests, varying levels of the organism of concern, e.g. *Listeria monocytogenes*, toxigenic *Escherichia coli*, *Salmonella* spp., enterotoxin-producing staphylococci, *Bacillus cereus* or spores of selected types of *Clostridium botulinum* are inoculated into packs of the food under consideration. These are stored together with un-inoculated control packs under controlled-temperature conditions and examined at various time intervals to determine whether the organism has grown, died or just survived.

Such challenge tests have been commonly used to verify processes for producing long life, ambient-stored, canned foods using a variety of spore-forming organisms in the inoculum but, new vacuum-packed, modified atmosphere-packed and other perishable chilled foods are being subjected to challenge tests. This is partly because of the commercial pressure to extend the assigned shelf lives of such foods and, although commercial refrigeration conditions are generally well-controlled, temperature control of domestic refrigerators is far more variable. As a product may be purchased by the consumer within a few days of production and stored in a domestic refrigerator for most of its shelf life (which may be several weeks) the product may be exposed to poor refrigeration conditions for a large proportion of its shelf life.

Hence, it is necessary to ensure that psychrotrophic strains of the spore-former, *Clostridium botulinum* or vegetative pathogens that may grow slowly at chill temperatures, e.g. *Listeria monocytogenes,* will not compromise the microbiological safety of the products during the entire period of the required shelf life. These tests are especially important for chilled products that are assigned a shelf life longer than 10 days and that are capable of supporting the growth of these organisms (Advisory Committee on the Microbiological Safety of Food, 1992).

Following outbreaks of food-borne illness, e.g. caused by *E. coli* O157 : H7 in apple juice and fermented meat products, *Listeria monocytogenes* in soft cheese, *Salmonella* and *E. coli* O157 : H7 in bean sprouts, challenge test studies are very often carried out by food research centres funded by trade bodies or government departments. In these cases, and occasionally for industry product development purposes, organisms may be inoculated at an early stage in a simulated manufacturing process rather than in the finished product, e.g. in a meat mix used for preparing salami products, so that the effects of the processing conditions on the inoculated organisms can be assessed. The results are frequently used to provide advice to food producers / manufacturers about methods and conditions for controlling the relevant organism both during the production process and in the finished product.

5.8 Trouble-shooting, crisis management and microbiology

Raw materials, in-process materials or finished products at the point of manufacture may occasionally fail to meet a required microbiological specification; an in-line process monitor may indicate a failure to reach a required temperature, pH value, etc.; or product once sold may fail to achieve its assigned shelf life through premature spoilage or may be suspected of having caused food-poisoning. All of these situations can trigger an action programme that requires the support of a microbiologist and microbiological information.

Trouble-shooting a microbiological problem in a process or product is a team task and for issues involving the manufacturing company's customer, e.g. retailer and / or consumers, the investigation may well also involve the customer's technical personnel and / or public health / environmental health personnel.

Of particular concern are any issues affecting public health such as suspected food-poisoning incidents implicating a specific food. For example, incidents involving *Salmonella* and ready-to-eat foods can lead to the closure of manufacturing sites with heavy commercial losses. The action taken by manufacturers in response to reports of *Salmonella* found in ready-to-eat foods as part of routine production monitoring or reported in relation to an incident of illness is always swift and thorough. It includes, in addition to checking all Critical Control Point monitoring records, a detailed microbiological investigation of raw materials, production environments, personnel and products produced on the same production lines, in the same area or incorporating common ingredients to the implicated product (Bell and Kyriakides, 2002). Fortunately, such incidents are uncommon; however, the active involvement of well-trained and experienced food microbiologists in such incident investigations, supported by appropriate laboratory facilities, methods and procedures, is both necessary and invaluable.

Whatever the area of application of microbiology, it must be understood that most microbiological test results are obtained some 18 hours to 5 days or more after the sample has been tested. This, however, does not detract from the value of a microbiological test that during the hazard analysis has been considered important for confirming the hygienic status of the product ingredients and the efficacy of the production process and intrinsic product safety factors, e.g. pH value, water activity, and extrinsic product safety factors, e.g. packaging atmosphere, storage temperature, and anticipated product shelf life, as well as the other areas already discussed.

Reliable microbiological information is essential in support of Hazard Analysis and in-line production control monitoring of the Critical Control Points in the process. The most reliable information will only be gained from the use of methods that are appropriately validated for the specific sample types to be examined and practices and procedures that are operated by well-trained staff in a properly constructed and equipped laboratory. These matters are discussed in the following chapters.

5.9   Further reading

Anon (1990, part revised 1997) *Evaluation of Shelf Life for Chilled Foods. Technical Manual No. 28.* Campden & Chorleywood Food Research Association, Chipping Campden, UK.

Anon (1997) *HACCP: A Practical Guide. Technical Manual No. 38,* 2nd edn. Campden & Chorleywood Food Research Association, Chipping Campden, UK.

Anon (2000) *An Introduction to the Practice of Microbiological Risk Assessment for Food Industry Applications, Guideline No. 28.* Campden & Chorleywood Food Research Association, Chipping Campden, UK.

Codex Alimentarius Commission (1997) *Food Hygiene Basic Texts.* Codex Alimentarius Commission, Rome, Italy.

Institute of Food Science and Technology (UK) (1999) *Development and Use of Microbiological Criteria for Foods.* Monograph of the Institute of Food Science and Technology, London, UK.

International Commission on Microbiological Specifications for Foods (ICMSF) (1988) *Microorganisms in Foods 4: Application of the Hazard Analysis Critical Control Point (HACCP) System to Ensure Microbiological Safety and Quality.* Blackwell Scientific Publications, Oxford, UK.

International Commission on Microbiological Specifications for Foods (ICMSF) (2002) *Microorganisms in Foods 7: Microbiological Testing in Food Safety Management.* Kluwer Academic / Plenum Publishers, New York, USA.

Mortimore, S.E. & Wallace, C. (1998) *HACCP-A Practical Approach.* 2nd edn Aspen Publishers Inc., Maryland, USA.

6 Laboratory Design and Equipment

6.1 Introduction

All food manufacturing companies either have an on-site microbiology laboratory or contract out microbiological testing to a company central laboratory or to an external, independent food microbiology laboratory. Wherever the location of the laboratory, it is important that the design of the laboratory is conducive to the work carried out in it. Appropriate consideration given to the design, construction and operation of a food-testing laboratory can deliver significant benefits in facilitating the maintenance of a clean environment and the production of reliable test results.

In addition, for on-site laboratories, it is imperative that through good laboratory design and operation, organisms cannot 'escape' from the laboratory and contaminate food production areas. For their own protection and the protection of others, it is important that food microbiology laboratory staff have some understanding of the standards and conditions required for microbiology laboratory design, construction and operation.

6.1.1 *Standards required for the design and construction of a microbiology laboratory*

Legislation in European Union countries requires employees to be protected from exposure to substances and biological agents that may be hazardous to health. In support of such legislation, biological agents have been categorised into hazard groups according to the health hazard as determined by expert committees.

In the UK, the Advisory Committee on Dangerous Pathogens (ACDP, 1995) has defined four hazard groups for biological agents, increasing in hazard from Group 1 to Group 4, based upon the infectivity of the organism and the severity of the disease it causes in humans. Except for toxigenic strains of *Escherichia coli*, e.g. *E. coli* O157 : H7 or *E. coli* O103 which are categorised in Hazard Group 3, all the other organisms and microbial toxins that form the work of a routine food microbiology laboratory are categorised in either Hazard Group 1 (a biological agent unlikely to cause human disease) or Hazard Group 2 (a biological agent that can cause human disease and may be a hazard to employees; it is unlikely to spread to the community and there is usually effective prophylaxis or effective treatment available). Examples of organisms categorised in Hazard Group 2 are *Bacillus cereus*, *Clostridium* spp., *Enterobacter* spp., *Escherichia coli* (non-toxigenic), *Klebsiella* spp., *Listeria monocytogenes*, *Salmonella* spp., *Staphylococcus aureus* and *Streptococcus* spp. Hazard Group 4 is reserved for

Plates 8, 9, 10, 11, 12, 13, 19 and 20 are located in the colour plate section at p. 100.
Plates 22 and 30 are located in the colour plate section at p. 228.

very highly infectious viruses such as the Ebola and Lassa fever viruses which are only handled in the most highly specialised and secure laboratories.

For each hazard group, minimum standards are described for the laboratory design and procedures required in order to handle the organisms in that group (Advisory Committee on Dangerous Pathogens, 1995 and 2001). Probably the most common basic microbiology tests carried out in most food industry laboratories are the 'total colony count' (TCC; see Chapter 9) and enumeration tests for coliforms (or Enterobacteriaceae), *Escherichia coli*, yeasts and moulds and *Staphylococcus aureus*. It is prudent to assume that, even if a laboratory only carries out 'simple' tests for TCC, incubated at 30°C, some Hazard Group 2 organisms will occasionally occur on the plates, especially if the samples examined are unprocessed raw food materials. For practical purposes, therefore, food microbiology laboratory design and construction takes account of the guidance given by the ACDP for the containment of Hazard Group 2 organisms (Table 6.1). Other national and international standards and recommendations relating to the design and construction of food microbiology laboratories include similar practical guidance, e.g. International Organisation for Standardisation (ISO) Standard 7218 : 1996, Downes and Ito (2001).

Some general 'common-sense' principles apply to the location, design and construction of a food microbiology laboratory. Adherence to these principles helps to ensure that the laboratory is not a source of microbial contamination to the laboratory personnel, food production areas, to other local businesses or to passers-by and that each area of the laboratory can easily be managed to prevent cross-contamination from extraneous sources to the samples under examination. The results obtained and reported from work carried out can then be more reliably associated with the sample, and the provision of reliable results is the main aim of any food microbiology laboratory.

The commentary on building structure, fittings, services and equipment in sections 6.2, 6.3 and 6.4 provides some indication of those areas of laboratory design and construction that require attention for assisting the safe and good practice of food microbiology.

6.2   The building

Adequate consideration and appropriate implementation of sound basic design and construction features for a laboratory building is necessary to provide the basis for developing a suitable environment for food microbiological work. Some aspects that need to be addressed are as follows:

(1)   The microbiology laboratory should be well isolated and separated from any food production areas, preferably being housed in a separate building.
(2)   The fabric of the building must be 'biosecure', i.e. there should be no holes / cracks etc. that would allow rodents, other animals or insects access to the laboratory areas.
(3)   The laboratory should be large enough to accommodate, adequately, all the necessary working areas, equipment, laboratory supply stores and staff, and to suit the expected workload.

Table 6.1 Some containment level requirements for organisms in Hazard Group 2 affecting laboratory structure and equipment, adapted from the Advisory Committee on Dangerous Pathogens (1995 and 2001).

Laboratory structure, services and equipment requirements
• There should be adequate space ($24\,m^3$) in the laboratory for each worker.
• The laboratory should contain a hand-wash basin located near the laboratory exit. Taps should be of a type that can be operated without being touched by hand.
• The laboratory should be easy to clean. Bench surfaces should be impervious to water and resistant to acids, alkalis, solvents and disinfectants.
• Procedures for routine disinfection and spillages should be specified.
• If the laboratory is mechanically ventilated, it must be maintained at an air pressure negative to atmosphere while work is in progress. In most laboratories operating at Containment Level 2 where there is mechanical ventilation simply to provide a comfortable working environment, it may not be practical to maintain an effective inward flow of air. The often constant traffic in and out of Containment Level 2 rooms may interfere significantly with attempts to establish satisfactory airflow patterns. However, where a laboratory is ventilated specifically to contain air-borne pathogens in the event of an accident, then engineering controls and working arrangements must be devised so as to counter the risk of air-borne transmission to other areas.
• There must be safe storage of biological agents (microorganisms).
• Personal protective equipment, including protective clothing, must be: (a) stored in a well-defined place;   (b) checked and cleaned at suitable intervals;   (c) when discovered to be defective, repaired or replaced before further use.
• Laboratory coats or gowns, which should be side or back fastening, should be worn and should be removed when leaving the laboratory suite. Separate storage facilities, e.g. pegs, apart from that provided for personal clothing, should be provided in the laboratory suite.
• Personal protective equipment which may be contaminated by biological agents must be: (a) removed on leaving the working area;   (b) kept apart from uncontaminated clothing;   (c) decontaminated and cleaned or, if necessary, destroyed.
• When undertaking procedures that are likely to give rise to infectious aerosols, a Class 1 microbiological safety cabinet (BS 5726:1992 or unit with equivalent protection factor or performance) should be used. Safety cabinets should exhaust to the outside air or to the laboratory air extract system. Some other types of equipment may provide adequate containment in their own right, but this should be verified.
• An autoclave for the sterilisation of waste materials should be readily accessible in the same building as the laboratory, preferably in the laboratory suite.
• Materials for autoclaving should be transported to the autoclave in robust and leak-proof containers without spillage.
• There should be a means for the safe collection, storage and disposal of contaminated waste.

(4) Drainage from the laboratory must be directed in such a way that it does not flow through / under food production areas. It should enter the main flow downstream of the factory drains.

(5) Separate areas to the main working laboratory area should be provided for:
- general stores
- sample reception and storage

- staff facilities (cloakroom and toilets)
- laboratory staff access point with laboratory coat change and hand-washing facilities
- media preparation and sterilisation tasks
- pathogen handling and culturing with dedicated coat-change and hand-washing facilities (for laboratories that handle pathogens such as *Salmonella*)
- equipment washing and waste decontamination tasks.

(6) The laboratory areas should be constructed so that a logical and practical work-flow can be achieved (see section 6.4).

(7) Windows should be designed so they cannot be opened and, if window ledges are present, these should be steeply sloping to prevent them from being used as storage areas. If windows require some form of shading, externally fitted blinds or solar protective film should be used. Internally fitted curtains or window blinds are not acceptable as these accumulate dust and debris that can become a source of contamination to the work in progress.

(8) Emergency exits should be provided at suitable points. Advice from local Health and Safety experts is usually sought to ensure that legal requirements are satisfied.

6.3 Internal structure, fittings and services

An appropriately designed and constructed building needs to be fitted with furniture and provided with services that are suitable for microbiological work. The following summarises the main areas requiring consideration. All the points indicated have some impact on the ability of laboratory staff to maintain a suitably clean working environment.

(1) Hot and cold water supplies and electric power points must be provided to the laboratory. Gas is usually required for Bunsen burners; however, alternatives to a permanent gas supply may include portable gas-powered Bunsen burners or 'burners' with electric heating elements.

(2) Dedicated hand-washing facilities with hot and cold water supplies (or mixer taps delivering adequately warm water) and supplied with bactericidal soap, nail brushes and paper towels must be provided in the laboratory coat changing area and in other relevant areas of the laboratory, e.g. the main laboratory, the pathogen laboratory, its access point and the washing up area. Taps must be non-hand operated and automatic liquid soap dispensers should be installed; dispensers may also be provided for the application of alcohol-based hand disinfectant after hand-washing. Hand-washing facilities must not be used to wash up equipment or utensils.

(3) A dedicated double sink and drainer with hot and cold water supplies and appropriate cleaning chemicals should be provided for equipment washing in the wash-up area. This installation must not be used for hand-washing.

(4) A suitable steam extraction and ventilation system should be provided for the autoclaves, glassware washers, hot air ovens and wash-up sinks in the equipment washing and autoclaving areas. Any extraction systems provided for the

laboratory must exit safely (if necessary, through filters) to the external environment and not into manufacturing areas or near air inlets to manufacturing areas.

(5) The internal laboratory areas should be designed and constructed to minimise traps where dust and debris can accumulate and to allow for easy cleaning, e.g. laboratory floors and walls should be constructed of smooth, impervious materials that are easy to clean and wall / floor joints should be fitted with sealed coving to assist cleaning.

(6) To facilitate cleaning, workbenches and cupboards must be either:
- built-in with impervious seals at the junctions with floor and walls, or
- easily removable, or
- of suspended construction allowing easy access underneath the benches.

(7) Bench work-surfaces must be level, smooth and impervious for easy cleaning and disinfection. Wood should be avoided for use as a work surface in microbiology laboratories because it can harbour microbial contamination and cannot be cleaned effectively; solid laminates or stainless steel are suitable materials for work-surface construction.

(8) Storage racking should be constructed of durable plastic or coated metal, not wood (for the reasons given in (7)).

(9) Chairs and laboratory stools should be constructed of durable plastic or coated metal, not made of wood or cloth-covered.

(10) Light fittings, e.g. fluorescent lighting tubes, should be protected by covers that can be removed for cleaning.

(11) A water treatment system, e.g. distillation unit, reverse osmosis, ion exchange with filtration, is required to provide high quality water for use in media preparation (Figure 6.1).

(12) Preferably, the laboratory should be temperature-controlled to facilitate incubator control as well as provide a pleasant working environment. Extractor fan type ventilators may assist where air-conditioning is not possible. Ventilation systems designed to bring air into the laboratory must be adequately filtered to prevent dust particles from becoming a source of contamination to the work in progress. It is strongly advisable to undertake a heat output calculation to ensure that the heat produced by autoclaves, incubators, refrigerators, other equipment and people does not overload the laboratory cooling system and environment; in summer, flat roofs can contribute significantly to the heat generated within a laboratory and it is easy to underestimate the heat output of a working food microbiology laboratory in mid-summer. **Such calculations are normally done by qualified engineers from companies supplying such environmental control systems**.

It is essential that laboratories operate regular maintenance and cleaning programmes to ensure that the working environment and equipment are maintained in undamaged and clean conditions.

Figure 6.1 Water purification and equipment (courtesy of Millipore (UK) Ltd, Water Purification Division).

Tap water may contain many impurities that may adversely affect the performance of microbiological growth media and hence bacterial growth; it therefore has to be purified before it can be used in microbiological test work, especially for bacteriology. As no single purification technology removes all types of contaminant to the levels required for all microbiological applications, a well designed water purification system for such purposes usually uses a combination of purification stages to achieve the final water quality required, e.g. distillation, reverse osmosis, ion exchange or electrodeionisation. Each purification technology must be used in an appropriate sequence to optimise its particular removal capabilities.

Distillation: Distillation is one of the oldest methods of water purification. Water is first heated to boiling point. The water vapour then rises to a condenser where circulating cool water lowers the temperature so that the vapour is condensed, collected and stored. Most contaminants remain behind in the original liquid phase vessel; however, volatile organic substances with boiling points lower than 100°C will not be removed efficiently.

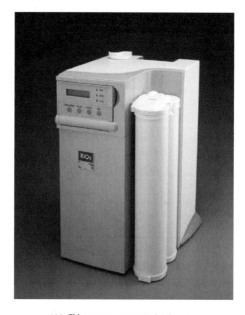

RiOs™ reverse osmosis (RO) system

Reverse Osmosis (RO): Natural osmosis occurs when solutions of two different concentrations are separated by a semi-permeable membrane. Osmotic pressure drives water through the membrane and this dilutes the more concentrated solution, bringing about an eventual equilibrium. If enough hydraulic pressure is applied to the more concentrated solution to counteract the osmotic pressure, pure water can be driven from a concentrated solution and collected downstream of the membrane as purified permeate water; this is reverse osmosis (RO). RO will remove a broad spectrum of contaminants in a single purification step. It is usually the most economical method of removing 90–99% of all contaminants. RO membranes can reject practically all particles, including bacteria, and organic compounds larger than 200 Daltons molecular weight by means of molecular sieving. RO also involves an ionic exclusion process; only solvent is allowed to pass through the semi-permeable RO membrane, while virtually all dissolved molecules, including salts and sugars, are retained. The RO membrane rejects salts (ions) by a charge phenomenon action; the greater the charge, the greater the rejection.

Ion exchange: The process of ion exchange involves water flowing through a matrix of spherical resin beads (ion exchange resins). Mineral ions in the water are exchanged for either hydrogen ions or hydroxyl ions attached to the resin beads. These ions then combine to form pure water. When all of the active sites on the resins are exhausted the resin needs to be either replaced or chemically regenerated using acid and base solutions.

Electrodeionisation (EDI): Electrodeionisation combines ion exchange and electrodialysis. This results in a process which effectively deionises water whilst the ion exchange resins are continuously regenerated by the electrical current in the unit. This electrochemical regeneration replaces the chemical regeneration used in conventional ion exchange systems.

6.4   Work flow

Given a well-designed, constructed and fitted laboratory, it is important to ensure that work is carried out in a logical flow. This will also facilitate good microbiological practice and assist in the prevention of cross-contamination of work in progress.

As examples Figure 6.2 shows a plan for a microbiology laboratory in which no specific bacterial pathogen work is undertaken and Figure 6.3 shows a plan for a microbiology laboratory that includes a food-borne pathogen-handling facility for organisms categorised in Hazard Group 2. The arrows indicate the access points for staff and materials. The position of each area in relation to each other facilitates a logical flow of work through the laboratory.

Figure 6.2   Plan for a microbiology laboratory in which no specific bacterial pathogen work is undertaken.

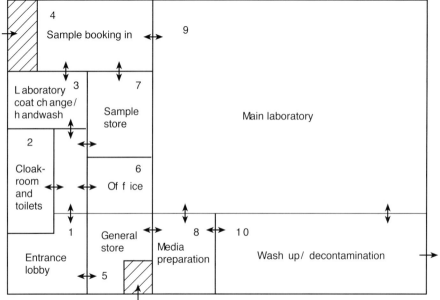

Notes: Arrows indicate work and people-flow access points from one area to another. The hygienic integrity of the laboratory should not be breached by staff accessing the laboratory directly through a store, e.g. from area 1, through area 7 to area 4, or from area 1 through areas 5 and 8 to areas 9 or 10.

Work benches and equipment (incubators, refrigerators, water-baths, etc.) in the main sample processing areas (Figures 6.2 and 6.3; areas 4 and 9) and the pathogen laboratory (Figure 6.3; area 12) as applicable, should be constructed and organised so that the work can progress in a logical flow from area 4 towards area 10 where the used media plates and tubes and used disposables etc. are decontaminated; disposables being sent to waste disposal and non-disposable equipment being washed and sterilised for re-use. Figure 6.4 summarises the key tasks associated with factory sample processing for microbiological examination that usually take place in specified laboratory areas (Figures 6.2 and 6.3) that are maintained to ensure a logical flow of work thus minimising the potential for cross-contamination of work to occur.

All laboratory staff must go through a coat change and hand-washing routine in the appropriate area (Figures 6.2 and 6.3; area 3) on entering and leaving the laboratory and no member of staff should be allowed to access the laboratory directly through a store, e.g. area 1, through area 7 to area 4, or area 1 through areas 5 and 8

Figure 6.3 Plan for a microbiology laboratory that includes a bacterial pathogen handling facility.

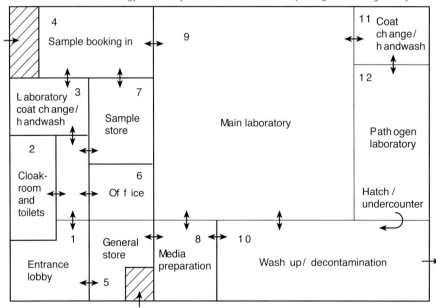

Notes: Arrows indicate work and people-flow access points from one area to another. The hygienic integrity of the laboratory should not be breached by staff accessing the laboratory directly through a store, e.g. from area 1, through area 7 to area 4, or from area 1 through areas 5 and 8 to areas 9 or 10.

to areas 9 or 10. The stores are made 'double entry' so that they can be stocked from outside the main laboratory working areas but accessed by laboratory staff from within the appropriate laboratory areas.

Clearly there may be variations of structure and operation of the access points to the areas within a laboratory and Figures 6.2 and 6.3 indicate by the 'hatched' areas, alternative means for delivering stores and samples into the laboratory whilst maintaining the integrity of the laboratory work and people-flow. In these cases, a secure 'counter' or hatch allows materials to be passed into the laboratory area while delivery personnel are kept 'outside' the laboratory. The key objective is to ensure that the laboratory is accessed only by authorised personnel via the coat-changing and hand-washing area in which suitable protective procedures are practised by all staff at all times.

6.5 Equipment

A considerable amount of capital equipment is required to support the work of a food microbiology laboratory. Table 6.2 indicates the basic capital equipment that should be found in different areas of the laboratory together with the furnishing and service infrastructure that should also be in place. In large laboratories, separate rooms are often allocated for incubators; these rooms should be fully air-conditioned. Central continuous incubator temperature monitoring systems with automatic high / low temperature alarms may also be found in large laboratories. Table 6.3 provides information about key items of laboratory equipment, their application, matters affecting their operation and relevant control parameters. Equipment should always be purchased against a clear specification appropriate for the purpose for which it is required and, where possible, appropriately accredited suppliers able to supply the

Plate 1 Some representatives of the Microbial Kingdoms.

(a) Gram-positive, spore-forming, rod-shaped bacteria; note the translucent spores in the centre of the cells.

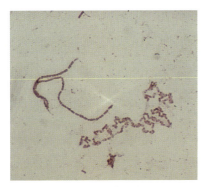

Magnification approximately ×700.

(b) Yeast cells; note budding.

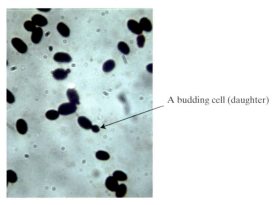

A budding cell (daughter)

Magnification approximately ×525.

(c) A typical food-borne mould, *Penicillium* sp., showing three levels of branching and spores (conidia).

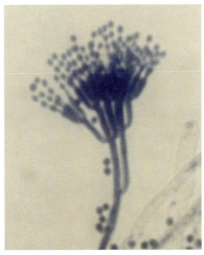

Magnification approximately ×700.

Plate 2 Examples of Gram-stained bacteria.

(a) Gram-positive, rod-shaped bacteria.

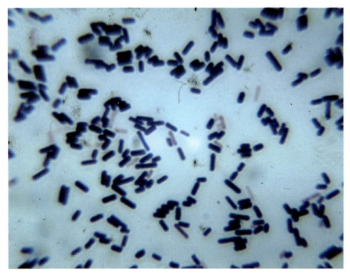

Magnification approximately ×1000.

(b) Gram-negative, rod-shaped bacteria.

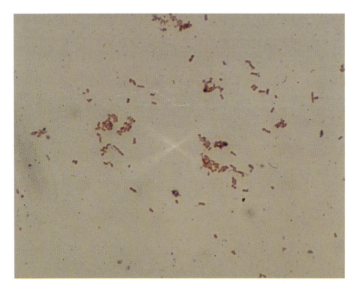

Magnification approximately ×1000.

Plate 3   Some typical bacterial cell morphologies.

(a) Chains of Gram-positive, rod-shaped bacteria.

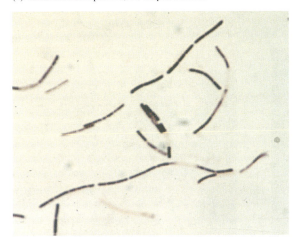

Magnification approximately ×900.

(b) Small clumps of Gram-positive coccoid bacteria.

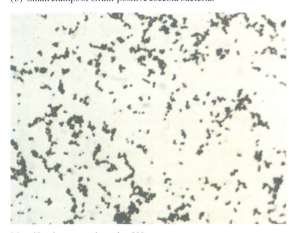

Magnification approximately ×900.

Plate 4   A phase-contrast micrograph of *Bacillus cereus* showing phase-bright endospores.

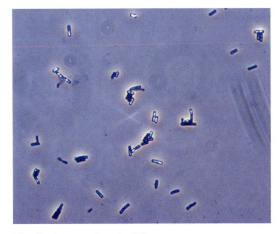

Magnification approximately ×900.

Note: See Chapter 11 for information concerning phase-contrast microscopy.

Plate 5 Visible mould on food.

(a) Various moulds on sausages – growth occurred as a result of moisture condensing within the pack.

(b) Mould colonies growing on the surface of pâté – mould had contaminated the pâté from garnish added after cooking and was allowed time to grow because the product had been kept beyond the end of its shelf life.

(c) Mould growing on the underside of the pastry lid of a meat pie – mould had entered the pie through a vent in the pastry and had grown in the moist interior of the pie.

(d) Mould growing on strawberries – natural mould contamination of the strawberries had been able to grow because of moisture condensing within the pack.

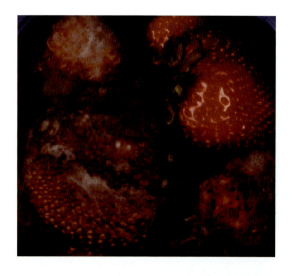

Plate 6 Some moulds of importance in food microbiology.

(a) *Penicillium* sp. – penicillia are widespread spoilage moulds of many refrigerated foods, e.g. pizza, cheese, pasta; also cause rots of citrus fruits and apples (see also Plate 1(c)).

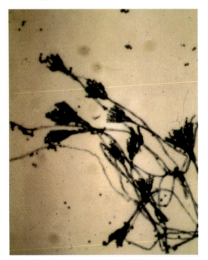

Magnification approximately ×280.

(b) *Aspergillus* sp. – aspergilli are common spoilage moulds of ambient-stored foods, especially low-water activity foods and those of tropical origin, e.g. jams, maize, peanuts.

Magnification approximately ×600.

Courtesy of Dr Ailsa D. Hocking, Food Science Australia, Sydney NSW.

(c) Spores (conidia) of *Cladosporium* sp. – cladosporia are common, frequently airborne, spoilage moulds of some fresh fruit and vegetables; also common on proteinaceous and fatty foods stored under chill conditions, such as raw meats, cheese, recipe dishes and reduced-fat spreads. Causes 'black spot' on deep-chilled (−2°C to −5°C) carcase meats.

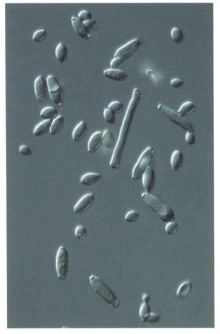

Magnification approximately ×800.

Courtesy of Dr Ailsa D. Hocking, Food Science Australia, Sydney NSW.

Plate 6 (*cont*)

(d) *Mucor* sp. – mucors are common spoilage moulds of bread and also some types of cheese, fruit and vegetables.

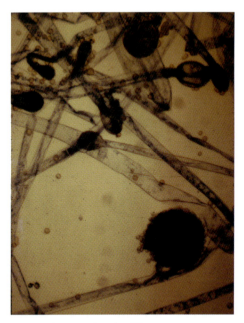

Magnification approximately ×400.

(e) *Geotrichum candidum* – known as 'machinery mould' and common on poorly-cleaned surfaces of processing equipment, particularly in dairies.

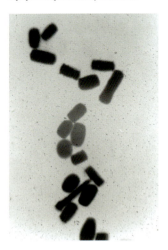

Magnification approximately ×800.

Plate 7 The Novasina instrument used to determine water activity.

Temperature-controlled chamber

Sensor cell containing food sample

Electronic measuring equipment and digital display

Plastic cups for holding the sample

Courtesy of Novatron Ltd.

Method of operation:
(1) A plastic cup is filled with the test sample material.
(2) The cup is immediately placed within the sensor cell compartment.
(3) The electrode unit is sealed on to the sample compartment.
(4) The lid is closed, the equipment is switched on and the temperature and humidity are allowed to stabilise and equilibrate respectively. This is observed when the reading of the digital display remains constant or the equipment can be programmed to detect stability automatically i.e. an audible / light indicator signalling that equilibrium stability has been reached.
(5) The a_w or ERH reading is then recorded either manually or, automatically by the printer.

Plate 8 Laboratory incubator (temperature controlled cabinets used for providing the correct temperature for the growth of specified organisms).

Courtesy of Norpath Laboratories Ltd.

Key points:
• Plates are stacked to allow good air circulation.
• Stacks of plates maximum six-high.
• Dedicated thermometer suspended in a 'heat-sink' liquid, e.g. glycerol.

Plate 9   Laboratory blender (Stomacher© 400 *Circulator*).
This instrument facilitates the removal of microbial populations from solid food materials into the liquid diluent.

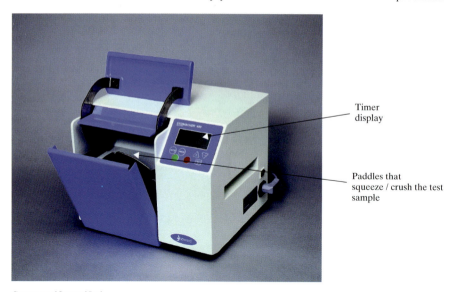

Timer
display

Paddles that
squeeze / crush the test
sample

Courtesy of Seward Ltd.

Key points:
• The instrument should be kept clean, particularly behind the paddles and the back of the door and ledges.
• The timer accuracy should be checked regularly.

Plate 10   Vortex mixer.
This instrument facilitates thorough mixing of a small inoculum volume in a larger diluent or medium volume.

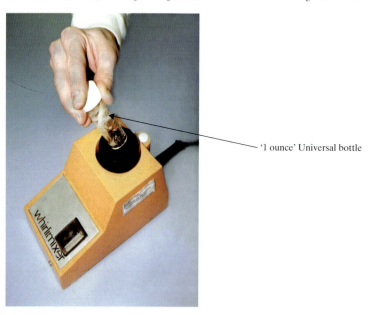

'1 ounce' Universal bottle

Courtesy of Norpath Laboratories Ltd.

Key points:
• The instrument should be kept clean.
• Damaged cups should be replaced.
• The tube / container should be held firmly to prevent it being dislodged from the hand. The top of the tube or bottle used should be sealed / capped to prevent spillage or aerosols from contaminating the laboratory.

Plate 11 Atmosphere control (anaerobe jars are used to incubate bacteria that require an oxygen-reduced environment).

The atmosphere within the jar is generated using cylinder gas or gas generating envelopes and an appropriate catalyst. Some will vent automatically if too much pressure is applied.

Courtesy of Don Whitley Scientific Ltd.

General method of use:
(1) Remove the lid.
(2) Place Petri dishes inside jar.
(3) Insert and activate gas generating envelope(s)* (see 'Key points' below).
(4) Secure the lid.
(5) Incubate for required time.

Key points:
• Ensure that the 'O' ring is in good condition and is seated correctly.
• Ensure the clamp on the jar is not over-tightened as excessive pressure on the clamps may distort the lid or 'O' ring and produce leaks.
• The method for generating the specific atmosphere required in the jar including the use of catalysts and indicators should be carried out strictly in accordance with the jar and / or gas generating kit manufacturer's instructions, as appropriate.
• Catalysts may require specific treatment for protection from chemical poisoning and information concerning the protection of catalysts from poisons should be obtained from the manufacturer / supplier.

Note:
Once the lid is removed, the atmospheric conditions in the jar are lost. If it is necessary to re-incubate cultures following an assessment out of the jar, then required atmospheric conditions must be re-established. The relevant laboratory standard operating procedure should be consulted to ensure the correct procedure is followed.
*For details of use with cylinder gas the specific jar manufacturer's instructions must be followed.

Plate 12 Disposable pipettes and pipetters (these are used for dispensing accurate volumes of liquid; in microbiology practice, volumes from 10 µl to 10 ml are commonly dispensed with these items).

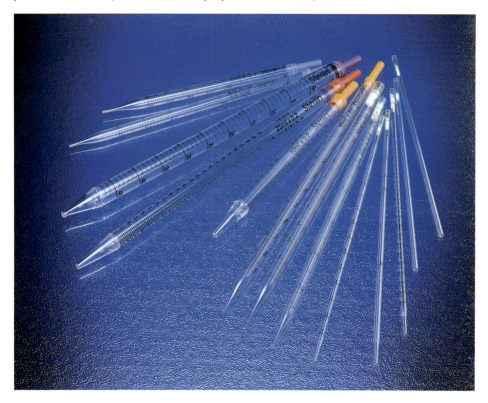

Courtesy of Bibby Sterilin Ltd.

Key points:
• Disposable pipettes or the tips of pipetters should be preferably single use only. In some circumstances it may be acceptable to use a single pipette more than once if the material to be dispensed is of the same concentration or a higher concentration than that pipetted on the first use.
• The laboratory standard operating procedure should indicate acceptable pipetting practice.

Plate 13   Plastic disposable equipment – loops, spreaders and Petri dishes (plates). Courtesy of Bibby Sterilin Ltd.

(a) Loops and spreaders.

Loops may be used for transferring very small volumes (1 μl–10 μl) of liquid culture to other culture media and for streaking cultures on plates. Spreaders are used to spread surface inoculated material or cultures over the surface of agar media in plates.

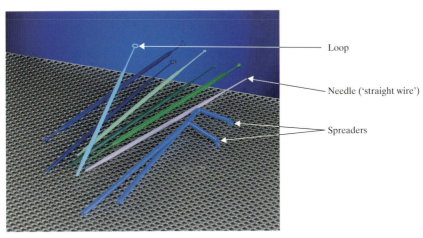

— Loop

— Needle ('straight wire')

— Spreaders

Key points:
• Plate spreading and streaking techniques must be practised with care to avoid contact with the edge of the plate which can lead to microbial growth that is difficult to assess and in turn lead to inaccurate results (see Figure 7.1).

(b) Petri dishes (plates).

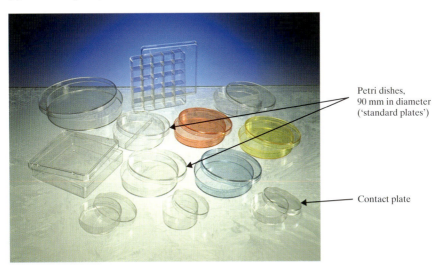

Petri dishes,
90 mm in diameter
('standard plates')

Contact plate

Key points:
• Most commonly used Petri dishes are 90 mm diameter triple-vented dishes (refers to the number of small raised plastic 'nodules' on the internal circumference of the lid).
• Single-vented dishes allow less atmosphere exchange giving minimal evaporation for long term incubation.
• Triple-vented dishes are particularly useful for any work for which good maintenance of the atmosphere to the cultures is required.
• Dishes with 2–3 compartments allow the use of 2–3 different media together or 2–3 cultures to be cultured per plate at the same time taking up less incubator space.
• Contact plates are small diameter plates which are carefully filled with an agar medium so that the surface of the medium forms a convex shape above the level of the edge of the plate; these are used specifically for microbiologically sampling flat surfaces by gently pressing the agar surface directly onto the surface to be sampled (the lid is then replaced, the dish labelled, incubated and assessed for microbial growth after incubation).

Plate 14 Example of good laboratory coat design and hair control.

 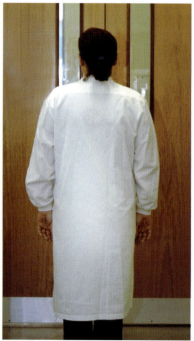

Courtesy of Norpath Laboratories Ltd.

Key points:
- Coat is fully buttoned and sleeves well-fitted to wrists.
- No personal clothing or jewellery is exposed.

Key points:
- Hair is tied back and tidy.

Plate 15 Illustration of the correct way to hold a pipette in use.

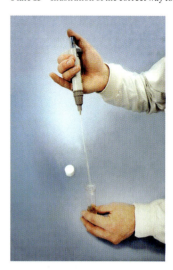

Courtesy of Norpath Laboratories Ltd.

Key points:
- Hold in a vertical or near vertical position.
- Keep the tip / straw / pipette away from the sides of the container from which material is being removed and the one to which it is being transferred. This will prevent extraneous material from contaminating the external surface of the tip / straw / pipette, and being transferred, thus contributing to inaccurate results.

Plate 16 Diagrammatic illustration of the difference in heat input to media in large volumes, e.g. ≥500 ml, autoclaved by a chamber versus a load-controlled heat process set to deliver a process of 121°C for 15 minutes.

(a) Heat input to media using chamber temperature to activate the process timer.

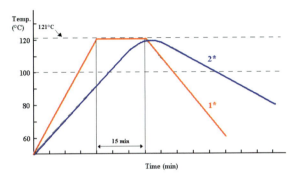

1* Chamber temperature
2* Load temperature

(b) Heat input to media using medium (load) temperature to activate the process timer.

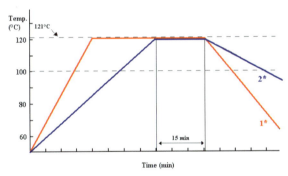

1* Chamber temperature
2* Load temperature

Notes:
• The area under curve (b2), between 100°C and 121°C, is far greater than that under curve (a2), between 100°C and 121°C, and so media heated in autoclaves in which processes are controlled by load-probes (temperature probes placed in a bottle of medium or 'dummy' load) have far more heat input. There may therefore be greater potential risks associated with load-probe controlled autoclaves with respect to heat-degradation of sensitive components and other changes in a medium that can occur because of excessive heat input.
• Clearly, mixed volume loads, e.g. 10 ml and 250 ml volumes, autoclaved in a load-probe controlled process in which the probe is placed in the larger volume can adversely affect the quality of medium in the smaller volumes.
• It is essential to understand how the laboratory's autoclave operates and is controlled to ensure that relevant process validations can be carried out so that media can be processed correctly.

Plate 17   A 'top-pan' laboratory balance.

'Top-pan' on which the material to be weighed is placed

Digital display of weight

Courtesy of Scientific Laboratory Supplies Ltd.

Key points:
• Balances should be sited on a level surface using the spirit level, if fitted, to check the levelling.
• Balances should be kept away from strong air movements; where a balance is used inside a dust extraction hood, care must be taken to ensure the air-flow does not affect the balance reading.
• Balances should be maintained in a clean condition at all times.
• The accuracy of all balances should be checked routinely (see Table 6.3).
• The balance should be tared (adjusted) to zero, before any materials are weighed.
• Balances should be serviced at regular intervals.

Plate 18   A typical hand-held pH meter.

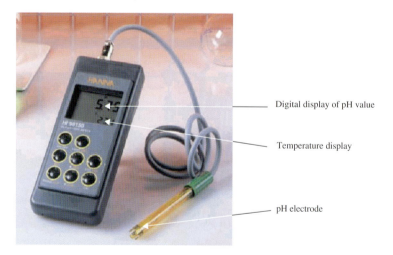

Digital display of pH value

Temperature display

pH electrode

Courtesy of Scientific Laboratory Supplies Ltd.

Key points:
• The pH electrode (probe) must be kept clean at all times.
• pH meters must be used with particular care to prevent probes from becoming damaged.
• pH meters should be calibrated before use with standard buffer solutions, preferably pH 4 and pH 7.
• After a pH reading has been taken, the end of the probe should be rinsed in purified water.
• When not in use, the meter should be left in 'standby' mode and the probe should be stored in purified water or pH 7 buffer solution.
• pH meters used for media and reagents should not be used to measure the pH value of test samples; separate probes should be kept and maintained for use with media and foods.
• Special cleaning procedures may be required to remove deposits of fat or protein if a pH meter is used with foods.

Plate 19   Colony counter. This instrument facilitates the accurate counting of colonies on or in agar media.

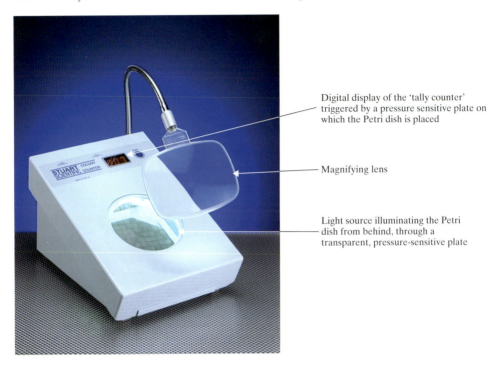

Digital display of the 'tally counter' triggered by a pressure sensitive plate on which the Petri dish is placed

Magnifying lens

Light source illuminating the Petri dish from behind, through a transparent, pressure-sensitive plate

Courtesy of Bibby Sterilin Ltd.

Note:
Colonies are counted by marking them individually using a marker pen or 'Chinagraph' pencil. Each mark operates the pressure-sensitive plate and increases the tally counter reading by 1.

Key points:
• The instrument should be kept clean at all times, particularly the light box plate and the magnifying lens.
• The light source should be checked to ensure it is working properly before use.
• The tally counter should be checked to ensure it counts sequentially and correctly, i.e. it does not 'stick' or does not move two or more digits at a time.
• A check should be made to confirm that each even, single point pressure applied to each colony registers only a single count.

Plate 20 Laminar air-flow and safety cabinets.

(a) Laminar air-flow cabinet.

Air flows through a HEPA filter towards the operator

Courtesy of Astec Microflow Ltd.

Key points:
• Designed to protect the test materials, e.g. food materials, media, from environmental contamination. The air flows through a high efficiency particulate air (HEPA) filter, horizontally, from the back of the cabinet towards the operator.
• Before and after each use, clean the work surface and sides of the cabinet with a non-corrosive disinfectant.
• Before working with materials inside the cabinet, but after cleaning, switch on the air flow and leave running for at least 20 minutes, or in accordance with the manufacturer's instructions. This flushes out dust and microorganisms from the filter and cabinet, and allows the laminar flow of air to become established creating the clean conditions for work.
• Always work at arms length, towards the back of the cabinet and keep the cabinet clear of unnecessary equipment and materials. This minimises air turbulence which can increase the risk of contamination.

(b) Class I safety cabinet.

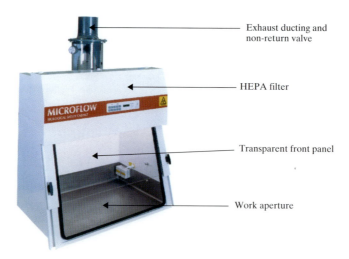

Exhaust ducting and non-return valve

HEPA filter

Transparent front panel

Work aperture

Courtesy of Astec Microflow Ltd.

Key points:
• Designed to protect the operator from the materials handled, e.g. cultures of microorganisms. The air flows from the operator into the cabinet and upwards through a high efficiency particulate air (HEPA) filter. The filtered air from the cabinet exhausts outside the laboratory into the external environment.
• Before and after each use, clean all internal surfaces of the cabinet with a non-corrosive disinfectant.
• Before working with materials inside the cabinet, but after cleaning, switch on the air flow and leave running for at least 10 minutes, or in accordance with the manufacturer's instructions; this allows the correct air velocity to establish. Before undertaking any work, check the controls and alarms to ensure that the operating conditions are satisfactory.
• Always work at arms length, towards the back of the cabinet and keep the cabinet clear of unnecessary equipment and materials. This minimises air turbulence which can increase the risk of contamination.

required equipment that also complies with a relevant National or International Standard (Table 6.3) should be favoured (see Chapter 7).

Because of the wide variation in types of equipment available to serve the same purpose, the manufacturer's manual and instructions should always be read thoroughly and used as the basis for producing an internal laboratory Standard Operating Procedure (SOP). Equipment manufacturers are a good source of information about the technology behind the items they supply, e.g. reverse osmosis, ion exchange columns, autoclaves, electrical impedance, enzyme linked immunosorbent assays etc., and can always be approached for help and training.

All equipment must be maintained and cleaned as appropriate to its function, frequency of use, requirements for safety, and as recommended by the manufacturer in order to ensure its continued best performance. Schedules for both maintenance and cleaning should be in place in all laboratories and it is essential that these are maintained and records kept of the tasks carried out.

Many laboratories employ, in addition to conventional methods, alternative microbiological methods (Table 6.4 and see Chapter 12), and the equipment for these may be allocated a dedicated laboratory or bench area.

A variety of 'consumable' materials, some of which are single-use (or 'disposable'), is required to support the work in a microbiology laboratory and would normally include the following:

- Sampling containers – sterile plastic bags or plastic containers
- Weighing containers – sterile plastic 'blender' bags or plastic containers
- Dilution containers – test tubes or small glass or plastic containers
- Media containers – glass 'Duran' bottles
- Labelling systems, e.g. 'Chinagraph' pencils, paper label 'guns'
- Selection of glassware – flasks, beakers, test tubes, measuring cylinders, bottles, funnels
- Pipettes – sterile serological glass, sterile plastic disposable, 'automatic' with sterile plastic tips (usually 100 μl and 1000 μl volumes)
- Petri dishes (plates) – sterile plastic
- Loops, straight wires and 'hockey stick' spreaders – sterile disposable or re-useable
- Filter membranes
- 'Tools' for handling samples, e.g. spoons, forks, scalpels, scissors, forceps, spatulas
- Microbiological media, chemicals, stains, dyes, etc.
- Disposable gloves
- Face masks
- Protective glasses
- Protective gloves
- Microscope slides and coverslips
- Autoclave bags
- Plastic sacks, bags and other containers for waste
- Alginate, cotton or sponge swabs
- Disinfectants
- Paper towels
- Autoclave temperature / process indicators
- Gas packs, catalysts and indicators for anaerobic jars
- Recording and reporting systems.

Figure 6.4 Summary of the key primary tasks in food sample processing related to laboratory area (see Figures 6.2 and 6.3).

Area 7	Area 4	Area 9	Area 8
Factory samples received and checked against customer's order	→ Sample booking-in*	→ Primary weighing to produce test sample	Dehydrated media, supplements etc. checked and recorded
Sample stored until required for testing		Weighed sample storage†	Weigh media ingredients
		Addition of diluent	Add water
		Blending / maceration	Disperse / dissolve media ingredients and dispense into required volumes
		Further dilutions made	Area 10 Sterilise media
		Plates and tubes of media inoculated → Discarded materials to Area 10	Area 9 Add supplements (if required)
		Media / inoculum mixed	Carry out quality control work and record results
		Plates and tubes incubated	Media for inoculation

Incubated plates and tubes assessed

Results recorded → Discarded materials to *Area 10*

Results reported

or

Confirmation tests carried out as required →

Results recorded → Discarded materials to *Area 10*

Results reported

* may be carried out as part of the sample receipt and checking process.
† most test samples are, and should be, processed immediately.

Table 6.2 Examples of key equipment and services that may be found in different areas of a food microbiology laboratory.

Area	Furnishing	Equipment	Services
Entrance lobby		Audible indicator of people entering the area	
General store	Shelving and cupboards	Maximum–minimum thermometer	
Cloakroom and toilets	Outdoor coat storage facility Hand-washing basins Toilets	Individual staff lockers	Hot and cold water supplies Soap and hand disinfectant dispensers Paper hand towels
Office	Desk(s) and chairs	Telephone Computer Filing system	Electrical power supplies
Laboratory coat change areas	Laboratory coat storage Hand-washing basins		Hot and cold water supplies Soap and hand disinfectant dispensers Paper hand towels
Sample store	Shelving and cupboards	Refrigerator / freezer Thermometer(s)	Electrical power supplies
Sample booking in	Laboratory furniture, e.g. benches, cupboards, chairs / stools	Laboratory information management system – can be paper based or electronic, i.e. computer	Electrical power supplies
Media preparation	Laboratory furniture, e.g. benches, cupboards, chairs Hand-washing basins	Water purification equipment Refrigerator Top-pan chemical balance pH meter Autoclave Water baths Medium dispenser Laminar air-flow cabinet Heat extraction equipment	Electrical and gas power supplies Hot and cold water supplies Soap and hand disinfectant dispensers Paper hand towels

Contd.

Table 6.2 *Contd.*

Area	Furnishing	Equipment	Services
Main laboratory	Laboratory furniture, e.g. benches, cupboards, chairs Hand-washing basins	Refrigerator / freezer Top-pan chemical balance to suit weight ranges required; usually, an accuracy of 0.1 g is satisfactory Gravimetric diluter Blender / homogeniser Vortex mixer Laminar air-flow cabinet Filtration apparatus Centrifuge Vacuum pump Water baths Anaerobic jars Bunsen burners Thermometers Timers Warm air incubators (30°C, 37°C and 55°C) Low-temperature incubator (22–25°C) for yeast and mould culture Illuminated colony counter Microscope with oil immersion lens	Electrical and gas power supplies Hot and cold water supplies Soap and hand disinfectant dispensers Paper hand towels Air-conditioning
Pathogen laboratory	Laboratory furniture, e.g. benches, cupboards, chairs Laboratory coat storage Hand-wash basin	Safety cabinet Refrigerator / freezer Water baths Warm-air incubators (30°C, 37°C) Microscope with oil immersion lens Thermometers Bunsen burners	Electrical and gas power supplies Hot and cold water supplies Soap and hand disinfectant dispensers Paper hand towels Air-conditioning
Wash-up and decontamination	Laboratory furniture, e.g. benches, cupboards, chairs Double sink and drainers Hand wash basin	Autoclave Glassware drying cabinet Sterilising oven (150–185°C) Thermometers Heat extraction equipment	Electrical and gas power supplies Hot and cold water supplies Soap and hand disinfectant dispensers Paper hand towels Cleaning chemicals

Table 6.3 Food microbiology laboratory equipment, function and control.

Equipment type	Application	Operational issues	Routine control parameters	Figure / plate and literature reference*
General				
Refrigerator / cold room	Provides appropriate conditions for: – prepared culture media storage – samples for examination – cultures for reference or further work – media reagents	Must never be overloaded Food samples should be kept separate from microbial cultures	Temperature 1–5°C unless otherwise specified	International Standard ISO 7218 (1996)
Freezer	Provides appropriate conditions for: – samples for examination – cultures for reference	Must never be overloaded Food samples should be kept separate from microbial cultures	Temperature below −18°C unless otherwise required	International Standard (ISO 7218 (1996)
Incubator / warm room	Provides consistently even and appropriate temperature conditions for microbial growth	Should be located in a room in which the environmental temperature is maintained at ≤ the temperature at which the incubator operates Must never be overloaded Petri dish stacks should be limited to a maximum of 6 high and allowed good air circulation	Temperature, usually ±1°C of the incubation temperature required, e.g. 25°C, 30°C, 37°C, or as otherwise specified	International Standard ISO 7218 (1996) (Plate 8)
Water bath – thermostatically controlled	Provides appropriate conditions for microbial growth *or* provides appropriate conditions for tempering and holding media in molten condition ready for use *or* for use in treating samples, e.g. heat treatment in tests for detecting / enumerating spore-forming bacteria	Level of water must be maintained above the level of the contents of the tubes / bottles in the bath Water must be changed and / or treated regularly to ensure it does not become a source of microbial contamination in the laboratory Aerosol control	Temperature, e.g. 44±0.5°C for incubating broth cultures of *E. coli*, 46±1°C for media tempering, 70–100°C for heat treatments (depends on type of organism to be detected)	International Standard ISO 7218 (1996)

Equipment	Function	Care/use	Control check	Reference
Blender (Stomacher, Pulsifier, homogeniser) with or without automatic timer	Disperses the food sample material and microorganisms in the liquid diluent	Must be kept clean	Time-clock or automatic timer used to time the mixing process for the specified period, e.g. 30 seconds, 1 or 2 minutes	(Plate 9)
Vortex mixer	Disperses an inoculum in diluent during the sample dilution process	Must be kept clean. Must be used in such a way as to prevent spillage from the tube		(Plate 10)
Incubation atmosphere control, e.g. anaerobic / microaerobic containers or cabinets	Provides appropriate conditions for microbial growth	Must be kept clean and disinfected. Use of a catalyst (where appropriate) must be in accordance with the manufacturer's instructions	Positive control culture used as a check that conditions are correct. Appropriate indicator systems must be used to confirm that the correct gas atmosphere has been obtained and is maintained	International Standard 7218 (1996) (Plate 11)
Horizontal laminar air-flow cabinet	Protects the work, e.g. sample, from extraneous contamination from the environment and the operator	Must be kept clean and disinfected. Air-flow rate. Filter integrity. Switch on the air supply for 20 minutes before use to set up clean air conditions	Exposure plates used during operation to check microbiological quality of air	International Standard ISO 7218 (1996) British Standard BS 5726: Part 4 (1992) (Plate 20(a))
Class 1 safety cabinet	Protects the operator and the laboratory environment from dust and aerosols arising from the work in the cabinet	Must be kept clean and disinfected. Air-flow rate. Filter integrity. Sterilise the cabinet before allowing access by maintenance engineers	Exposure plates used during operation to check microbiological quality of air	International Standard ISO 7218 (1996) British Standard BS 5726: Part 4 (1992) British Standard BS EN 12469 (2000) Kennedy & Collins (2000) (Plate 20(b))

Contd.

Table 6.3 *Contd.*

Equipment type	Application	Operational issues	Routine control parameters	Figure / plate and literature reference*
Class II safety cabinet	Protects the work from extraneous contamination *and* the operator and environment from the work, e.g. cultures being handled	Must be kept clean and disinfected Air-flow rate Filter integrity Sterilise the cabinet before allowing access by maintenance engineers	Exposure plates used during operation to check microbiological quality of air	International Standard ISO 7218 (1996) British Standard BS 5726:Part 4 (1992) British Standard BS EN 12469 (2000) Kennedy & Collins (2000)
Class III safety cabinet	Used for handling highly dangerous organisms, i.e. Hazard Group 4. Such cabinets are not usually found in routine food industry microbiology laboratories	Special instruction is required in the operation and maintenance of these cabinets		
Volumetric dispensers	Used to dispense accurate volumes of liquid	Must be kept clean	Multiple weight checks (3 or more) on volumes dispensed and records kept	British Standard BS EN ISO 8655 (2002)
Water treatment / purification system (distillation, reverse osmosis, ion exchange and filtration)	Provides water free from nutrient and toxic elements for use in media preparation	Water should be produced fresh for use, i.e. minimise storage time for treated water Storage in polythene or soda glass containers must be avoided as these can cause pH drift Metal should be excluded from pipework and containers as metal ions can leach into the water Containers for purified water may need to be disinfected periodically	Conductivity check to target <10 micro Siemens/cm (μS/cm) Aerobic colony counts on the water produced Records of all checks must be kept	Kyriakides *et al.* (1996) (Figure 6.1)

Equipment	Purpose	Condition	Requirements	Reference
Colony counter	Provides a magnifying and suitable background with illumination, to permit accurate counting of colonies on plates	Must be kept clean		(Plate 19)
Drying cabinet	Thermostatically controlled heating cabinet for drying washed glassware	Must be kept clean	Temperature $\pm 2°C$ of the temperature required	
Sterilising oven	Thermostatically controlled hot oven for sterilising washed glassware	Must be kept clean	Temperature usually $> 170°C$ controlled to $\pm 2°C$ of the temperature required Chemical or biological indicators used to check the process Records of all checks must be kept	
Autoclave	For sterilising microbiological test media *and* for decontaminating laboratory waste For all but the smallest laboratories, separate machines are preferred for processing media and waste	Must be of adequate size for the volumes of materials required to be autoclaved Specific autoclaves should preferably be dedicated for the different tasks Autoclaves with a built-in thermocouple and chart recorder are preferable	Each load process should be monitored and checked using a calibrated maximum thermometer and / or chemical or biological indicators Records of all checks must be kept	British Standard BS 2646 : Part 4 (1991), Parts 1, 3 and 5 (1993) Anon (1999) (2) (Plate 22(a), (b))
Light (optical) microscope and accessories	For using high powers of magnification (oil-immersion lens) to observe directly the characteristics of microorganisms (size, shape, staining type) Low power magnification plate microscopes may be used to examine colonies on or in agar media	Must be kept clean, particularly the condenser, eye-pieces and objective lenses Glassware accessories (slides and coverslips) must also be kept optically clean	A microscope must be properly set up before use. This may require adjustment of the iris diaphragm, condenser focus and light power (see Chapter 11)	British Standard BS 7012-1 (1998) (Figure 11.2)

Contd.

Table 6.3 Contd.

Equipment type	Application	Operational issues	Routine control parameters	Figure / plate and literature reference*
Centrifuge	Used in sample preparation for some proprietary tests, e.g. ELISA tests for bacterial toxins. The high speed of rotation achieved in a centrifuge creates a centrifugal force outwards that is far greater than gravity and suspensions in tubes settle out very quickly	Must be of adequate size and rotation speed (measured in revolutions per minute – rpm) for the tests required (The 'power' of a centrifuge is measured in terms of the increase in the force of gravity (g) exerted upon the sample and is expressed in units of relative centrifugal force (rcf), e.g. rcf $4000 \times g$) Specifications for centrifuges quote both the maximum rpm and rcf values achievable by the machine May need a rotor that can be made biologically secure, e.g. if live cultures are to be centrifuged Must be kept clean	A rotor must be properly balanced and the chamber properly secured before the centrifugation operation is commenced Manufacturer's instructions for the safe operation and routine maintenance of the centrifuge must be followed	British Standard BS EN 12547 (1999)
Monitoring / measuring				
Balance	For weighing samples for testing *or* microbiological medium components for media preparation	Top-pan balances are preferred for practicality Balances must be kept clean and level Balances used should adequately cover the range and sensitivity required for the work, e.g. for food samples up to 25 g, a balance sensitivity of 0.1 g with a 100 g range is adequate; for small quantities of media components (antibiotics, dyes), an analytical balance with a sensitivity of 1 mg and a 10 g range may be required	Routine accuracy checks are required using calibrated weights traceable to the national standard, e.g. to within 0.01 g of the target weight for food sample balances and 0.0001 g for analytical balances Check zero before each weighing Records of all checks should be kept	International Standard ISO 7218 (1996) Anon (1999) (3)

Equipment	Description	Checks/Calibration	Reference
Gravimetric diluter	Automatically dispenses diluent to achieve the correct sample dilution factor The machine weighs the sample, calculates the weight of diluent to be added to achieve the correct dilution factor then adds diluent to that weight	Balance and test sample / diluent container support must be kept clean	Routine accuracy checks required using calibrated weights traceable to the national standard, e.g. to within 0.01 g of the target weight Check zero before each weighing Records of all checks should be kept	International Standard ISO 7218 (1996) (Plate 30)
Conductivity meter	For measuring the conductivity of the water produced from the water treatment / purification system	Electrode must be kept clean	Standardisation against a solution of known conductivity within the range of the machine, e.g. 1–100 µS/cm	
pH meter	Measures the pH value of food samples or media	Electrode must be kept clean Separate electrodes should be used for media and foods; electrodes used for foods may need periodic specialised cleaning to remove fat or protein deposits	Calibration against a minimum of 2 standard buffer solutions, e.g. pH 4.0 and 7.0 to within ± 0.1 pH unit, making appropriate compensation allowance for temperature	International Standard ISO 7218 (1996) British Standard BS 3145 (1978) Anon (1999) (9)
Timer	For timing different laboratory operations, e.g. blending time		Check accuracy against the national time signal	
Temperature measuring devices, e.g. mercury in glass thermometer; thermocouple / electronic devices	For routine monitoring of the temperature of all temperature-controlled units, e.g. incubators, refrigerators	Temperature-measuring devices should be capable of measuring temperature to the accuracy required, e.g. $\pm 0.5°C$ Total immersion thermometers should not be used if only partially immersed or vice-versa The specification of the thermometer should be suitable to its application	All temperature-monitoring devices should be calibrated (using several different temperatures in the range of use) against a reference thermometer that is traceable to a national standard	British Standard BS 593 (1989) Anon (1999) (7)

Contd.

Table 6.3 *Contd.*

Equipment type	Application	Operational issues	Routine control parameters	Figure / plate and literature reference*
Pipetters, automatic or plastic disposable	For transferring specific volumes of liquid between containers	The barrels of automatic pipetters must be kept clean	For automatic pipetters, multiple weight checks (3 or more) on volumes dispensed and records kept For plastic disposable pipettes, supplier quality assurance certificate for each batch	British Standard BS 700–1 and 3 (1982) British Standard BS 5732 (1985) British Standard BS 6706 (1986) British Standard BS 1132 (1987) British Standard BS EN ISO 8655 (2002) Anon (1999) (12) (Plate 12)

*International and National Standards are regularly reviewed and updated and it is always advisable to ensure the latest relevant Standards are available in the laboratory. Current Standards are listed on the website of the British Standards Institution at www.bsi-global.com and International Standards on the International Organisation for Standardisation website at www.iso.ch

Table 6.4 Examples of alternative microbiological methods that require capital equipment for their operation.

Technique / Method	Examples
Electrical	Bactometer®, bioMérieux UK Ltd
	Malthus System™, IDG (UK) Ltd
	RABIT, Don Whitley Scientific Ltd, UK
Enzyme Linked Immunosorbent Assay (ELISA)	VIDAS® *Salmonella*, bioMérieux SA, France
	Salmonella -Tek™, bioMérieux UK Ltd
	TECRA® *Salmonella* Visual Assay, TECRA Diagnostics UK Ltd
	EiaFOSS, Foss UK Ltd
Nucleic Acid Hybridisation Probe	Gene Trak® *Salmonella* Assay, Gene Trak Systems, USA
Polymerase Chain Reaction (PCR)	TaqMan™ for *Salmonella*, PE Applied Biosystems, USA
	Probelia™, Sanofi Diagnostics Pasteur, France
	BAX™, Qualicon, USA

Wherever possible, to facilitate good aseptic practices, sterile plastic disposable equipment should be used, e.g. pipettes, containers for sampling, weighing and dilution and Petri dishes (Plates 12 and 13b). Equally, where laboratory equipment and tools are used for sampling within food production areas, it is important that glass items are avoided. If this is not possible, glass items should be listed on the factory 'glass register'.

Most laboratories have individual preferences in their ways of working and use different types of equipment, methods and procedures; it is therefore necessary for all new staff to be fully and effectively trained to ensure they become competent in the methods and operations of the particular laboratory in which they are working (see Chapter 7).

6.6   Further reading

Advisory Committee on Dangerous Pathogens (ACDP) (1995) *Categorisation of Biological Agents according to Hazard and Categories of Containment,* 4th edn. HSE Books, Sudbury, UK.

Advisory Committee on Dangerous Pathogens (ACDP) (1998) Supplement to: *Categorisation of Biological Agents according to Hazard and Categories of Containment,* 4th edn. HSE Books, Sudbury, UK.

Advisory Committee on Dangerous Pathogens (ACDP) (2001) *The Management, Design and Operation of Microbiological Containment Laboratories.* HSE Books, Sudbury, UK.

Anon (1990) Council Directive 90/679/EEC on the protection of workers from risks related to exposure to biological agents at work. *Official Journal* No. L374, 31/12/90. p. 1.

Anon (1994) *Health and Safety. The Control of Substances Hazardous to Health (COSHH) Regulations 1994. Statutory Instrument No. 3246.* HMSO, London, UK.

Brown, K.L. (ed.) (1994) *Guidelines for the Design and Safety of Food Microbiology Laboratories,* Technical Manual No. 42. Campden Food & Drink Research Association, Chipping Campden, UK.

Downes, K.P. and Ito, K. (2001) *Compendium of Methods for the Microbiological Examination of Foods.* 4th edn. Compiled by the APHA Technical Committee on Microbiological Methods for Foods. American Public Health Association, Washington D.C., USA.

International Standard ISO 7218 : 1996 *Methods for Microbiological Examination of Food and Animal Feeding Stuffs. Part 0. General Laboratory Practices.* International Organisation for Standardisation, Geneva, Switzerland.

7 Laboratory Operation and Practice

7.1 Introduction

In addition to an understanding of the standards and conditions necessary in the design and construction of microbiology laboratories, it is essential that laboratory staff have a good understanding of the practices and procedures that should be adopted if microbiology is to be practised safely and the results of the work carried out are to be reliable.

As discussed in the 'Introduction' to Chapter 6, for practical purposes, microbiology laboratory design and construction should take account of the guidance given by the Advisory Committee on Dangerous Pathogens (ACDP, 1995 and 2001) for the containment of infectious agents categorised as Hazard Group 2. The same ACDP guidance level should also be applied to laboratory operations. Table 7.1 outlines some of the requirements indicated for organisms in Hazard Group 2 that affect food microbiology laboratory operation and procedures as given by the Advisory Committee on Dangerous Pathogens (ACDP, 1995 and 2001); some of these are discussed in more detail in the following sections.

All laboratory methods, practices, operations and procedures should be fully documented in clear language and unambiguous style for use as a reference source for staff and for training of new staff (see Chapter 8).

7.2 Standard operating procedures

A 'Standard Operating Procedure' (SOP) is a procedure that has been determined, usually by senior laboratory staff, as the most appropriate means for carrying out a specified activity in the laboratory. SOPs may relate to any aspect of the laboratory operation e.g. equipment operation, routine equipment checks, housekeeping schedules / requirements, staff practices. All SOPs should be documented and these documents signed and dated by the author and / or an appropriate senior member of the laboratory staff. The text within an SOP should be kept focused on the topic and written in an easy to read and search manner, e.g. using bullet points and flow diagrams where possible (see Chapter 8 and Tables 8.3 and 8.4).

Plates 9, 14, 15 and 16 are located in the colour plate section at p. 100.

Table 7.1 Some containment level requirements for organisms in Hazard Group 2 that affect laboratory operation and procedures, adapted from the Advisory Committee on Dangerous Pathogens (1995 and 2001).

Requirements for laboratory procedures and practice
(1) Access to the laboratory is to be restricted to authorised persons.
(2) Effective disinfectants should be available for immediate use in the event of spillage. There must be specified disinfection procedures.
(3) The laboratory door should be closed when work is in progress.
(4) Hands should be decontaminated immediately when contamination is suspected, after handling infective materials and before leaving the laboratory. When gloves are worn, these should be washed or preferably changed before handling items likely to be touched by others not wearing gloves, for example, telephones, paperwork. Computer keyboards and, where practicable, equipment controls should be protected by a removable flexible cover that can be disinfected.
(5) Bench surfaces should be regularly decontaminated according to the pattern of work.
(6) Mouth pipetting should be forbidden.
(7) Used laboratory glassware and other materials awaiting sterilisation before recycling should be stored in a safe manner. Pipettes, if placed in disinfectant, should be totally immersed.
(8) All waste material, if not to be incinerated, should be disposed of safely by other appropriate means.
(9) All accidents and incidents should be immediately reported to and recorded by the person responsible for the work or other delegated person.

7.3   Laboratory staff and personal practices

The practices and behaviour of staff in a microbiology laboratory directly affect the quality of the work carried out and, hence, the reliability of the results obtained. It is therefore important that all staff adhere to common standards of good personal practice and behaviour designed to ensure there is no adverse effect on work and the quality of results that can be attributable to the laboratory staff. The following indicates the commonly accepted basic standards and 'rules' relating to laboratory staff and staff practices in food microbiology laboratories.

(1)   All staff must be trained to the standard required for carrying out designated tasks. Training records must be kept to demonstrate that the standard has been attained and is maintained. Each laboratory designs and operates its own staff training system and the systems for recording the training done; a training record should be in place for each member of staff and every task for which they are trained should be assessed, against pre-determined criteria, by a competent and authorised person who should comment on and sign the training record when the standard for each task is met. Where possible, objective evidence should be obtained and recorded to demonstrate competence in a task, e.g. by comparing balance check-weight results, or results from test sample dilution series that have been plated and counted both by the trainee and by a trained person. Table 7.2 shows an example of what training topics may be contained in a basic training record for a member of staff new to a microbiology laboratory.

Table 7.2 Example of a basic training record for staff new to a microbiology laboratory.

Name:
Date employment commenced:
Position / grade:

Task	Date commenced	Assessment criteria and record*	Date completed	Assessor's (authorised person) signature
Induction training The laboratory's role and staff Laboratory layout and equipment Health and Safety policy First aid in the laboratory Security policy Reporting structure Personal clothing procedure Laboratory clothing procedure Personal hygiene / hand-washing procedures Requirement for confidentiality Use of telephone / message taking and reporting				
Introduction to laboratory quality systems: Laboratory's Quality Policy Equipment files Standard Operating Procedures Equipment calibration and monitoring Accreditation standard External quality assessment standard Record systems				

Contd.

Table 7.2 *Contd.*

Basic laboratory operations†: Stores receipts – booking-in, quality control, unique identifiers Disinfection chemicals and procedures Use of balances ⎫ Use of pipettes ⎬ including checks and records Use of pH meter ⎭ Use of incubators, water baths, refrigerators and freezers and associated temperature monitoring systems and records Use of laminar air-flow cabinet		
Basic microbiological procedures†: Aseptic procedures Diluents, dilutions and dilution series Incubation conditions and equipment Colony counting		
Media preparation†: Microbiological media – types and sources Water purification system including checks and records Preparation procedures Sterilisation procedures and equipment Quality control systems Unique identifiers applied to prepared media Record systems		

Sample preparation†: Booking-in of factory samples and unique identifiers applied Storage conditions Sampling equipment Weighing test sample / dilution / blending Further dilutions Pipetting – equipment and practices		
Media inoculation procedures and practices†: Pour plates Spread plates Liquid media Streaking Sub-culture techniques Incubation of test materials		

* Assessment criteria for each area of training must be established and documented. Assessment may be by observation of the trainee performing the task, by comparing results obtained by the trainee and a trained person performing the same task or by some other objective and documented means. Where possible, a cross-reference to the trainee's results obtained during training should be made in the training record so that these can be inspected at a later date if necessary.

† As part of the training schedule, the Standard Operating Procedures associated with each practice, procedure or method must be read and understood by the trainee.

Note that the reading, interpretation and recording of test results after incubation are critical tasks to be undertaken by more experienced members of staff only.

(2)   Only laboratory staff should be authorised to enter the laboratory and generally all non-laboratory personnel must be discouraged from entering the laboratory and should be prohibited in the pathogen handling area. Certain persons, e.g. auditors, repair and maintenance personnel, may be authorised to enter the laboratory for specific purposes but should always be accompanied and supervised by laboratory staff.

(3)   (a) Access to the laboratory for all staff must only be through a dedicated coat-changing and hand-washing facility (see Figure 6.2).
(b) All laboratory staff must adhere to the laboratory's coat-changing and hand-washing policy, which should be specified in a Standard Operating Procedure. In any case, all personnel should wash and dry their hands thoroughly each time on entering (before donning laboratory coat) and leaving (after removing laboratory coat) the laboratory. Disposable paper towels should be used in preference to hot air hand dryers; cloth hand towels should be prohibited.

(4)   (a) Laboratory coats of good design, e.g. with wrap-over button-neck and elasticated wrists, ('Howie style'), must be worn at all times in the laboratory (Plate 14, see colour plate sections for all plates). Coats should be maintained in a clean condition and in a state of good repair. Sufficient numbers of clean coats should be available for each member of staff to ensure regular changes can be made and for any additional changes required in the event of unexpected or accidental coat contamination.
(b) Coats used in the microbiology laboratory must never be worn outside the laboratory.
(c) Coats should never be laundered at home by staff; a dedicated laundering service should be employed that can handle and effectively clean such coats safely. Soiled coats must be adequately protected during transport to the laundry, e.g. using dedicated laundry bags that are cleaned together with the coats or using water soluble bags that are placed, with their contents, directly into the washer.

(5)   Long hair should be tied back in such a way that it will not fall forward during work (Plate 14).

(6)   Good personal hygiene practices must be observed at all times. Eating, drinking, smoking, spitting, nail-biting and the application of anything to the face or mouth must be prohibited in the laboratory.

(7)   Pot plants, pictures, personal belongings and decorative material must not be kept in the laboratory.

Some companies that maintain a food microbiology laboratory, particularly those on food manufacturing sites, operate a specific health policy for the staff, especially those working with pathogenic microorganisms, in which they are routinely monitored by stool testing, e.g. annually, for organisms such as *Salmonella* to ensure that carrier states are not established. Visitors may have to sign a health declaration.

7.4   Laboratory housekeeping

The standards of environmental tidiness and cleanliness operated and maintained in the laboratory can also affect the quality of work carried out in the laboratory. Attention paid to the following points is important for maintaining acceptable environmental and housekeeping standards in the laboratory.

7.4.1   *General*

(1)   The laboratory must be kept tidy at all times. To facilitate cleaning, boxes of 'disposables' etc. should be stored off the floor, and levels of stocks held in the laboratory should be kept to a minimum. Bulk stocks should be held in a suitable dedicated microbiology store (see Figure 6.2).

(2)   A good disinfectant, for example hypochlorite based, must always be available; the use of hypochlorite is especially important when there is a need to inactivate bacterial spores, e.g. if the laboratory routinely handles dried foods. Benches should be disinfected at the end of each working period and as necessary to maintain a safe working environment. Disinfectants must be used at recommended strengths and adequate contact time must be allowed. Procedures covering routine laboratory surface disinfection and spillage containment and disinfection should be written into the laboratory's Standard Operating Procedures.

(3)   A regular laboratory cleaning and disinfection programme must be maintained to prevent floors, walls, ceilings, fitments and equipment such as water baths and the interior of incubators and refrigerators from becoming sources of infection for personnel, samples and media. Cleaning staff must have clear instructions and should not be allowed to handle or clean-up hazardous materials or attend to equipment such as incubators.

(4)   Soiled laboratory coats should preferably be autoclaved before being sent for laundering or a reliable method employed that will protect any handlers of the coats from inadvertent contamination during the transporting and laundering of the coat prior to return to the laboratory. A laboratory Standard Operating Procedure should specify the procedures to be employed for handling soiled coats to be laundered.

7.4.2   *Laboratory waste disposal*

(1)   Safe waste disposal systems must be provided, understood and used effectively by all staff. Waste must never be taken from the laboratory through food manufacturing areas.

(2)   Used pipettes, tips for pipetters and microscope slides should be discarded into containers of disinfectant until they can be autoclaved prior to disposal. Pipettes and tips for pipetters should be **fully immersed** in the disinfectant.

(3) Samples entering the laboratory for examination must be disposed of through the laboratory and never returned to the factory. Unused food samples that have entered the laboratory should not be consumed. Samples in the laboratory must be kept secure to prevent unauthorised removal.

(4) Autoclave processes for decontaminating waste should normally achieve a minimum of 121°C for at least 30 minutes and preferably 45 minutes, or an equivalent heat process. Some laboratories employ external incineration services to destroy waste.

7.4.3 *Environmental monitoring*

Because the frequent handling of food samples and microbiological cultures can generate dust and aerosols that can spread around the laboratory environment, contaminating work surfaces etc., it is important to ensure the laboratory environment and equipment are routinely monitored for microbiological contamination. A schedule should be drawn up covering the key areas to be examined, frequency of assessment and relevant test organisms, and this must be reviewed regularly. Bench surfaces used for sample and culture processing, hand-wash basins, taps, handles and door seals, overhead structures, laminar air-flow cabinets, balances, laboratory blenders, (see Plate 9), and any other relevant areas must be identified and suitable means used to monitor their hygienic status, e.g. visual assessment, swabs, contact plates, settle plates or other sampling devices for monitoring air quality. Records must be kept of the results of all such tests.

7.5   Laboratory quality standards and procedures

7.5.1 *General*

To achieve and maintain consistently high performance standards, food microbiology laboratories are dependent on a variety of external services and supplies. Equipment, e.g. balances, autoclaves, incubators, etc., and consumable supplies, e.g. media, plastic disposables, etc., should always be obtained from reputable suppliers who have attained an appropriate standard of quality management system, e.g. International Standard BS EN 9001 : 2000 or an equivalent Standard. All suppliers should be expected to provide a reliable technical support service as well as technical data and quality assurance data in relation to their products.

   It is also important to ensure that all equipment required to support microbiological testing work is operating to a consistently high standard. This requires the implementation and maintenance, by laboratory staff, of adequate equipment calibration regimes and daily recorded checks to ensure the correct functioning of all key laboratory equipment, e.g. water baths, autoclave(s), incubators, refrigerators, pH meters, volumetric equipment (pipetters, gravimetric diluters, dispensers), balances, laminar air flow equipment and automatic counting systems (see Table 6.3).

7.5.2   *Laboratory assessment / accreditation and external quality assessment (proficiency testing)*

As a means of providing some assurance to the staff of the laboratory, the laboratory's employers and the employers' customers (as applicable) that appropriate standards are being maintained in the laboratory, many laboratories now subscribe to externally operated, independent quality management schemes (assessment / accreditation bodies) for assessing good laboratory procedures and practices. In these schemes, an annual inspection of the laboratory's quality systems, procedures and practices is carried out and reported on by independent inspectors. This can help the laboratory staff to maintain and develop the systems required to support the quality of work carried out and provides a certificate of accreditation / assessment to laboratories that comply with the agreed standards (see Chapter 8).

In addition to an annual audit by an independent body, laboratories also participate in one or more External Quality Assessment (EQA) (or Proficiency Testing) schemes. Through these schemes, the laboratory is provided with materials containing known organisms and numbers of organisms for examination; the microbiological content of the materials supplied is not revealed to the laboratory until after the test results have been returned to the EQA provider. These tests provide a challenge to the whole laboratory operation and procedures including:

- sample receipt,
- media preparation and quality control,
- sample handling procedures and traceability,
- staff practices, and
- method efficacy.

EQA providers, who are independent of the laboratory, are contracted to provide, at an agreed frequency (commonly every two months), samples of specified microbiological composition for the laboratory to detect and / or enumerate. In some schemes, the sample instructions indicate which tests should be undertaken whilst, in others, the laboratory must decide which tests are appropriate from a microbiological history invented and supplied by the EQA provider. Results are returned to the EQA provider for assessment against both the anticipated result and the results from other scheme participants. Following a statistical analysis of all results returned from participants, the laboratory is informed of the expected result and scored for its own result against the expected result. Reports from each of these challenges can be used by the laboratory staff to assess their ongoing performance so it is important for all staff to participate e.g. twice per year, in the examination of EQA samples for all the tests for which they are trained. Some laboratories also use these samples for staff training purposes.

Subscription by the laboratory to these schemes (whereby the overall operating and quality standards of the laboratory including supplies, equipment maintenance and operation, test methods used and staff performance throughout sample handling and processing, can be assessed for all the types of test carried out and involving all trained staff) can provide a useful method for helping to demonstrate the credible operation of the laboratory.

7.5.3 *Practices*

The aim of carrying out any microbiological test is to obtain the most accurate assessment of a specified microbial population in the test sample. In order to achieve this, relevant validated methods must be used (see Chapters 8 and 12) and these must be operated using consistently high standards of practice and procedure.

All sample handling and associated test sample manipulations must be carried out using high standards of aseptic practice that are operated uniformly by all staff. These include the operations of weighing, pipetting and sub-culturing by spreading and streaking, in all of which the sample material, sample dilutions or cultures are exposed to the laboratory environment. Poor staff practices are a hazard to the work (samples, cultures and results) in the laboratory, the laboratory environment and the people in the laboratory, and can perpetuate sources and routes of cross-contamination to the work.

There are a number of stages in the procedures for processing factory and test samples and cultures at which considerable care must be taken to minimise reliably the possibility of cross-contamination. This is achieved by the operation of high standards of aseptic practice. Table 7.3 indicates those steps in sample processing that require the specific application of aseptic practice and the following are some fundamental requirements for achieving good aseptic and general microbiological practices and ultimately, reliable results:

- All containers, tools, media and other test equipment / materials used for directly handling / processing the factory or test sample must be sterile. Preferably, and wherever possible, single use, sterile, plastic, disposable tools / implements should be used. However, if reusable implements are used, they must be properly cleaned before sterilisation. An autoclave process is the preferred means of sterilisation. Where an alcohol flaming method is used to sterilise implements, they **must** be adequately cooled prior to use otherwise the microorganisms being sought may be killed and a false result will be obtained. In addition, aerosols may be liberated into the environment if the hot implement contacts food material, broth or agar.
- Spreading and streaking techniques used for dispersing an inoculum over the surface of an agar in a Petri dish (plate) must be undertaken with care to avoid contact with the edge of the plate which can lead to microbial growth that is difficult to assess and can lead to inaccurate results (Figure 7.1).
- Mixing procedures for each inoculum with an agar, in the pour plate technique (see Chapters 9 and 10), must be undertaken with care to avoid splashing the sides and lid of the plates. Contamination and confluent growth, where colonies grow and spread into each other so that individual colonies cannot be distinguished, can result from over-vigorous mixing. This makes incubated plates difficult to assess and leads to inaccurate results.
- When a water bath is used for tempering bottles of molten agar, the bottles must be properly dried using disposable paper towels prior to pouring the agar into Petri dishes. This prevents un-sterile water from the water bath from contaminating the inoculum, which would otherwise add microorganisms to the test sample under examination and lead to inaccurate results.

Table 7.3 Sample processing procedure and the steps at which specific aseptic practice is required.

Process steps	Aseptic handling specifically required (Yes) Good microbiological practice required (GMP)
Receipt of factory sample and storage of factory sample	GMP
Weighing of test sample	Yes
Addition of diluent	Yes
Blending / homogenisation	GMP
Further dilution(s)	Yes
Inoculation (media / petri dish)	Yes
Addition of molten media	Yes
Mixing / dispersion	Yes
Incubation	GMP
Assessment of growth	GMP
Sub-culture for confirmation	Yes
Incubation	GMP
Assessment of results	GMP

7.5.4 *General procedures*

The following are some of the generally accepted and therefore common areas where specific procedures and controls should be applied to the variety of the routine operations in many food microbiology laboratories. Failure to operate the correct procedures affects the accuracy of the results obtained. Although individual laboratories inevitably have their own variations of 'standard procedures', the actual procedure used should be documented and, where possible, a reference / justification should be given in support of the in-house procedure. Staff in each laboratory must understand the quality systems specific to the laboratory and ensure that they are implemented correctly at all times.

Test sample preparation and handling

The terminology relating to sampling can be confusing. Figure 7.2 shows the common stages of sampling and sample processing that are encountered and the terms used to describe the sample material at each stage.

(1) (a) All factory samples taken for microbiological tests should be examined within a timescale appropriate to the sample, so that changes in the microbial composition of the sample between sample collection and testing are minimised; this is important to ensure that the interpretation of the results is relevant. For example, a chilled product taken from a manufacturing line should be

Figure 7.1 Diagram of plates illustrating correct positioning of streaks and spreads.

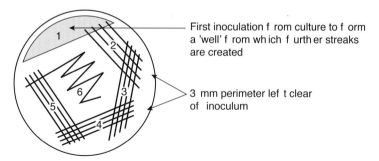

First inoculation f rom culture to f orm a 'well' f rom wh ich f urth er streaks are created

3 mm perimeter lef t clear of inoculum

(a) A streak plate. The loop should be flamed (or replaced if disposable) after step [1] so that only the inoculum on the plate is carried forward in step [2]. This is important to obtain satisfactory reduction of the inoculum and thus adequate separation of cells, especially if the initial inoculum was taken from a colony which may contain a very high number of cells, e.g. 10^{10}–10^{11}. The streak is a 6-streak form with minimal overlap between 1 (well) and 2 (first streak), and streaks 2/3, 3/4, 4/5 and 5/6.

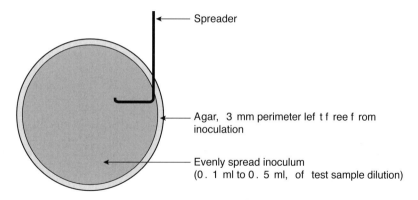

Spreader

Agar, 3 mm perimeter lef t f ree f rom inoculation

Evenly spread inoculum (0. 1 ml to 0. 5 ml, of test sample dilution)

(b) A spread plate. The spread is even and completely covers the surface of the agar except that approximately 3 mm around the edge of the agar must be left clear.

examined ideally within 1 hour of sample collection to ensure the sample's condition adequately reflects the manufacturing circumstances at the time of sampling; for the same reason, swabs should be processed within 30 minutes of being taken. Many food processors, however, now use external contract testing laboratory services, especially for pathogen testing, and factory sample transport conditions and timescales must therefore be carefully designed and operated to ensure that the test results obtained are meaningful.

(b) All samples must be packed cleanly for transport taking care to keep raw or 'dirty' samples e.g. dusty and environmental samples well-separated from clean samples. This segregation of sample types must be maintained during subsequent storage and handling in the laboratory. For chilled factory samples, the main priority is to ensure that effective refrigerated conditions are maintained; ideally, this is accomplished by means of a refrigerated vehicle but, where this cannot be arranged, insulated boxes containing ice packs are often used. Once

Figure 7.2 Sampling and sample terminology.

Sampling and testing stage	Notes / consideration
*Factory sample** ↓ Laboratory ↓	Can be taken from raw materials / production line / finished products / environments, etc.
Weigh *test sample***, e.g. 10 g, 25 g	Appropriate quantity according to type of test / specification
For detection tests **For enumeration tests**	
Add pre-enrichment / enrichment medium Add primary diluent	Usually 9 times the weight of test sample
↓ ↓ Blend to produce test sample suspension Blend to produce test sample suspension	Minimum 30 seconds
↓ ↓ Incubate (1) Further dilutions as necessary (2)	(1) Temperature / time (2) (See Figure 10.1)
↓ ↓ Sub-culture Inoculation of medium	Pour plate / spread plate / broth / streak plate
↓ ↓ Incubate Incubate	Temperature / time
↓ ↓ Assess growth[†] Assess growth[†]	Calculate and report result
↓ ↓ Confirmatory tests as required Confirmatory tests as required	Calculate and report result

Notes:

* When a factory sample is delivered to the laboratory it may also be known as the laboratory sample. Factory samples may be essentially of 3 types:

(a) A number of small quantities of material removed from different parts of a bulk load, e.g. frozen meat blocks, powders packed in drums (egg, milk powder), herbs, spices, etc., or production line samples, e.g. from processing stages such as slicing, mincing, mixing, cooking, extrusion, that are bulked together to form one larger sample. Aseptic procedures are required and such sampling must be carried out by properly trained staff.

(b) Swabs or samples of gross debris for environmental monitoring purposes. Aseptic procedures are required and such sampling must be carried out by properly trained staff.

(c) Sealed units of food materials, e.g. finished product from production lines, for direct delivery to the laboratory. Aseptic procedures are not usually required but care is required not to contaminate the outer packaging.

** 'Test sample' is the term used to describe the portion or quantity used in the microbiological examination to which the primary diluent or initial growth medium is added.

† A further subculture and incubation stage may be required depending on the method used.

in the laboratory, the factory samples must be transferred quickly to an appropriate refrigerator.

(c) When microbiological testing is sub-contracted to an external laboratory, it is inevitable that the 'ideal' timescales mentioned above are almost always exceeded; however, wherever possible, 'same day' testing should be achieved for all factory samples held in refrigerated storage, and especially for swabs. Microbiological examination should never commence later than 24 hours after the factory samples are taken unless they are required to be held by the receiving laboratory under specified conditions for shelf life testing or any other clearly defined purpose. The only types of sample that may be held for longer periods without incurring significant changes to their microbiological composition are frozen and ambient-stable foods but, even for these types, storage conditions must be maintained as appropriate to the material type, and testing must be carried out within the timescale required by the customer and should not be delayed without very good reason.

(d) A specified quantity of test sample e.g. 10 g or 25 g is removed from the factory sample using sterile implements and good aseptic practices. It may be necessary to sanitise the factory sample packaging prior to opening and removal of the test sample and there may be a requirement to remove the test sample from a particular area of the pack e.g. an area away from the point of opening, to minimise the possibility of extraneous contamination of the test sample. All requirements relating to test sample removal, weighing and dilution should be documented for each type of factory sample.

(2) (a) For solid food materials, test sample treatment must ensure good dispersion of the sample in the diluent, e.g. using a laboratory blender and commonly, a minimum blending time of 30 seconds, maximum 1 minute. A timer should be provided and used to ensure that the required time is achieved.

(b) Samples should normally be blended singly, but where more than one sample is blended, the bags should be staggered in the blender so that the contents of each bag are presented to the paddles. Care must be taken not to exceed the capacity of the blender.

(c) Heat (35–40°C) and / or surfactant are used where necessary to aid dispersion, e.g. for high fat ($> 30\%$) samples; however, heat should be used with caution where samples are to be examined for psychrotrophic bacteria that may be very heat-sensitive.

(d) Care is required in test sample treatment to ensure that there is minimum shock to organisms and maximum recovery by using, for example,

- resuscitation techniques where necessary, e.g. dried products should be re-hydrated gently using a method appropriate for the sample type
- extra dilution, quenching agents or special diluents to inactivate inhibitory substances such as sanitisers or essential oils naturally present in foods such as onion, garlic and some herbs and spices, as well as some components of cocoa materials alone, or in chocolate.

A laboratory Standard Operating Procedure should detail the required approach to be used for each test sample type and care needs to be taken by laboratory staff to understand and implement these.

(e) When using techniques involving heat treatment of the test sample or dilutions of the test sample, e.g. for spore enumeration, sufficient time must be allowed to ensure that the whole sample receives the required heat treatment. Where possible, sample dilutions should be prepared in '1 ounce' Universal bottles fitted with tightly fitting metal caps, carefully sealed with waterproof tape, that can then be completely submerged in the water bath for the heat treatment. Otherwise, care must be taken to ensure the level of test sample suspension in the bottle / tube is at least 1 cm below the hot water level. The time taken to reach the heat treatment temperature varies according to the volume and density of test sample suspension and the size of container used. Care must be taken to ensure that the heating characteristics of each test material volume / container are known and used appropriately. Preferably, to avoid mistakes, only one system should be used in a laboratory, e.g. 10 ml of 1 in 10 test sample suspension in a '1 ounce' McCartney bottle or test tube.

Note: A 'dummy' tube (containing a small volume of the test sample suspension and a thermometer) should be used to check the adequacy of the heat treatment, and, when sub-culturing from heat-treated test samples, care must be taken to ensure there is no carry-over of untreated sample material from around the neck of the container.

A laboratory Standard Operating Procedure should detail the particular procedures required that are relevant for each test sample type and these must be understood and properly implemented by all staff.

(3) Diluted test samples must be inoculated into growth media within 20 minutes of their initial dilution and, preferably, should be re-mixed immediately before being used to inoculate media.

(4) Where diluents for primary dilution are pre-prepared in specific volumes, e.g. 90 ml and 180 ml, care must be taken to ensure the accuracy of these volumes, as substantial losses can occur due to evaporation during autoclaving. Inaccurate volumes can contribute to inaccurate results being attributed to the test sample. Weighing out the diluent, either manually or using commercially available automated equipment e.g. gravimetric diluter, from a bulk container onto the weighed test sample in the sample container, e.g. a blender bag, will overcome this problem and improve the accuracy of the test sample dilution.

Particularly large volume changes (losses) can occur during the autoclaving and subsequent storage of diluents made up in 9 ml volumes used to make secondary dilutions of the test sample, and this can markedly influence the accuracy of colony counts. For these diluents, it is essential to undertake weight or volume checks, after autoclaving, and to minimise the storage time before use, especially if tightly fitting, screw-capped containers are not used.

(5) A laboratory should have suitable control systems to ensure that contaminated diluents or media (made in-house or bought in ready-prepared) will be recognised e.g. sterility checks of incubated uninoculated plates. All staff must understand what these systems are and ensure that they are fully implemented.

(6)   (a) In order to avoid sample-to-sample cross-contamination, pipetting practices must ensure the pipetting unit is kept free from contamination. The barrel of a pipette unit, in particular, can become contaminated with sample material or diluent and care must be taken to prevent this from occurring. Pipetting units and pipettes should be held in the vertical or near vertical position for use (Plate 15) and always stored in dry conditions with the barrel clear of any surface that may be contaminated. Where tips are used with automatic pipetting units, the plunger should always be released gently to prevent the liquid from 'jumping' and contaminating the barrel. Pipetting units should be inspected frequently and it is good practice periodically to dispense a volume of sterile diluent to a plate of non-selective agar (for pour- or spread-plate technique) which should be incubated at a mesophilic temperature. Any growth at all on these plates indicates a need to investigate staff pipetting practices.
(b) Disposable pipettes, pipette tips and 'straws' should each be used for volume transfers once only. This will help minimise cross-contamination and improve the consistency and accuracy of the volumes transferred.
(c) Pipetters must be cleaned and sanitised regularly to assist aseptic practice and be calibrated routinely to ensure accuracy. After cleaning, they must be dried adequately and kept dry when in use.

(7)   (a) Surface plating media and broth media must not be used directly from the refrigerator. Media should be allowed to equilibrate to room temperature (or as specified in the laboratory Standard Operating Procedure) before use.
(b) Surfaces of agar in plates for spreading must be smooth, level, entire and adequately dried just prior to use. The spread plate inoculum volume should be limited to a maximum of 0.5 ml for a 90 mm diameter Petri dish.
(c) Broths to be used for anaerobes must be freshly steamed and cooled before use, to expel dissolved oxygen.

(8)   As a general rule, all prepared 'pour plate', 'spread plate' and 'streak plate' Petri dishes should be incubated in an inverted position to prevent condensation from wetting the agar surface and causing colonies to merge, forming confluent growth that cannot be used to obtain a reliable result; however, the main exceptions to this are:
  • plates used to grow mould colonies, where mould spores may be dislodged by turning the plates over and over,
  • plates containing membrane filters,
  • plates in anaerobe jars where the air is removed by vacuum, and
  • plates containing pieces of food on an agar surface or filter paper.
    When plates are prepared for high temperature incubation, e.g. 55°C, or for long incubation times, e.g. >3 days, precautions must be taken to minimise evaporation and consequent drying out of media, e.g. by increasing the volume of agar used from the normal 12–15 ml to 20–25 ml, and humidifying the incubator with containers of water or by loosely wrapping the plates in 'Petri dish sleeves' or blender bags. Where containers of water are used, these must be changed regularly and kept clean and sanitised to prevent them from becoming a source of microbial contamination in the environment. In the case of some tests for moulds that require incubation for >7 days, the Petri dish sleeves or

blender bags must be perforated to allow the diffusion of oxygen into the atmosphere above the agar surface.

(b) Adequate incubation time must be given to all inoculated media. If test samples are processed in the afternoon and the media require 24 hours incubation, results of growth on the media must be assessed the following afternoon and not in the morning.

(c) After incubation, all media (agar and broths) must be thoroughly examined to extract all useful information and any unusual observations should be recorded with the results, e.g. colony descriptions if different from usual. Notes concerning the colony colour, shape, size, topography, moist or dry growth and any other distinguishing characteristics, can all help in tracing organisms that may have become a problem to a food production process. Good observation and descriptive skills alongside clear notes of the observations made can prove invaluable in solving a production microbiological problem.

(d) If after incubation there is difficulty in distinguishing colonies from debris, or if real colonies are very small, plates should be incubated for a further 24 hours before counting and / or a low power microscope or hand lens should be used to assist in colony recognition and counting. A common mistake, for example, is to confuse particles of cooked egg yolk with small colonies of microorganisms resulting in an incorrect high count being recorded.

(9) Unless otherwise required by the laboratory's customer, all microbiological counts are usually expressed per gram or millilitre of the test sample unless a detection (presence/absence) test is carried out when the quantity of test sample examined must be given. The results of enumeration tests ('counts') to be reported are calculated from the number of colonies growing on the agar medium derived from a known volume of a known dilution of a test sample (see Chapter 10). Confusion sometimes exists in expressing swab results as, often, swabs are dispersed in a volume of diluent, only a small quantity of which is plated. In this case, the number of colony-forming units per ml of the primary swab dilution is only relevant when calculating the results (for which the dilution factor should be taken into account) of the number of colony-forming units per swab (or, per unit area swabbed) (see Chapter 10).

(10) Organisms that may have a particular significance to the safety of a product, e.g. *Salmonella* spp., or those that represent unusual findings for the product examined may be isolated occasionally. The cultures isolated on these occasions should always be kept, stored under appropriate conditions, to provide the basis for further investigation.

Media preparation and use

The standards of media preparation and use that are operated in the food microbiology laboratory can affect the quality of test results obtained by that laboratory. Careful attention to and maintenance of high standards in the following areas are important for achieving acceptable standards of media quality for use in the laboratory.

(1)   Before handling any dehydrated medium or medium supplements, the manufacturer's safety information should be consulted and, where specific precautions are recommended, then these should always be observed. In any case, it is good practice to use a dust extraction hood for dispensing the dehydrated powdered material, taking care to ensure the extraction air-flow does not adversely affect the operation of the balance. If there is any doubt at all concerning appropriate handling procedures, then the supplier of the medium should be consulted.

(2)   Media must always be prepared and used according to the instructions of the manufacturer or author (where the medium formulation is taken from a publication).

(3)   Dehydrated powders / granules must be weighed accurately using a balance that is calibrated and checked routinely (see Table 6.3).

(4)   The quality of the water used for making growth media can affect the performance of microbiological media. Tap water contains many chemical impurities, including calcium, magnesium, chlorine, fluorine and, occasionally, toxic metal ions such as copper. These can adversely affect the characteristics of selective media designed for isolating specific bacteria from mixed cultures. Tap water should, therefore, never be used to make media for bacteriological use. Media for the isolation and identification of moulds, however, must always contain trace elements in order to permit the development of the coloured structures and spores by which these organisms are characterised and identified. This is normally achieved by the addition of a trace element solution, containing zinc and copper, to the normal growth medium prepared with purified water (see Chapter 9) but, if a trace element solution is not available, tap water usually makes a workable alternative that is preferable to the use of purified water alone. Traditionally, it was not necessary to add trace elements to mycological media as they were found in sufficient quantities as impurities in the formulations produced by the media manufacturer; as a result, they are rarely mentioned in the formulations for mycological media. However, the media produced by some suppliers are now so chemically pure that good growth of mould colonies does not occur unless trace elements have been added.

Water purified by distillation, reverse osmosis or deionisation is suitable for dissolving dehydrated microbiological media and should be used freshly prepared. If purified water has to be stored, it should be in lidded, clear glass containers, not in polythene or soda glass storage containers which can cause a change in its pH value. Metal pipework and containers must also be avoided to prevent toxic metal ions from leaching into the water. Water storage containers and equipment should be cleaned regularly to prevent the development of biofilms and, especially, the growth of algae.

(5)   Water purification equipment is available commercially (see Figure 6.1), and the manufacturer's information and instructions should be clearly understood and adhered to in order to ensure the consistent production and availability of good quality water, particularly for use in media preparation. Regular checks

should be made of the conductivity of purified water and the results should be recorded (see Table 6.3).

(6) All media must be properly dissolved in the appropriate volume of purified water and dispensed into suitable quantities before sterilisation. Most media are made up in large volumes and the specified, weighed quantity of dehydrated medium must be added to the required full volume of purified water to ensure that the correct concentration of media constituents in the final medium is obtained.
Note: Most media manufacturers' instructions require the addition of the specified quantity of dehydrated medium to 1 litre of water. It is important to ensure that a full 1 litre of water is used, as adding water to the pre-weighed dehydrated medium and making the volume up to 1 litre would lead to a higher concentration of medium constituents than that required and may adversely affect the performance of the medium. It is important, therefore, to ensure that the media manufacturer's instructions are followed precisely. Some high-sugar media, e.g. 50% glucose malt extract agar, specify that the sugar concentration be 'weight for weight' (w/w). In this case, the medium would, for example, be prepared by weighing out 500 g glucose plus the minor ingredients and then making up to 1000 g by **weighing** in the required amount of water.
  Although most media are formulated to be readily soluble in water, the application of gentle heat and stirring using a hot-plate / magnetic heater-stirrer unit can greatly assist the effective dispersion and dissolution of the dehydrated material.

(7) After the medium has been dissolved completely it may need to be dispensed into smaller volumes prior to sterilisation. Media are normally dispensed into volumes of up to 500 ml or 1 litre for autoclaving; to obtain rapid heat penetration of each container during autoclaving, maximum volumes of 500 ml are usually recommended. The laboratory's Standard Operating Procedure for media preparation should detail the method for weighing the ingredients and water addition, dispersion, dissolution and dispensing to be used for each medium type.

(8) (a) Almost all microbiological media are heat processed / sterilised prior to use. Different media require different levels of heat process that are determined by the heat sensitivity of their ingredients and the need to inactivate bacterial spores, e.g. violet red bile lactose agar (VRBLA) used for the enumeration of coliforms is processed by boiling because its formulation prevents the growth of bacterial spores which do not therefore need to be destroyed. Rappaport-Vassiliadis (RV) Enrichment Broth, used as a selective enrichment medium in detection tests for *Salmonella* spp., is sterilised by autoclaving at 115°C for 15 minutes whilst Plate Count Agar used for assessing the 'total colony count' of samples, is sterilised by autoclaving at 121°C for 15 minutes. This latter temperature–time combination is a very common medium sterilisation process that is applied to many microbiological media. It is essential, however, that the correct process is applied to each medium since over-processing by heat can destroy the required characteristics of the medium. The medium manufacturer's instruction for the medium sterilisation process to be used for each

product is found on the product pack labels and often in their catalogue of products and must always be adhered to.

(b) Medium sterilisation processes requiring temperatures > 100°C are carried out in autoclaves or medium preparators. These are, in effect, large pressure-cookers, i.e. vessels that can be pressurised to achieve the high temperatures, e.g. 121°C, required to inactivate heat-resistant bacterial spores under controlled conditions.

(c) Medium preparators are usually used to prepare large volumes of a single medium formulation. The pre-dispersed and dissolved medium can be placed into the preparator for sterilisation, or the water and dehydrated medium can be added directly to the preparator container in which it is stirred and heated to disperse and dissolve prior to the sterilisation process being applied.

(d) Where autoclaves are used, these are loaded with containers of dissolved media and the sterilisation process is applied. Different media requiring the same sterilisation process can be sterilised together in an autoclave. The factors affecting the loading of the autoclave, selection of the process to be applied and monitoring of autoclave processes need to be understood and considered carefully before media are subjected to an autoclave process. Table 7.4 indicates some of the factors and considerations that should be taken into account.

(e) Suitable time / temperature recorders should be used for checking the correct functioning of the autoclave or media preparator. The thermal profile (temperature achieved in different parts of the autoclave) must be understood for media (in all sizes of container used in the laboratory) and for waste loads.

(f) The 'come-up' and 'come-down' time / temperature profile of a process contributes to the total heat received by the load (Plate 16). Due consideration must be given to these contributions to prevent over-processing of media.

(g) Every load of media should be checked for adequate processing using in-load monitors placed in containers of medium distributed throughout the load (Table 7.4). A laboratory Standard Operating Procedure should specify in detail how the laboratory autoclaves and media preparators should be used to achieve the correct media sterilisation processes. Technical advice can also be obtained from the manufacturer of the autoclave or medium preparator.

(9)   The required pH value of media prepared for use is specified by the manufacturer and is normally contained in the instructions for preparation on the product pack label. It is important to ensure that the pH value of prepared diluents and growth media is correct at the point of use. Liquid and solid media can be tested for pH value (usually at ambient temperature, 25°C) using appropriate probes (immersion and surface respectively), and the equipment manufacturer's instructions for calibration and use of the pH meter and probes should be followed implicitly. Checking media pH values should form part of the normal media quality control procedures carried out by the laboratory (Table 7.5) all of which should be explained in the laboratory's SOPs.

(10)   (a) Prepared media storage and use conditions, including the conditions for tempering molten agar, and the 'life' under the specified conditions must be applied according to manufacturer's / author's recommendations. The labor-

Table 7.4 Considerations to be taken into account when autoclaving media.

Area	Considerations
Loading of the autoclave (not applicable to a media preparator)	Whenever practically possible, medium volumes of similar size, e.g. 500 ml, should be processed together in a single load, to ensure that all the containers receive the same process and, therefore, minimise the potential for over- or under-processing that could occur with media of mixed volumes. Containers of media should be loaded into the autoclave in a pattern that allows free circulation of the steam and effective heat penetration of the load. For this purpose, round bottles are preferred to 'medical flats'.
Process monitoring	Modern autoclaves are fitted with automatic, electronic temperature- and time-controllers. Once the required process is programmed into the autoclave, the process operates automatically and a chart recorder linked to an independent thermocouple probe(s) inserted into the chamber (and load) provides a printed record of the process achieved. These records must be inspected immediately after each process to ensure the correct process was actually achieved. In addition to electronic monitors, independent physical monitors (maximum thermometer) or chemically based monitors (in tubes or on paper strips) are placed in critical positions in each load to provide additional rapid information about the process achieved. A limitation of some of these monitors is that they may not indicate that over-processing has occurred, i.e. too high a temperature was reached or the length of time at the correct temperature was too long. This is generally not a problem for waste loads or equipment sterilisation but can affect the quality of microbiological media. It is important to understand what information is actually given by the specific types of monitor so that an appropriate one can be selected for use.
Process applied	The timing of the heat process has traditionally been commenced when the autoclave chamber or drain temperature has reached the required temperature, e.g. 121°C. In this case, the process is controlled by the chamber temperature probe. However, in many modern autoclaves, the process timer is linked to a temperature probe located within the load (load-controlled) and for some media types, this can lead to over-processing because of the extended time that the load is exposed to high temperatures (Plate 16). Because autoclave electronics are now very sophisticated, many process variations are available. Autoclave processes applied to microbiological media therefore require careful validation to ensure that over-processing and under-processing do not occur. Prepared media quality control procedures must be designed to show that the functional properties of each medium are correct after heat processing and that the medium is sterile.

Table 7.5 Some prepared media quality control procedures routinely carried out in laboratories*.

Parameter[†]	Method / comment
Colour and clarity	Visual check
pH value	pH meter should be calibrated daily or prior to each use Temperature at which measurement is carried out should not exceed 25°C Appropriate pH probe must be used for liquids and gels pH probes used for foods should not be used for media
Sterility check	Incubation of un-inoculated medium under the appropriate conditions and assessment for growth of contaminants. No growth should be visible.
Efficacy check / positive control (quantitative / qualitative)	Particularly useful for selective-diagnostic media to ensure the 'target' organisms, e.g. *Salmonella*, can grow well and produce the relevant diagnostic colony characteristics Reference cultures of the relevant organisms are used
Efficacy check / negative control	These, again, are particularly useful for selective-diagnostic media to ensure that competitor organisms are either inhibited or produce colonies that are distinguishable from the target organism. Reference cultures of organisms related to the target organism are used, e.g. *E. coli* on selective culture media for *Salmonella*. Media manufacturers often recommend specific species or strains of organisms for use as positive and negative controls

* Each laboratory's specific media quality control requirements should be documented in relevant standard operating procedures.
[†] Results of all checks should be recorded.

atory's media preparation procedures and schedules must be designed to ensure that microbiological media are as fresh as possible at the time of use. Old, over-cooked media must not be used; also, after melting, media must not be re-set for re-use. Special care is necessary to limit the shelf life of media containing acids, dyes and / or antibiotics as their efficacy can deteriorate quickly in a short time (a few days).

(b) Media should be protected to ensure that moisture loss is minimised during preparation and storage. Reduction in volumes of broth media and over-drying of agar media by evaporation can result in a concentration of selective agents or a reduction in water activity to the extent that growth of the target organisms is compromised. Broth media are usually dispensed in 10 ml quantities and agar media poured into 90 mm diameter Petri dishes in 12–15 ml quantities. For some specific purposes, agar quantities may need to be increased to 20–25 ml, e.g. incubation at temperatures > 40°C, or incubation for extended times, e.g. > 3 days. The laboratory's operating procedure for the specific method to be employed should indicate the required volume of medium to be used for each test type. As plate-pouring is generally a manual task, it is worthwhile practising to ensure the correct volume of medium can be poured consistently to minimise the potential for this stage of the method to contribute to incorrect results.

(11) When, as a result of quality control tests, a batch of medium is considered to be of unsatisfactory quality or found not to perform correctly, then records and procedures should be checked to determine the cause and rectify the problem. Table 7.6 indicates some of the more common faults found in laboratory prepared media and some possible causes of these problems.

(12) Agars must be properly tempered to 45–46°C before pouring. This should be achieved using temperature and time monitoring systems based on data from media cooling studies for each medium volume used, **not** by subjective hand-ling of bottles. To minimise deterioration of media, molten agars for pour plates should be held at 46°C and used within 3 hours; molten media not used within the specified time should be discarded. The 44°C water bath used for *Escherichia coli* confirmation must not be used for agar tempering as, besides being at slightly too low a temperature, the constant addition and removal of media containers from the water bath produces unacceptable variations in the water temperature invalidating the *E. coli* test result. In addition, it is poor practice to place sterile tempering media in close proximity to incubating cultures.

Table 7.6 Some media quality faults and their causes, adapted from Bridson, 1998; Kyriakides *et al.*, 1996.

Faults	Possible cause(s)
Incorrect pH value	Deterioration of dehydrated medium due to medium being stored incorrectly or out-of-shelf life Poor quality water used to make up the medium Chemical contamination e.g. from glassware Medium powder not properly dissolved Medium overheated – in sterilisation process – during re-melting – during molten holding time pH test carried out above 25°C Faulty pH probe / meter pH meter incorrectly calibrated
Abnormal precipitation, turbidity (The medium manufacturer's description of the prepared medium should be checked to ensure that the normal medium does not produce precipitates)	Deterioration of dehydrated medium Poor quality water used to make up the medium Dirty glassware Medium powder not properly dissolved Medium overheated – in sterilisation process – during re-melting – during molten holding time Supplement added when the medium was too hot (if applicable) Incorrect pH value

Contd.

Table 7.6   *Contd.*

Abnormal darkening	Medium powder not properly dissolved Medium overheated – in sterilisation process – during re-melting – during molten holding time Incorrect pH value
Abnormal colour	Deterioration of dehydrated medium Poor quality water used to make up the medium Dirty glassware Medium powder not properly dissolved Incorrect heat process Incorrect pH value
Black flecks in the medium	Charred agar or sugars: can occur when heating media over a Bunsen flame
Clear flecks in the medium	Small pieces of set agar due to overcooling of agar which sets on the sides of the container and disperses on agitation Seen clearly in media to which opaque supplements have been added
Coagulation	Supplements added to a medium that is too hot
Soft gel in agar medium	Error in media preparation leading to insufficient concentration of agar in the medium, e.g. under-weighing of dehydrated medium, addition of too much water or supplements (if applicable) Poor mixing during dissolution leading to agar not being completely dissolved Overheating at low pH values
Poor growth of cultures or poor differential characteristics of colonies growing on diagnostic agar	Deterioration of dehydrated medium Poor water quality; not purified and contains inhibitory substances from contaminated glassware or tubing used to dispense media Error in media preparation leading to insufficient concentration of medium constituents in the prepared medium, e.g. under-weighing of dehydrated medium, addition of too much water Supplements added to a medium that is too hot Incorrect quantity of supplement Mould media prepared without trace elements Medium powder not properly dissolved, insufficient levels of nutrient components in the medium Excessive heating of medium Over-drying of media in plates Prepared selective media stored too long Media stored in direct sunlight producing toxic products by photo-oxidation of components, e.g. the dye rose bengal in rose-bengal chloramphenicol agar

7.5.5 *Reference cultures*

Reference cultures (reference organisms) are cultures of microorganisms that have a fully documented provenance. They are obtained from Culture Collections that are usually maintained as national resources (Table 7.7) although some research organisations have collections derived from work carried out over the years and from which cultures have been saved and maintained. Some larger commercial food laboratories may also have a small collection of cultures obtained from the routine work of the laboratory or from investigative work during problem solving, e.g. an organism found in a spoiled product.

Table 7.7 Examples of National Culture Collections.

Culture collection	Contact information
United Kingdom National Culture Collections (UKNCC): *Information about the following UK collections can be accessed via the UKNCC website:*	www.ukncc.co.uk
National Collection of Type Cultures (NCTC)	PHLS Central Public Health Laboratory 61 Colindale Avenue London NW9 5HT UK
National Collection of Yeast Cultures (NCYC)	Institute of Food Research Norwich Laboratory Colney Lane Norwich NR4 7UA UK
National Collection of Industrial and Marine Bacteria Ltd (includes National Collection of Food Bacteria)	23 St Machar Drive Aberdeen AB2 1RY Scotland
CABI Bioscience Genetic Resource Collection (includes the UK national collection of filamentous fungi)	CABI Bioscience UK Centre Bakeham Lane Egham Surrey TW20 9TY UK
Other national collections: American Type Culture Collection (ATCC)	10801 University Boulevard Manassas Virginia 20110-2209 USA www.atcc.org/
Institut Pasteur Collection	25–28 rue du docteur Roux 75724 Paris Cedex 15 FRANCE www.pasteur.fr/externe

Reference cultures are used in the routine work of food microbiology laboratories to check the quality and performance characteristics of media prepared in the laboratory; in particular, selective media that should be tested with known, relevant organisms at regular intervals to ensure their efficacy and also, to maintain technician experience of specific colony morphologies and diagnostic characteristics of the target and non-target organisms on the medium used. This regular experience assists the ongoing correct interpretation of the results of growth. Reference cultures are also used as positive and negative reaction controls when specific confirmatory tests are carried out, e.g. biochemical tests, serology, enzyme-linked immunosorbent assays.

Reference cultures are also used to check the efficacy of a test method, i.e. verify the ability of a test method to detect an organism in a test sample artificially inoculated with the reference organism; in this procedure, the reference organism is passed through all stages of a microbiological method. Method efficacy tests usually form part of the laboratory's Internal Quality Assessment (IQA) system set up to regularly monitor staff proficiency and the overall performance of laboratory operations. An outline of the conduct of a method efficacy / verification test is shown in Figure 7.3. Thus, the ability of a test method to recover selected organisms from a particular foodstuff is tested by deliberately inoculating the food with known levels of pure cultures of known organisms (reference organisms) and then using the specified recovery techniques, media and test methods (including any confirmatory tests) to recover the inoculated organisms. Such tests should be carried out regularly to verify / re-check methods to ensure full method efficacy and are certainly necessary when 'new' sample foodstuffs are introduced to a laboratory, e.g. for the detection of pathogens in chocolate, cheese, powders and highly spiced or preserved foods.

Great care must be exercised to ensure that reference cultures do not contaminate the laboratory environment or any of the other work in progress, so the highest standards of aseptic practice are required at all times throughout the handling and culturing of reference organisms. As the risks of cross-contamination can be high, even in the best run laboratories, and the consequences of cross-contamination could be severe and dramatic for both the laboratory and its customers, these cultures should only be handled by fully trained and experienced personnel to ensure that the potential for extraneous contamination caused by the cultures is minimised. Additional precautions may be applied to the handling of these cultures, e.g. use of a separate room with dedicated facilities, use of a dedicated bench in the main laboratory area or dedication of a specific time to reference culture handling when other work is not in progress.

Reference cultures should preferably be obtained from a recognised culture collection (Table 7.7) or be fully traceable to such a collection, e.g. the same strains obtained as commercially, pre-prepared cultures on loops, paper discs, 'sticks', etc.

The laboratory's procedure for storing and handling reference cultures and preparing, storing and using working cultures of the reference culture should be documented in a Standard Operating Procedure. It is usual however, to limit the number of sub-culture steps between the 'Reference Stock' culture and the 'Working Culture' to minimise the possibility of any extraneous contamination of the culture (Figure 7.4(A), (B)). Where there is any doubt as to the purity, identity, age or handling history of the working culture, it should be discarded and a new working culture used.

Laboratory cross-contamination from reference cultures and other cultures to

Figure 7.3 Example of a method efficacy (verification) test procedure for bacteria by controlled product inoculation. Only properly trained staff should be allowed to conduct these tests. Aseptic handling procedures must be used at all times.

Step	Operation	Notes
A	Inoculate, using a loop, 10 ml of broth medium (appropriate for the test organism) with a small quantity of a working culture of the test organism (Figure 7.4)	Nutrient Broth is suitable for most aerobic bacteria, Malt Extract Broth for yeasts Reinforced Clostridial Medium (RCM) – freshly de-aerated, for *Clostridium* spp. and Alkaline Peptone Water (pH 8.6) + sodium chloride (3%w/v) for *Vibrio parahaemolyticus*
B	Incubate the inoculated broth medium until growth is visible, i.e. the broth becomes cloudy	The defined conditions of atmosphere, temperature and time appropriate for the test organism should be used Normally overnight incubation is sufficient, e.g. for *E. coli* and *Salmonella* spp., but may need to be longer for other organisms such as lactic acid bacteria, *Campylobacter* spp. and yeasts
C	Using a diluent suited to the requirements of the test organism make serial dilutions of the incubated broth culture down to 10^{-8}	Maximum Recovery Diluent (MRD) is suitable for most organisms but specific diluents may be required for some organisms, e.g. freshly de-aerated RCM for *Clostridium* spp.
D	Using the spread plate technique, inoculate 0.5 ml volumes of the 10^{-5} to 10^{-8} dilutions of the broth culture onto two plates each of an appropriate selective and a non-selective agar	(Figure 7.1)
E	Weigh and dilute **5** replicate test samples from a factory sample e.g. for *Salmonella* spp., weigh 5×25 g test samples and add 225 ml Buffered Peptone Water to each 25 g test sample; for *Listeria monocytogenes*, weigh 5×25 g test samples and add 225 ml half Fraser broth to each test sample; for *E. coli*, weigh 5×10 g test samples and add 90g / ml MRD to each test sample	The laboratory's standard method / procedure should be used for the sample types under test for the specified organism
F	Add 1 ml of the 10^{-5} broth culture dilution to one of the diluted test samples Repeat for the 10^{-6}, 10^{-7} and 10^{-8} broth culture dilutions using another 3 of the diluted test samples	The remaining diluted test sample is the un-inoculated control
G	Blend all 5 diluted, inoculated / un-inoculated test samples from F above to produce test sample suspensions	
H	Using the spread plate technique, inoculate 0.5 ml volumes from each of the diluted and blended, inoculated / un-inoculated test sample suspensions from G above onto two plates each of an appropriate selective and a non-selective agar for the test organism	

Contd.

Figure 7.3 *Contd.*

Step	Operation	Notes
I	Incubate plates from D and H and, for organisms such as *Salmonella* and *Listeria monocytogenes*, the blended, inoculated / un-inoculated test sample suspensions from step G	The defined conditions of atmosphere, temperature and time appropriate for the test organism should be used
J	After incubation, select plates from D and H that show an acceptable range of colony numbers, e.g. 15–150 colonies per plate. Count the colonies and use the count to calculate the number of colony forming units (cfu) per ml of broth culture and also of the inoculated test sample suspensions	**Full records of all raw data (counts) obtained and the calculated cfu/ml must be kept**
K	For enumeration tests, e.g. *E. coli, Staph. aureus, Bacillus cereus, Clostridium perfringens*, the laboratory's normal confirmation tests should be carried out on colonies on plates obtained at step J. For detection tests, e.g. *Salmonella* or *Listeria monocytogenes*, continue the normal testing procedure for the now incubated test sample suspensions from G, incubated at step I	The laboratory's normal detection or enumeration methods must be followed including (where relevant) the full confirmation tests on the required numbers of suspect colonies, e.g. for *Salmonella* spp., *Listeria* spp.
L	**Record and report *all* results obtained including the dilutions used for inoculation, colony counts obtained, biochemical reactions (where appropriate), number of colonies taken for confirmation and results of confirmation tests for each isolate (where appropriate) and the final results obtained**	Results obtained are assessed against the performance criteria defined in the laboratory's own procedural standards* and the overall results of the assessment are recorded
	If the method efficacy tests fail to detect or enumerate the inoculated organism correctly, the test should be repeated and the cause of the failure investigated	

* (1) For detection methods, the method used must be able to recover low numbers of the test organism, e.g. *Salmonella* spp. and *Listeria monocytogenes* should be detectable at inoculated levels of <10 cfu/ 25 g test sample.
 (2) For enumeration methods, the method should produce countable plates with numbers of colony forming units in the region of the normal specifications for the foods examined, e.g. coagulase producing staphylococci, should be distinguishable and countable (and confirmed) at levels 20–100 cfu/g test sample.
 (3) Results obtained for the un-inoculated test sample should be within the normal factory sample specification requirements, e.g. *Salmonella* 'not detected' in 25 g test sample, *E. coli* <100 cfu/g test sample, coagulase producing staphylococci <20 cfu/g test sample.
 (4) The closeness of agreement between the counts obtained from step D (actual numbers inoculated into the samples) and step H (counts of the target organism obtained from the inoculated test samples) should be checked against the laboratory's own standard of requirement for agreement, e.g. within ±0.5 log in count or two standard deviations from the mean.

Contd.

Figure 7.3 *Contd.*

Note:
The method efficacy test should be designed to ensure that the numbers of the test organism used reflect the numbers that are targeted by the method applied to normal test samples, e.g. specifications for *Salmonella* spp. are frequently 'not detected' in 25 g so the numbers of the test organism used in a method efficacy test should preferably be <10 cfu *Salmonella* /25 g test sample.

Having determined what numbers of the test organism are required per gram of test sample, it is necessary to calculate the culture dilution level necessary to provide those numbers of test organism, taking into account the quantity of test sample to be examined.

Examples
(1) For *E. coli* or coagulase-producing staphylococci, if 100 cfu of the test organism/g test sample is required, and the broth culture (B) contains 1×10^8 cfu of the test organism/ml (obtained from steps C, D, H, I and J), then:
 – the method requires 10 g of test sample.
 – a 'count' of 100 cfu/g therefore requires 1000 organisms (cfu) in total.
 – a culture containing 10^8 cfu/ml therefore must be diluted to 1 in 100 000 (10^{-5}) to provide a suspension containing 1000 cfu/ml.
 – adding 1 ml of the 1 in 100 000 (10^{-5}) culture dilution to the diluted test sample (10g + 90ml MRD) therefore means that 1000 cfu have been added to the 10 g test sample suspension, i.e. providing a 'count' of 100 cfu/g test sample*.
(2) For *Salmonella* spp. or *Listeria* spp., if 10 cfu of the test organism/25 g test sample is required, and the broth culture (B) contains 1×10^8 cfu (obtained from steps C, D, H, I and J) of the test organism/ml, then:
 – the method requires 25 g of test sample.
 – a 'count' of 10 cfu/25 g therefore requires 10 organisms (cfu) in total.
 – a culture containing 10^8 cfu/ml therefore must be diluted to 1 in 10 000 000 (10^{-7}) to provide a suspension containing 10 cfu/ml.
 – adding 1 ml of the 1 in 10 000 000 (10^{-7}) culture dilution to the diluted test sample (25g + 225ml broth) therefore means that 10 cfu have been added to the 25 g of test sample, i.e. providing an inoculum level of 10 cfu/25g test sample*.

* The final figures are approximate because of the very small added dilution effect of the culture inoculum, however, 'rounded off', the counts remain as 100 cfu/g and 10 cfu/25 g test sample respectively.

samples and media being processed has occurred in many laboratories and, it is likely, will still occur from time to time even in well-controlled laboratories. In these events, false positive information can be produced, i.e. results are reported to indicate that a food sample is contaminated with the pathogen when it is not. The consequences of such an occurrence when this involves *Salmonella* for example, can include cessation of production, product recalls and closure of the manufacturing line or factory. As protection against the possibility that a reference culture, particularly of bacterial pathogens such as *Salmonella* spp., *Campylobacter jejuni*, *Listeria monocytogenes*, may cross-contaminate work in the laboratory without detection, laboratories should obtain strains or serotypes that are encountered only rarely in foods and reported rarely in public health statistics, yet still produce the typical diagnostic characteristics of the genus / species as appropriate. If necessary, these can then be more readily identified as being of laboratory and not food origin.

In addition, a means of rapidly and reliably identifying the laboratory reference strains, e.g. specific diagnostic antiserum for the *Salmonella* serotype used, should be kept available so that the laboratory reference strain can be quickly eliminated as a laboratory contaminant in the event of confirmed positive cultures apparently being isolated from food samples (see Chapter 11). Each laboratory should have an established policy in relation to this matter and pre-prepared procedures and lines of communication when informing customers. Advice on the suitability of cultures for use and reliable methods for their rapid identification can be obtained from the culture suppliers.

The laboratory's policy on sources of reference cultures, listing of specific organisms used for defined purposes, specific procedures for handling reference cultures

Note: In order to minimise the potential for contamination of laboratory reference cultures through numerous sub-culturing steps, it is preferable to limit the number of sub-cultures between the primary culture source (reference culture collection or commercial preparation) and the laboratory working culture. Many laboratories limit this number of sub-cultures to a maximum of 3. However, for practical purposes, a larger number of sub-cultures may be specified in the laboratory's Standard Operating Procedures and, in any case, **it may require 2 or more sub-cultures to ensure that the organism's characteristics are fully restored** prior to use in quality control procedures. Laboratories should have procedures in place to establish and check the resuscitation and sub-culturing methods required for each reference organism; this is to ensure that the intended characteristics of each reference organism from each source (where multiple sources are used) are obtained and maintained.

Figure 7.4   Examples of procedures for the preparation and use of Reference Cultures.

A

Culture source and preparation	Notes
Culture obtained from a **Reference Culture Collection**, e.g. freeze dried culture in an ampoule.	Anon (1999)(6)
First culture step ↓ Re-constitute to produce the **Laboratory Reference Stocks**. Store in suitable conditions for preservation, e.g. deep frozen beads (−20°C).	Preparation of the **Laboratory Reference Stocks** should be preferably only a single culture step from the Culture Collection culture (see Note above). The highest standards of aseptic technique are vital as contaminants may grow during incubation as quickly as, or more quickly than, the **Reference Culture** itself.
Second culture step ↓ Use the **Laboratory Reference Stock** to prepare several **Working Cultures*** in broth or on agar slopes or plates, e.g. enough for a one month supply.	Preparation of the **Working Cultures** should be preferably only a single culture step from the **Laboratory Reference Stocks** (see Note above). Before all **Working Cultures** have been used up, a new set should be prepared from the **Laboratory Reference Stocks**. **Working cultures** preferably should not be sub-cultured to produce further **Working Cultures**. **Working Cultures** must *never* be used to replace **Laboratory Reference Stocks**.
↓ After incubation, store **Working Cultures** in a dedicated refrigerator (or as otherwise advised).	Once use is commenced, use the **Working Culture** for a maximum of one week, after which it must be discarded. It should be discarded sooner if it becomes contaminated or dies.

Contd.

Figure 7.4 *Contd.*

B

Culture source and preparation	Notes and requirements
Culture obtained from a **Commercial Preparation**, e.g. loops, paper discs, sticks.	Anon (1999)(6)
↓ Store as recommended by the manufacturer and use directly as the **Laboratory Reference Stocks**.	
First culture step ↓ Use the **Laboratory Reference Stocks** to prepare several **Working Cultures*** in broth or on agar slopes or plates, e.g. enough for a one month supply. Discard the **Laboratory Reference Stock** unit (loop, stick etc.) used to prepare the **Working Cultures**	Preparation of the **Working Cultures** preferably should be only a single culture step from the **Laboratory Reference Stocks** (see Note above). Before all **Working Cultures** have been used up, a new set should be prepared from the **Laboratory Reference Stocks**. **Working Cultures** preferably should not be sub-cultured to produce further **Working Cultures**.
After incubation, store **Working Cultures** in a dedicated refrigerator (or as otherwise advised)	Once use is commenced, use the **Working Culture** for a maximum of one week after which, it must be discarded. It should be discarded sooner if it becomes contaminated or dies.

* Regular purity and identity checks (Gram stain, biochemical profile and serological checks, as appropriate)

and preparing and using laboratory reference stocks and working cultures should all be fully documented, understood and carried out by laboratory staff.

7.5.6 *'Uncertainty of Measurement' of microbiological test results*

Despite the best of efforts, no measurement is absolutely accurate and this is particularly true of measurements in food microbiology, e.g. enumeration (count) tests, because of the number and variety of potential errors that may occur and the difficulties in precisely reproducing the high quality of practice required for the manual operations performed during the course of any particular test (see Figure 6.4). This means there is some 'Uncertainty of Measurement' associated with the result of any microbiological test and the staff in a laboratory need to be aware of the size of the uncertainty associated with their own test results obtained from their particular sample types when examined using the laboratory's specific methods.

In practical food microbiology, although there are some influences on 'Uncertainty of Measurement' that can be assessed readily, e.g. physical measurements such as weight and volume, there are others, unfortunately, that cannot, e.g. microorganisms may not be distributed evenly throughout the sample or there may be unpredictable growth behaviour during the test (Table 7.8). Therefore, any assessment of the 'Uncertainty of Measurement' for microbiological test results is unlikely to be as precise as that achievable in physics or mathematics. This does not mean that efforts made to estimate this 'uncertainty' would be wasted. If the reliability (accuracy) of results can be improved at all by controlling or minimising any adverse influences, i.e.

Table 7.8 Some examples of influences and their control in practical food microbiology on the 'Uncertainty of Measurement' of the test result.

Influence	Is it controllable?
Measurement equipment for:	
weight	Yes
volume	Yes
temperature	Yes
pH value	Yes
Samples:	
Factory sample representative of the total quantity of the batch from which it was taken	Unlikely, unless the bulk is an homogenous liquid, e.g. milk
Test sample representative of the factory sample	Unlikely unless the sample is an homogenous liquid, e.g. milk
Handling conditions (transport, storage)	Influence can be minimised by controlling the temperature of storage and of any transport used, and the time between initial sampling in the factory and examination in the laboratory
Microorganisms:	
Even distribution in the factory sample and test sample	Unlikely unless the sample is an homogenous liquid, e.g. milk
All organisms of interest in the test sample in a physiological condition suited to growth under the test conditions	No
Methods	Yes
Procedures	No, influence can only be minimised
Practices	No, influence can only be minimised by training
Human error	No, influence can only be minimised by training and careful practices

reducing the level of 'Uncertainty of Measurement' in a test result, then action should be taken.

Because a large number of individual influences on 'uncertainty' in microbiological test results cannot be measured, many laboratories operate a practical approach and endeavour to assess the effect of the total laboratory operation (staff practices, methods, procedures, etc.) on the results obtained. This is done by applying a statistical test (analysis of variance) to the results from duplicate counts obtained from examinations of selected test samples and this is carried out routinely as part of the laboratory's quality assurance procedures (Anon, 1999 (13)). The results obtained from the analysis allow the laboratory staff to assess the reliability of procedures and practices in place that are designed to maximise the reliability of test results. It is important that laboratory staff understand their own laboratory's approach to, and method of making, some estimate of 'uncertainty' in test results as well as those areas in the laboratory that could make significant contributions to the level of 'uncertainty' if inadequately controlled.

It is not practical to estimate or demonstrate the 'uncertainty' in test results relating to qualitative microbiological tests such as detection (presence / absence) tests for specific pathogens, e.g. *Salmonella* or *Listeria monocytogenes*, in a specified quantity of food sample, e.g. 25 g, 50 g, 100 g because 'detection' is not a measurement and the test result does not provide data that can be analysed statistically. However, a practical approach to demonstrating competency in handling the low numbers of target organisms that these tests aim to detect is to use the duplicate colony count results obtained during method efficacy studies (Figure 7.3) in a statistical analysis similar to that used for estimating 'uncertainty' for enumeration tests. This will show how reliably repeatable the serial dilutions are for use in generating the very low inoculum numbers required for the tests.

A laboratory Standard Operating Procedure should detail the approach to be taken for determining the laboratory's own 'uncertainties' in the microbiological test results obtained for test samples.

7.6 Further reading

Anon (1993) *Guide to the Expression of Uncertainty in Measurement*. International Organization for Standardization, Geneva, Switzerland.

Baylis, C.L., Jewell, K., Oscroft, C.A. & Brookes, F.L. (eds) (2001) *Guidelines for Establishing the Suitability of Food Microbiology Methods. Guideline No. 29*. Campden & Chorleywood Food Research Association, Chipping Campden, UK.

Bridson, E.Y. (1998) *The Oxoid Manual*, 8th edn. Oxoid Ltd., Basingstoke, UK.

British Standard BS 5763 (1996) *Microbiological Examination of Food and Animal Feeding Stuffs, Part 0 – General laboratory practices*. British Standards Institution, London, UK.

Corry, J.E.L., Curtis, G.D.W. and Baird, R.M. (2003) *Handbook of Culture Media for Food Microbiology. Progress in Industrial Microbiology Volume 37*. Elsevier Science, B.V., Amsterdam, The Netherlands.

Downes, K.P. and Ito, K. (eds) (2001) *Compendium of Methods for the Microbiological Examination of Foods*. Compiled by the APHA Technical Committee on Microbiological Methods for Foods. American Public Health Association, Washington D.C., USA.

Environment Agency (2002) *The Microbiology of Drinking Water – Parts 1–3 – Water Quality and Public Health. Methods for the Examination of Waters and Associated Materials*. Environment Agency, Nottingham, UK.

European Standard BS EN ISO/IEC 17025 (2000) *General Requirements for the Competence of Testing and Calibration Laboratories*. European Committee for Standardization, Brussels, Belgium.

Jewell, K. (ed.) (2004) *Microbiological Measurement Uncertainty: A Practical Guide, Guideline No. 47*. Campden & Chorleywood Food Research Association, Chipping Campden, UK.

Kyriakides, A., Bell, C. & Jones, K. (eds) (1996) *A Code of Practice for Microbiology Laboratories Handling Food Samples – (Incorporating Guidelines for the Preparation, Storage and Handling of Microbiological Media), Guideline No. 9*. Campden & Chorleywood Food Research Association, Chipping Campden, UK.

Snell, J.J.S., Brown, D.F.J. and Roberts, C. (1999) *Quality Assurance – Principles and Practice in the Microbiology Laboratory*. Public Health Laboratory Service, London, UK.

UKAS Publication ref: LAB12, October 2000. *The Expression of Uncertainty in Testing*. United Kingdom Accreditation Service, London, UK.

Voysey, P.A. & Jewell, K. (1999) *Uncertainty Associated with Microbiological Measurement*. CCFRA Review No. 15. Campden & Chorleywood Food Research Association, Chipping Campden, UK.

8 Laboratory Standards of Operation: Accreditation and Documentation

8.1 Introduction

Commencing from 1985, most microbiology laboratories serving the UK food industry have either committed themselves to operating according to the requirements of a laboratory accreditation standard or are in the process of doing so. Driven by the need to demonstrate to their customers a certain level of operational and technical competence, the assessment and / or accreditation of laboratories by 'third party' bodies, has become a normal requirement of laboratories that serve major food manufacturers and retailers. Indeed, third party accreditation of microbiology laboratories is also now widespread in the public health and public analyst laboratory sectors. This third party accreditation is supplemented and supported by regular internal audits of all aspects of the laboratory operation carried out by the laboratory staff themselves. It is important that all staff are aware of, and understand, the key elements comprising their own laboratory's approach to meeting the particular quality standards subscribed to.

8.1.1 *Laboratory accreditation: assessment criteria*

Laboratories should have a declared policy or 'mission statement' that defines the laboratory's main objectives. In addition to providing a microbiological service to the company or companies to which the laboratory is bound, mission statements normally state, in more or less words, the general objective: 'to provide a professionally high standard of microbiological work and service that will ensure that test results are consistently reliable for use in support of the production of safe and high quality foods.' The standards of work and service may also be linked with compliance to current industry best practices and / or particular accreditation criteria, e.g. the United Kingdom Accreditation Service (UKAS).

In the context of laboratories, accreditation is the activity of 'setting them forth as credible', or 'vouching for their credibility'. Laboratory accreditation or a certificate of compliance to particular assessment criteria is conferred by independent bodies and may be national schemes or commercially operated industry schemes (Table 8.1). Internationally, the European Co-operation on Accreditation (EA) and the International Laboratory Accreditation Co-operation (ILAC) maintain lists of national accreditation bodies for countries in Europe, North, Central and South America, Asia, The Pacific, India and Africa. These can be found on the respective internet websites shown in Table 8.1. Schemes for assessing and accrediting / certificating food industry microbiology laboratories have mainly only been in operation since the mid to late 1980s when it was recognised that procedures and practices in laboratories

Table 8.1   Examples of some laboratory assessment / accreditation bodies.

UK schemes for testing laboratories United Kingdom Accreditation Service (UKAS) 21–47 High Street Feltham Middlesex TW13 4UN Campden Laboratory Accreditation Scheme (CLAS) Campden & Chorleywood Food Research Association Chipping Campden Gloucestershire GL55 6LD LABCRED Law Laboratories Ltd Blakelands House 400 Aldridge Road Great Barr Birmingham B44 8BH
Examples of other national accreditation bodies for testing laboratories* Australia – National Association of Testing Laboratories, NATA Belgium – Ministry of Economic Affairs, BELTEST Denmark – Danish Accreditation – DANAK France – Comité Français d'Accréditation, COFRAC Germany – Deutscher Akkreditierungsrat, DAR Spain – Entidad Nacional de Acreditación, ENAC United States of America – American Association for Laboratory Accreditation, A2LA

* The names, addresses and contact telephone and fax numbers of nationally operated accreditation
 schemes for testing laboratories can be found on the websites of the organisations of:
 – European Accreditation (EA) at www.european-accreditation.org, and
 – International Laboratory Accreditation Cooperation (ILAC) at www.ilac.org

needed to be improved. Although each scheme has its own specific requirements and
means of assessing competence, the general methods of operation of accreditation
schemes are broadly similar and include the following:

(1)   The laboratory is supplied, by the scheme operator, with information and the
      criteria that must be fulfilled in order to achieve accreditation.
(2)   The laboratory assesses the scheme's requirements and implements systems to
      meet the criteria as appropriate.
(3)   The laboratory formally applies for assessment.
(4)   A pre-assessment visit to inspect the laboratory's documentation, procedures
      and practices may be made by the scheme operator to determine if there are any
      major omissions in the quality system assessed against the criteria.
(5)   The laboratory reviews and amends, as appropriate, its system in light of the
      pre-assessment visit report.
(6)   The scheme operator formally assesses the laboratory's quality systems under
      its normal operating conditions against the scheme's requirements / criteria.

(7)   A report is given to the laboratory indicating any non-compliances against the scheme's requirements and a recommendation is made in respect of accreditation.
(8)   The laboratory corrects the non-compliances and reports the corrective actions to the scheme operator, providing documentary evidence where appropriate.
(9)   The scheme operator assesses the corrective actions (which may require a further inspection). If satisfactory, accreditation is awarded.
(10)   Re-assessment inspections are made at regular intervals, usually annually and sometimes these are made unannounced.

Steps 4 and 5 above are only usually included in the assessment procedures of national scheme operators, e.g. UKAS.

The European Standard *General Requirements for the Competence of Testing and Calibration Laboratories* (BS EN ISO / IEC 17025, 2000) is the current standard against which accreditation scheme operators judge the competence of testing laboratories, including food microbiology laboratories. This is, in general, quite a readable document and any person expected to work in a testing laboratory meeting the standards of an accreditation body, whether it is a national scheme or not, is advised to read this document to gain an overall view of the scope covered by the standard, particularly in the areas relating to the quality system and general technical requirements.

Table 8.2 gives a summary of the key areas addressed by third party accreditation bodies and the associated general requirements. This provides a logical and systematic framework around which to build and operate the quality systems of any food microbiology laboratory. Although all laboratories are, in many ways, different one from another, e.g. structure and layout, staff levels, work-load, sample types examined, methods used; the essential operations carried out are very similar in principle, i.e. media preparation and sample processing, and therefore the controls and checks implemented at key points are necessarily similar (Figures 8.1 and 8.2).

Clearly, in the space of a single inspection of a working laboratory (which usually involves a one-day visit by an assessor) in which the work may vary from day to day, it is not possible to assess, in detail, every aspect of the laboratory's activities. It is by a combination of assessment of the documentation (standard operating procedures, methods, records etc.) and observation of selected tasks, i.e. by 'sampling' different aspects of the laboratory's operation, that the assessor determines the overall competence of the laboratory to carry out selected tasks and the extent to which the quality systems and procedures comply with the accreditation scheme's criteria. However, to be of value, the assessment must take place, as far as practically possible, under the routine operating conditions of the laboratory.

In between these third party assessments of laboratory operations, the laboratory staff have a duty to maintain high standards of procedures and practices. One means of achieving this is by instituting and maintaining regular 'internal' audits of the different activities of the staff of the laboratory (as applicable to their experience and training). Such audits should be scheduled and documented. They should be carried out at a frequency that ensures that all operating procedures and practices are examined at least twice per year.

Table 8.2 *Some key areas addressed by accreditation bodies.*

Area addressed	General requirements	Sources of further information
Management	A clear management and technical staff structure should be in place whose responsibility, authority and interrelationships are defined and who are provided with the necessary resources to ensure that the required quality of laboratory operations can be maintained The integrity of the laboratory personnel should not be allowed to be influenced by external pressures, e.g. from commercial or financial sources or as a result of conflicting interests, such as to affect their judgement or compliance with standards	European Standard BS EN ISO / IEC 17025 (2000) Documentation of individual accreditation bodies
Accommodation and environment	The laboratory should be designed, structured and operated (good housekeeping standards) to ensure that environmental conditions will not adversely affect the quality of the work done and results obtained Environmental conditions should be monitored using appropriate methods / tests and records kept of the results	Brown (1994) Chapter 6 *ibid*
Equipment	The laboratory should be provided with all the necessary equipment of appropriate specification to facilitate correct sampling, measurement, sample preparation, processing and testing procedures Records should be kept up to date for each item of equipment including the laboratory's equipment unique identifier and location, a full description and method of operation, manufacturer's manual / instructions and any reports relating to maintenance, repairs and calibration Designated staff should be trained in the correct operation, maintenance and monitoring requirements for the equipment Appropriate calibration and monitoring procedures should be established, documented and operated Clear, written instructions on the use and operation of all equipment relevant to the tests carried out must be readily available	Kyriakides *et al.* (1996) Anon (1999) (4) Chapter 6 *ibid*

Contd.

Table 8.2 *Contd.*

Area addressed	General requirements	Sources of further information
Equipment calibration	Certain equipment, e.g. thermometers, temperature probes, autoclaves, weight-measuring equipment, in a microbiology laboratory may need to be subjected to a periodic, e.g. annual, calibration check against a relevant reference material as a measurement standard. Any discrepancies noted between the reference standard and the laboratory item must be taken into account in future use, e.g. a laboratory thermometer tested in its normal application range, say for incubators at 30–37°C, that reads 30.1°C against 30°C on the reference thermometer should be labelled as reading 0.1°C high and the appropriate adjustment made when using the thermometer for routine laboratory checks All measurement standard materials, e.g. weights, thermometers, must be traceable to a primary standard of the International System of Units of Measurement, e.g. a National Standard; this is normally done through a relevant calibration service provider Schedules of calibrations required and records of results and any actions taken should be maintained	European Standard BS EN ISO / IEC 17025 (2000) Documentation of individual accreditation bodies Anon (1999 (2) (3) (7) (8))
Staff	Staff employed should be suitably qualified and experienced, then trained and assessed for competence to perform the specific tasks required within the laboratory Staff undergoing training must be properly supervised and should be required to demonstrate the necessary skills and competence to perform a specified task before being authorised to carry out the task unsupervised. A laboratory policy and procedure should be in place to identify staff training needs and provide appropriate internal and external training systems and resources Records should be kept of all training undertaken by staff, the results of such training and they should identify the personnel involved in conducting / supervising and assessing the training activity	Chapter 7 *ibid*
Quality system	The laboratory should have documentation covering its policies, operational systems, methods, procedures and instructions and this should be readily available to, communicated to, understood by, and implemented by, all appropriate personnel	European Standard BS EN ISO / IEC 17025 (2000) Documentation of individual accreditation bodies

Contd.

Table 8.2 *Contd.*

Area addressed	General requirements	Sources of further information
Document control	All documentation forming part of the quality system should be uniquely identified and issued only under the authority of designated personnel Documentation should be reviewed regularly to ensure it is kept up to date and any amendments should be identified clearly, appropriately authorised and communicated to relevant personnel	European Standard BS EN ISO / IEC 17025 (2000) Documentation of individual accreditation bodies
Requests, tenders, contracts and sub-contracting	The laboratory is obliged to ensure that the appropriate capability, resources and methods are available to support any work undertaken and that the necessary requirements are communicated to all appropriate staff If any work is sub-contracted, then the laboratory is obliged to ensure that the sub-contractor is competent to carry out the work to the required standard	European Standard BS EN ISO / IEC 17025 (2000) Documentation of individual accreditation bodies
Purchases of services and supplies	The laboratory should operate a supplier evaluation and approval system in respect of all services and supplies that are bought-in, particularly for those that supply equipment, consumables, services, etc. that are critical to the quality of the testing work carried out in the laboratory. Supplier companies that are registered to the ISO 9001 : 2000 standard or an equivalent standard should be operating appropriate controls and record systems that can provide laboratories with a high degree of confidence in ongoing product quality Where relevant, purchase orders should indicate the particular quality / specification of the item / material required	European Standard BS EN ISO / IEC 17025 (2000) Documentation of individual accreditation bodies Anon (1999 (4))
Documentation	For training and reference purposes, all key procedures relating to and affecting the quality of work in the laboratory should be clearly documented	Wilson and Weir (1995) Snell *et al.* (1999)
Records	All quality and technical records made in the laboratory must be legible and complete, must identify the responsible person(s) associated with the record and be readily retrievable	European Standard BS EN ISO / IEC 17025 (2000) Documentation of individual accreditation bodies

Contd.

Table 8.2 *Contd.*

Area addressed	General requirements	Sources of further information
Internal audits	The quality system and all laboratory procedures and practices should be audited by designated staff against a written schedule Audit findings and corrective actions should be recorded and personnel involved identified on the record	European Standard BS EN ISO / IEC 17025 (2000) Documentation of individual accreditation bodies
Microbiological methods	Microbiological methods used by the laboratory should preferably be selected from those published in international, national or regional standards, by reputable technical organisations or in peer-reviewed journals. Where e.g. a commercial test kit is used, the manufacturer's instructions should be followed All methods, efficacy verification studies and relevant references to the methods must be documented, authorised by the designated person and kept up to date	European Standard BS EN ISO / IEC 17025 (2000) Documentation of individual accreditation bodies
Method validation	If methods are used in the laboratory that are not standard or published methods, then they must be demonstrated as suitable for the intended purpose by a structured validation process often undertaken by the manufacturer. This usually involves comparison of results from the proposed new / alternative method and a standard method obtained using a large number of samples comprising different sample types, and / or an inter-laboratory study where two or more laboratories examine a number of samples by agreed methods and compare results. A selection of deliberately inoculated (spiked) samples using specific organisms at levels normally expected to be detected / enumerated by the test are often included in method efficacy validation studies. For each validation study, the procedure, results and conclusion as to suitability of the method for the intended purpose should be recorded	European Standard BS EN ISO / IEC 17025 (2000) Baylis *et al.* (2001) Chapter 7 *ibid*
Quality control (QC)	In addition to sourcing supplies from reputable, ISO standard registered companies, regular internal quality control and checks should be undertaken to monitor the day-to-day efficacy of tasks performed, e.g. equipment checks; media preparation QC; routine replicate sample examinations to check consistency of the methods; procedures and practices; cross-checking of plate-counting and examination of incubated plates to check the quality of plate preparation including depth of agar poured, quality of plate pouring, distribution of colonies, edge contamination etc.	Chapter 7 *ibid* Anon (1999 (1) (1A) (2) (3) (7) (9) (10) (11) (12) (13))

Contd.

Table 8.2 *Contd.*

Area addressed	General requirements	Sources of further information
Proficiency testing	Laboratories should participate in an externally managed proficiency testing scheme covering all of the organisms for which tests are carried out in the laboratory and involving all relevantly trained staff. Poor results in such tests must be investigated and any problems resolved. All such investigative actions must be recorded.	Chapter 7 *ibid* European Standard BS EN ISO / IEC 17025 (2000) Documentation of individual accreditation bodies

8.2   Standard Operating Procedures

As already indicated in Chapter 7, a 'Standard Operating Procedure' (SOP) is a procedure that has been determined usually by the senior microbiology staff of the laboratory, as the most appropriate means for achieving a specified activity in the laboratory correctly. All key laboratory operations should be identified and documented in 'Standard Operating Procedures' or other relevant documentation established for use in staff training and for day-to-day reference purposes. This helps to ensure that a common and correct approach is taken by all staff to each of the key tasks. Table 8.3 lists some of the key documentation, including SOPs, normally required in a microbiology laboratory and gives an indication of the topics that should be covered in such documentation. Table 8.4 shows examples of the headings and expected contents of two different types of SOP.

As old equipment is renewed, or new methods, procedures and equipment are introduced to the laboratory, relevant accompanying documentation including SOPs should be developed to ensure that all laboratory documentation is maintained relevant and up to date.

As part of training in any particular task, laboratory staff must read and ensure they understand any Standard Operating Procedures relevant to the job to be done. This important part of training should be included on each individual's training record.

8.3   Sample processing documentation

In order to be able to demonstrate full control of all laboratory activities and, if necessary, properly investigate and resolve any quality issues or, more rarely, complaints, it is important to be able to clearly and reliably link and trace, as far as practically possible, all aspects of work related to all the tests carried out on each factory sample received by the laboratory.

Records relating to stores receipts and associated quality checks on laboratory consumables, receipt of factory samples, handling, examination (testing), culturing, confirmation, test results, and all media preparation records should all be made in such a way as to allow an unbroken chain of traceability (links) from any final,

Figure 8.1 Some quality **check points** and **considerations** relating to media preparation.

Check point	Consideration
Dehydrated media, supplements, water	Dehydrated media and supplement supplies should be used within the shelf life prescribed by the manufacturers Physical condition should be checked, e.g. aggregated lumps should be absent from powdered products Purified water quality should be checked and results recorded
↓	
Weigh media ingredients*	Manufacturers' instructions should be checked for the weight per volume required Balance should be checked for accuracy and results recorded
↓	
Dispense water volume*	Containers used should be clean and of appropriate size, e.g. double the volume of the medium volume being prepared Volumetric equipment used to dispense water should be checked for accuracy and results recorded
↓	
Disperse / dissolve ingredients	Manufacturers' instructions should be checked for the appropriate procedure All components of the medium must be fully dispersed and dissolved
↓	
Sterilise medium	Label with a medium unique identifier All equipment used must be in good condition and properly calibrated Manufacturer's instructions should be checked for the appropriate heat process to be used, e.g. boiling, autoclaving temperature / time Heat processes should be monitored to ensure that overheating is avoided. Autoclave processes should be monitored using time–temperature recorders and / or independent physical (maximum thermometer) or chemically based monitors and results recorded
↓	
Add supplements (if required)	Manufacturers' instructions should be checked for quantity per medium volume required and method of addition Where appropriate, autoclaved basal media should be allowed to cool before adding supplements
↓	
Quality assurance	Checks on prepared media should be carried out e.g. pH value, sterility, efficacy (qualitative / quantitative) using reference cultures and results recorded
↓	
Media use	Media storage conditions should ensure that no moisture is lost and there is no opportunity for deterioration of components, e.g. through photo-oxidation Agar to be re-melted should not be overheated during melting and storage of the molten media after tempering (at 45–47°C). Molten media should generally be used within 3 hours of melting or as specified by the manufacturer Drying of agar surfaces to be used as spread or streak plates should be accurately controlled to prevent over-drying

* Media manufacturers' instructions should be checked carefully to determine the correct water volume required, e.g. 'add 1000 ml water', 'make up to 1000 ml' or 'make up to 1000 g'

Figure 8.2 Some quality **check points** and **considerations** relating to factory sample handling and test sample preparation and examination.

Factory sample receipt	→	Factory sample storage until required for testing	→	Factory sample booking-in	→	Primary weighing to obtain test sample	→	Weighed test sample storage	→	Addition of diluent	→	Blending / maceration	→	
Check that sample description is correct *Check sample condition and transport history* *Record in laboratory system*		*Check that storage conditions (physical protection, temperature, humidity, time) are correctly applied* *Monitor daily and record*		*Check sample description* *Apply a sample unique identifier*		*Check that correct tools and equipment are to hand* *Check that conditions and method of opening sample container are correct, e.g. disinfect bench area, disinfect packaging; and type of sample to remove (component or composite etc.)* *Check balance accuracy* *Use sample unique identifier on container labels*		*Check that storage conditions (physical protection, temperature, humidity, time) are correctly applied* *Monitor and record*		*Check that diluent type is correct* *Check that diluent volume is correct*		*Check type of blending / maceration required* *Check time required*		

Contd.

Figure 8.2 Contd.

Process step		Records required
Further dilutions made	→	*Check that diluent type is correct* *Check that diluent volume is correct* *Check that adequate mixing is achieved* *Maintain sample unique identifier*
Plates and tubes inoculated, media added to plates (as appropriate for tests required)	→	*Check that media prepared are correct (type and volume) for the tests required* *Record media unique identifiers* *Check agar temperature (not too hot causing death of organisms in test sample inoculum and not too cool causing premature gelling of agar)* *Check that agar container is properly dried on the outside before pouring agar medium* *Maintain sample unique identifier on all plates and tubes*
	→	

		Records required	
Medium / inoculum mixed	→	Check that mixing process yields a clean plate (no splashes, spills) Check that inoculum is evenly distributed in the medium Check that the correct volume / depth of agar is poured	
Plates and tubes incubated	→	Check that correct incubation conditions / temperature are used and the incubator is functional Check that incubators / water baths are not overloaded	
Incubated plates and tubes assessed	→	Check that correct incubation time has been applied Check the colony morphology and media reactions to be assessed for each medium Re-check plate quality to ensure that an accurate assessment of growth can be made reliably	

Contd.

Figure 8.2 Contd.

Results reported	→	**Results recorded**
Check the reporting structure and format of reports required based on the sample unique identifier	←	*Check that test results are recorded for the correct sample using the sample unique identifier* *Ensure that the results are recorded clearly and that the assessor is identified* *Check if any follow-up is required*
		↑
Confirmation tests carried out as required and assessed	→	**Confirmatory tests if required**
Records required	←	
		↑
Check that correct tests are performed by trained and authorised persons *Check that any media or test kit unique identifiers are recorded along with the sample unique identifier for traceability purposes* *Check that test assessment criteria are correctly applied*		

Results recorded	→	Check that test results are recorded for the correct sample using the sample unique identifier Ensure that the results are recorded clearly and that the assessor is identified Check if any follow-up is required
	↓	
Results reported	→	Check the reporting structure and format of reports required, based on sample unique identifier

Table 8.3 Some key documentation required in an accredited microbiology laboratory and an indication of the topics that are covered.

Documents	Coverage
Laboratory policy document	Accreditation standard and mission statement Organisation and reporting structure Staff job descriptions Authorised signatories Supplies purchasing policy including standards of goods required Plan of laboratory Conditions applying to sub-contracting of work Complaints procedure
Laboratory manual	Approved laboratory practices Safety procedures (emergencies, culture spillage) First aid, security, authorised personnel Sample and supplies receipt procedures and unique identification systems Reference cultures (list, source, applications, handling procedures) Media preparation procedures, unique identification system and quality control requirements Microbiology methods (approved methods, validation procedures, efficacy checks) Approved microbiological techniques, practices and procedures and monitoring systems including responsible staff External proficiency and internal proficiency testing systems Calculation of results Records required
Laboratory housekeeping – Standard Operating Procedure	Requirements and means of ensuring that all areas of the laboratory environment and equipment are kept as clean and tidy as appropriate to the function of a food microbiology laboratory Complete list of areas and equipment* to be included in the housekeeping schedule Responsible staff and rota operated Frequency of cleaning and method of cleaning including chemicals and chemical strengths to be used Health and Safety requirements Laboratory environmental microbiological checks, e.g. swabs, settle plates, to be carried out, including frequency, location, test types required Records required
Equipment files	Purchase documentation and manufacturers' operating manuals / instructions Maintenance and repair records Calibration records Monitoring records
Equipment – Standard Operating Procedures	Description including make / model / serial number Location in laboratory and laboratory equipment unique identifier Description of function and applications Service and maintenance requirements Calibration and monitoring requirements Responsible staff and method of correct operation 'Out of specification' reaction procedure Records required

Contd.

Table 8.3   *Contd.*

Documents	Coverage
Staff training manual	Training subjects and schedules Training methods Assessment criteria for each subject Approved internal and external courses Responsible staff (trainers, assessors) Records required
Training records	Individual files for each member of staff containing full training record Evidence that assessment criteria for each subject have been achieved Each training subject signed off by the trainer / assessor and the trainee

* Any computers used in the laboratory must be included in these lists and schedules.

Table 8.4   Examples of headings and associated content of Standard Operating Procedures.

Standard Operating Procedure (Laboratory Reference Number)
Date and signature of responsible person

	Examples of information that should be included	
	Equipment SOP	**Task SOP**
Title	Balance	Housekeeping
Aim	To describe the function, applications and requirement for correct operation of the laboratory's balances	To describe the requirements and means for ensuring that all areas of the laboratory environment and equipment are kept as clean and tidy as appropriate to the function of a food microbiology laboratory
Equipment (name / identity number(s))	Make / model / serial number Location / laboratory equipment unique identifier	Description and location of any specific equipment and chemicals required to aid housekeeping tasks
Person(s) responsible for operation	Senior technicians and technicians trained to specified level	Laboratory staff trained to specified level Explanation of staff rota for specified housekeeping responsibilities
Use / purpose / function	Accurate measurement of the weight of test materials, microbiological media and media supplements	To maintain the laboratory environment and equipment in an appropriate hygienic state for the microbiological examination of food and related samples

Contd.

Table 8.4 *Contd.*

	Examples of information that should be included	
	Equipment SOP	**Task SOP**
Method / procedure / practice / safety requirements	Method of keeping the equipment clean Need for levelling the balance before use Method of use	Complete list of areas and equipment to be included in the housekeeping schedule Frequency of cleaning specified Method of cleaning specified, including chemicals and chemical strengths to be used Health and Safety requirements indicated
Calibration / monitoring	Frequency of calibration Calibration service provider used Location of calibration information Frequency and method of monitoring Responsible person(s) for both monitoring and initiating action to out-of-specification results Weights (masses) to be used and method of use Records to be kept	Responsibility for visual inspection Environmental microbiological checks, e.g. swabs, settle plates, to be carried out, including frequency, location, test types required Action required to out-of-specification results Records to be kept
Action in the event of failure (including documentation to be completed)	Equipment labelling system, e.g. "Do Not Use" Responsible person to be notified Documentation requiring completion	Responsible person to be notified in the event the housekeeping schedule is not maintained Records to be kept
Relevant related documents (for cross-reference purposes)	Other SOPs or files in which information concerning balances may be found	Other SOPs or files in which information concerning housekeeping requirements and related information may be found

reported result, back through all the operational processes to e.g. the batch of dehydrated medium used and, to the original factory sample receipt record. Figure 8.3 gives examples of the types of records and identifiers, i.e. '*Laboratory Unique Identifier*', *Laboratory Equipment Unique Identifier* '*Sample Unique Identifier*' and '*Medium Unique Identifier*', that can help achieve practical traceability for a microbiological enumeration test carried out using plating media.

Different laboratories have different ways of recording the types of information shown in the tables in Figure 8.3, but the key point is to ensure that all information relevant to each factory sample and each test carried out on the sample is recorded in such a way as to allow the practical traceability of the test results. Thus, starting from the 'Final Result' in Table G or H (Figure 8.3), a path can be readily traced back through the records to identify all staff involved in the sample handling, testing and media preparation as well as the batch numbers of media and other key materials used and the dates / times of key activities such as start and end of

Figure 8.3 Example of records for sample processing and medium preparation in relation to tests for the enumeration of microorganisms showing the links that ensure traceability from the test sample result back to the factory sample received and to the test materials used.

Table A

Stores receipts record and quality checks for media and laboratory supplies*

Date / Time received	Supplier	Description including any codes / dates and condition of materials received	Supplier's Certificate of Quality Assurance received	Laboratory (in-house) QC tests carried out	Results	**Laboratory Unique Identifier[†]**	Materials passed / date and time released for use Additional comments	Signature of authorised person**

* Separate record systems are usually maintained for plastic disposables and microbiological media but the general record type and information recorded is similar.
† Assigned by the laboratory

Table B

Factory sample receipt record

Date / Time received	Customer	Description including any codes / dates	**Sample Unique Identifier***	Tests required / methods to be used	Date to be tested	Additional comments, e.g. condition of sample, temperature	Signature of authorised person**

* Assigned by the laboratory.

Contd.

Figure 8.3 *Contd.*

Table C

Media preparation

Date / Time	Medium	Supplier	Supplier Batch Code / **Laboratory Unique Identifier***	Quantity of dehydrated medium weighed (g)	Volume prepared	Volumes dispensed	Signature of authorised person**	Heat process Boiling† / autoclave Required	**Medium Unique Identifier***	Additional comments	Signature(s) of authorised person**

* Assigned by the laboratory. See also Table A and Figure 8.1.
† If medium is boiled, signature of operator performing this task is required in this column.

Table D

Autoclave records

Date	Time	Autoclave number (laboratory equipment unique identifier)	Load content of media using **Medium Unique Identifiers**	Required process	Commercial monitor used (supplier, type and batch code / **Laboratory Unique Identifier**)	Process records checked* (chart, chemical record, chemical monitor, maximum thermometer) Satisfactory / Unsatisfactory	Action (taken in relation to unsatisfactory process)	Additional comments	Signature of authorised person**

* Where possible, the actual result should be recorded, e.g. temperature read on the thermometer, colour change on the chemical monitor.
Note: Chart records for each process should be dated (including time) and signed by the checker.

Table E

Supplements record

Date / Time	Medium Unique Identifier	Supplements used	Supplement supplier	Supplement batch code / Laboratory Unique Identifier	Quantity of supplement added per volume of medium	Additional comments	Signature of authorised person**

Note: Supplements are often added to the basal medium on the day of medium preparation and heat-processing, in which case the information relating to supplements may be included in Table C. In cases where supplements are added to the re-melted base medium on the day of use, a separate supplement record may be needed.

Table F

Media quality control

Date	Medium Unique Identifier	Physical appearance (Satisfactory / Unsatisfactory)	pH value	Signature of authorised person**	Reference cultures used (date code etc.)	Incubation temperature and Incubator / water bath number†	Time into incubation	Signature of authorised person**	Date / Time removed from incubation	Growth control results			Signature of authorised person**	Pass / Fail and Comments	Action / Date / responsibility and document / report cross-references	Signature of authorised person**
										Sterility	Positive control	Negative control				

† Laboratory equipment unique identifier.

Contd.

Figure 8.3 *Contd.*

Table G

Test sample record for enumeration tests

Date / Time test started	Sample Unique Identifier	Test	Test weight and primary dilution		Dilutions required	Volume plated	**Medium Unique Identifier**	Incubation temperature and Incubator number*	Time into incubation	Signature of authorised person**	Date removed from incubation	Assessment of plate quality	Counts made on each duplicate plate at each dilution		Calculated count per gram of test sample / **Final Result**	Additional comments / confirmatory tests required	Signature of authorised person**
			Weight	Primary dilution	-1-2-3 etc.								-1 -2 -3				

* Laboratory equipment unique identifier.

Table H

Test sample record for biochemical and serological confirmatory tests

Date / Time test started	Sample Unique Identifier	Test	Laboratory Unique Identifier (test kits / reagents) and / or **Medium Unique Identifier**	Incubation temperature and Incubator number*	Time into incubation	Signature of authorised person**	Date removed from incubation	Assessment of confirmatory test results	**Final Result**	Additional comments / confirmatory tests required	Signature of authorised person**

* Laboratory equipment unique identifier

** In large laboratories in which a number of different people may contribute to different aspects of the sample examination process or media preparation, it is important that each individual indicates their involvement / responsibility by initialling / signing the appropriate record.

incubation, i.e. a complete laboratory history of the test result, which is the aim of such a record system.

Unambiguous sample and / or medium unique identifiers, as applicable, should be applied either directly (by writing on them) or using labels, to all containers, plates and broth bottles or tubes associated with the test sample. The base of the Petri dish and the body of a culture bottle should be used for labelling (not the lid of the Petri dish or the cap of a container as these can be mixed up easily).

Where further confirmation work or identification of cultures grown is required (see Chapter 11), the sample unique identifier should continue to be transferred to any test kits or additional media required to complete the work (Figure 8.3, Table H).

Also essential, are the records that relate to the laboratory's routine monitoring checks made of equipment performance and general laboratory operations (Table 8.5) because these are required:

(1) as part of the work traceability and quality record
(2) to demonstrate the correct performance of the equipment (Table 8.6) referred to in the preparation and processing records (Figure 8.3) for the time they are used.

Table 8.5 Routine laboratory operating and monitoring records required for commonly used equipment and routine tasks in direct support of the traceability of work and results.

General laboratory operation records*	Daily equipment check records*
Stores receipts† 1. Media (date of receipt, supplier, description, quantity, batch code, shelf life, supplier's QC records) 2. Sterile plastic disposables (date of receipt, supplier, quantity, batch code, supplier's quality control records and / or quality assurance certificate) 3. Reference control cultures (date of receipt, supplier, description, quantity, batch code, shelf life, supplier's quality control records) Laboratory's own quality control records on any stores items received (item, date of receipt, batch code, number tested, parameter tested, results) Equipment calibration records	Balance Water conductivity of purified water Incubator temperatures Water bath temperatures Refrigerator temperatures Freezer temperatures pH meter Pipetter volume Autoclave Gravimetric diluter

* All records must be signed and dated by the person carrying out the check / making the record.
† Different batch codes of disposables received should be kept separately and used in sequence of date of receipt. Store records should be kept to indicate the date and time at which each batch code is released into the laboratory for use. Where the same batch code is received on different dates, then an alphabetic or numeric system should be applied to distinguish the different deliveries and assist traceability of disposable equipment or media usage in the laboratory.

Table 8.6 Example of the layout and headings of an equipment check record.

Equipment name and laboratory equipment unique identifier

Date / Time	Parameter checked*	Standards used (if applicable)** Laboratory unique identifiers	Number of checks carried out	Results	Satisfactory / Unsatisfactory	Action / Additional comments	Signature of authorised person

* For example, temperature, pH value, weight, volume.
** For example, checkweights (identifying number), standard pH or conductivity solutions.

Some general requirements applicable to the completion of records are as follows:

- All records must be kept up-to-date, complete and accurate.
- All staff working on any process that affects the test sample results produced must sign / initial the appropriate record and date it where appropriate.
- All records must be clearly readable and it should not be possible for staff to mis-interpret any record (dittos and blanket statements are not acceptable; '–', 'X' and 'OK' might mean different things to different members of staff).
- The need for copying information from one record to another should be minimised to avoid mistakes on transfer.

8.4 Summary

It can seem a daunting prospect to be required to develop, maintain and operate such a quantity of documentation alongside the practical work involved in examining food and food-related samples for their microbiological content. However, it is a useful training exercise to draw up a step-by-step flow diagram of your own laboratory's media and sample processing procedures and consider for yourself the documentation and record systems that would logically and usefully support them and that would facilitate full traceability of test results (for guidance see Figures 8.1, 8.2 and 8.3). It then becomes clear that a certain minimum number of records is required.

A simple and clear but thorough, logical and complete documentation and record system, relating to all laboratory operations, maintained and operated correctly by all laboratory staff alongside good standards of practical microbiology (see Chapter 7) can provide laboratory staff, their employers and their customers with a high degree of confidence in the reliability of test results produced by the laboratory. This is the ultimate aim of all those practising food microbiology and the requirement of those who use the results in decision-making.

8.5 Further reading

Anon (2002) *EA-04/10 : Accreditation for Microbiological Laboratories.* European Cooperation for Accreditation, Brussels, Belgium.

European Standard BS EN ISO/IEC 17025 (2000) *General Requirements for the Competence of Testing and Calibration Laboratories.* European Committee for Standardization, Brussels, Belgium.

Snell, J.J.S., Brown, D.F.J. and Roberts, C. (1999) *Quality Assurance – Principles and Practices in the Microbiology Laboratory.* Public Health Laboratory Service, London, UK.

Wilson, S. and Weir, G. (1995) *Food and Drink Laboratory Accreditation – a Practical Approach.* Chapman & Hall, London, UK.

9 Conventional Microbiological Methods I: Equipment, Basic Techniques and Obtaining Samples

9.1 Introduction

Conventional food microbiological methods require the operation of the same fundamental aseptic practices and sub-culture techniques that are used in almost all branches of microbiology; however, in food microbiology, special procedures and detection methods have been developed that, in many respects, are quite different from those used for example, in medical microbiology. This is because the microflora of food is very different from that of clinical specimens, not just in terms of the types of microorganism encountered and of interest or importance to foods, but also in terms of the numbers of the target organism present, the numbers of competitor organisms present and the degree of sub-lethal damage or stress caused to the organisms by food processes and environments.

If bacterial pathogens contaminate a food sample, they are often present in low numbers and may be in competition with much higher numbers of harmless microorganisms; also, food processing often (intentionally) creates a harsh environment, e.g. low pH value or high salt concentration, which stresses the microorganisms, so that, although they remain viable, it is difficult to grow them on some microbiological media without prior resuscitation. In contrast, clinical specimens are likely to contain high numbers of the target organism that are healthy, actively growing and often with few competitors. Conventional methods in food microbiology, therefore, tend to be more complex, involving more sub-culture steps / manipulations than those in medical microbiology and take longer to obtain a reliable result; such methods are, however, essential if false negative results are to be minimised.

There are two important questions that have to be answered when a factory sample arrives in a food microbiology laboratory. These are 'What organisms should I be looking for in this sample?' and 'How many of these organisms am I expecting to find?', because the answers to these questions determine the general test method to be used and also whether and how the method needs to be 'adapted' for the particular factory sample. Where a laboratory routinely tests many similar food sample types, e.g. daily factory samples taken from a single production line, then these decisions may have to be made only once (unless the product type changes). If, however, the laboratory examines factory samples that have been stored for some time since they left the production line, e.g. shelf life samples or samples associated with customer complaints, or if samples are received from many different sources, then, even if the

Plates 10, 11, 12, 13 and 16–20 are located in the colour plate section at p. 100.
Plates 21, 22 and 25 are located in the colour plate section at p. 228.

sample types from the different sources appear similar, carefully considered decisions must be made for each sample before testing can begin.

Discussions of some of the considerations that need to be taken account of in such decision making can be found in texts by Bell and Kyriakides (1998 and 2002) and Institute of Food Science and Technology (1999).

The basic techniques and practices used in the food microbiology laboratory to examine food and food-related samples are described in this chapter, as well as some aspects of the sampling of foods for microbiological examination. Some microbiological methods that are specific to the examination of foods are discussed in Chapter 10 whilst culture confirmation and identification tests are described and discussed in Chapter 11. Further guidance and information concerning some of the considerations involved in the selection of appropriate tests for different food sample types can be found in a publication of the Institute of Food Science and Technology (UK), *Development and Use of Microbiological Criteria for Foods* (IFST, 1999).

9.2   Basic tools of the food microbiologist

The types of factory sample encountered in a food microbiology laboratory can vary widely and often do not arrive neatly packaged in sample containers; indeed, the staff of many factory laboratories are responsible for collecting food and environmental materials for examination directly from production areas. It is not uncommon for food microbiology laboratories to be presented with factory samples weighing several kilograms, such as whole pies or hams, cheeses or frozen joints of meat, from which smaller quantities must be aseptically removed and placed in an appropriately labelled, sterile sample container before discarding the remainder. A wide range of equipment may therefore be required for collecting factory samples, for sub-sampling foods into smaller, more manageable aliquots (pieces or volumes) and for weighing out the amount to be tested to provide the test sample.

9.2.1   *Sampling equipment and terminology*

A number of different types of factory sample may be received by a laboratory and these may or may not require some form of preparation or sub-sampling; for enumeration tests, in particular, following preparation steps, often only a small quantity of the primary dilution of the test sample i.e. containing 0.1 g or less of the test sample, is actually inoculated into the microbiological growth medium, the remainder being disposed of. The term 'sample', may be interpreted in different ways by different people and it is important that laboratory staff understand what is meant by this term. For some descriptions of the different types of sample see Chapter 7 and Figure 7.2, while a more detailed discussion of sampling techniques and their significance can be found in this chapter (see '9.7 Obtaining and handling samples').

With a few exceptions, it is not essential for the food microbiologist to use specialist sampling equipment and some laboratories use plain, domestic, stainless steel cutlery or solid steel scalpels i.e. the blade is not removable, and spatulas are often used for soft and semi-solid samples. Wherever possible, liquids should be sampled using a pipette rather than by pouring, as the latter can generate aerosols and result in contamination of both the sample and the surrounding environment. Whatever equipment is used, it must be capable of being thoroughly cleaned and sterilised, unless commercially pre-sterilised, single use, disposable tools such as scalpels and pipettes (plastic or glass) are used; sterile disposable plastic knives and spoons are sometimes used for soft samples and powders.

Sufficient stocks of all sampling equipment should be available, especially of those items that need to be washed and autoclaved before re-use. For re-usable stainless steel tools, sterilisation by autoclaving after thorough washing is preferable to flaming with industrial methylated spirit, for two reasons:

(1) items such as scalpels with removable blades have 'nooks and crannies' that can harbour food particles and may not be cleaned and sterilised effectively by cursory rinsing 'at the bench' and then flaming;
(2) containers of industrial methylated spirit can become heavily contaminated with food particles in which bacterial spores may survive not only in the container but also, when protected, during the flaming operation.

Both of these areas of potential failure in aseptic practice can cause a cross-contamination hazard to materials being sampled.

A wide variety of containers, of different shapes and sizes, may be used for factory sample transport and storage. Traditionally, these have been made of glass or similar material that is re-used, after cleaning and re-sterilisation. This is time-consuming and must be done properly; therefore, pre-sterilised (irradiated), plastic disposables are generally preferred. Another consideration is that breakable materials such as glass should never be taken into food factories (unless they are for food product packaging) because if there is a breakage they present a foreign body hazard to food materials being processed and hence a potential danger to the consumer.

Different sizes of container are required, ranging from 10 ml to 1 litre, though 100 ml and 250 ml sizes are likely to be those most commonly used; sometimes, however, other types of container may be useful. Sterile blender ('Stomacher') bags, for example, can be used for the 'inverted bag technique' of sampling where the trained operative holds the outside of the bottom of the bag and folds back the opening, without touching the inside, so that the bag forms an inverted, sterile 'glove' over the hand; the operative then picks up and holds the sample material using the 'gloved hand' and refolds the bag so that the sample remains inside the bag. This is a convenient means of sampling materials that are not packaged, such as bulk material at an intermediate stage of processing, e.g. minced (ground) meat, or unwrapped finished products, e.g. bakery goods, or samples of debris from the production environment (the technique has even been used for sampling cattle faeces!).

Blender bags are also useful when taking swabs, especially the large sponge swabs that are sometimes used for sampling production environments for the presence of pathogens such as *Salmonella*. After swabbing the selected environmental area, the swab is placed in a blender bag (if several swabs are to be combined for testing, they may be placed in the same bag) and transported to the laboratory where a pre-enrichment broth or enrichment medium is added directly into the bag. This is then sealed and the material collected on the sponge swab(s) is dispersed, by hand massage or blender, in the medium. The bag and its contents are then placed directly into an incubator, ensuring that the bag is adequately supported e.g. in a beaker so that its contents cannot spill during incubation.

Whatever type of container is used for sampling, it is important that, when a sample is collected, the container is adequately labelled with all the sample details; for containers with removable lids, the container itself should be labelled, not the lid, as lids can become mixed up easily when several containers are opened simultaneously in the laboratory.

9.2.2 *Pipettes*

Pipettes are manufactured in a range of different designs and volumes (Plate 12, see colour plate sections for all plates). Graduated pipettes, the most common type, are used for measured volumes, often 1 ml or 10 ml. Some graduated pipettes have a wide aperture for taking up particulate suspensions. If made of glass, graduated pipettes are generally re-useable; they usually have a cotton wool plug at the 'bulb' end which needs to be removed before the pipette can be washed and re-sterilised. More commonly, however, plastic or glass, disposable pipettes are used in food laboratories. Pasteur pipettes have a fine tip and are used for transferring small quantities of liquids; they are almost always disposable. Traditionally they are made of glass but plastic disposable versions with an integral bulb are often used nowadays. It is **never** permitted to pipette by mouth in the microbiology laboratory; a bulb or equivalent dispensing aid must **always** be used.

Many laboratories now use pipetters with disposable tips, especially for transferring small volumes, i.e. 1 ml or less. These are often convenient for dispensing small quantities of reagents such as selective agents, especially if the material dispensed is to be sterilised subsequently. When sterile tips are used to dispense dilutions of test samples, very great care is needed because, if used carelessly, the material being dispensed can contaminate the pipetter unit and barrel and then cross-contaminate subsequent materials being pipetted (see 'Sample preparation and handling' in Chapter 7). Where these units are used, sterile (irradiated) pipette tips must be handled carefully when being attached to the pipetter as they can become contaminated easily by microorganisms from the environment and the operator's fingers and likewise, great care must be taken to prevent contamination of the external surfaces and barrel of the pipetter during use. Effective cleaning and sanitising regimes described in the laboratory's own Standard Operating Procedures must be adhered to.

9.2.3 *Loops, wires and spreaders*

Loops are pieces of wire with a ring-shaped end (loop); the end opposite the loop is fitted and clamped into a handle. Loops are used to transfer very small quantities of reagents, liquid cultures or colonies from one place to another. In particular they are used for 'streaking' an agar plate so that microbial cultures can be purified and isolated (see Figure 7.1). Wires are straight-ended and used to inoculate cultures into agar, e.g. by 'stabbing' (see below), or to inoculate cultures, especially moulds, as single points of inoculation onto the surface of agar; plastic, disposable 'wires' are often known as needles (as are the inflexible dissecting needles often used to transfer mould growth). Loops and wires are used for qualitative tests, i.e. when an accurate volume of material is not required, whilst spreaders are L-shaped rods, also known as 'hockey sticks', that are used for spreading a known quantity of a test sample dilution evenly over an agar surface, so that the colonies that develop during incubation can be counted accurately.

Traditionally, loops and wires were made of platinum but now are usually made of metal such as 'nichrome' which is an inexpensive alternative to platinum. Both of these metals, made into wires, are reasonably flexible, so that they do not damage soft agar gels when used properly. Spreaders are traditionally made of glass. All of these items are therefore re-useable and need to be sterilised, both before and immediately after use. Metal loops and wires are sterilised by heating in an oxidising Bunsen flame, i.e. a flame that is colourless or blue, but not yellow, until the entire wire portion is red hot. Heating the loop or wire from the 'clamp' end towards the tip minimises the creation of aerosols from any remaining material. They must then be allowed to cool properly in the air before being used to sub-culture live microorganisms, to prevent killing any cells; to avoid excessive time delays when sub-culturing many samples, two loops are often used alternately so that one is cooling whilst the other is in use. Glass spreaders are sterilised by dipping in industrial methylated spirit then passing through a non-oxidising (yellow) Bunsen flame to ignite the alcohol, allowing it to burn until the flame extinguishes naturally. The spreader must then be allowed to cool in air before making contact with the test sample suspension on the agar, again to prevent killing any cells in the inoculum. **Spreaders must never be put back into a container of alcohol unless it is certain that the flame has extinguished** because of the danger of igniting the alcohol.

Nowadays, laboratories often use pre-sterilised, plastic, disposable loops, wires (needles) and spreaders. Using these can save time and therefore labour costs, but are expensive to buy when large quantities are needed and they must be disposed of, which may not be considered 'environmentally friendly'. However, a major advantage to their use is that no flame sterilising is required and thus the potential for hot tools to be used in contact with cultures and the consequent adverse effect on results is avoided. Because pre-sterilised plastic tools do not require flame sterilisation with alcohol, which, in the authors' opinion is not a consistently well-practised technique, they are highly preferable in support of good practice by comparison with their metal and glass counterparts. Some examples of plastic loops, wires and spreaders are shown in Plate 13a.

9.2.4   *Other laboratory equipment*

Petri dishes, commonly known as plates, are used in all laboratories for growing microorganisms on or in agar media. Many laboratories use sterile, plastic disposable Petri dishes (Plate 13b) but the traditional glass dishes are still used in some laboratories. Used glass dishes, however, require careful procedures for primary decontamination, washing, drying and re-sterilisation prior to re-use and the disposable plastic types are safer in general use. A variety of other items of equipment is used in most food microbiology laboratories, ranging from disposables to items of capital equipment (see Chapter 6).

Bulk quantities e.g. 500 ml, 1 litre, of prepared microbiological growth media and diluents are best stored in screw-capped, glass bottles; a common type used in food microbiology laboratories is the 'Duran' bottle. 'Duran' bottles are fitted with a plastic 'pouring ring' around the lip of the bottle so that the contents do not drip down the outside of the bottle when pouring from the bottle, as this can result in contamination of water baths or the laboratory environment. It is important that the pouring ring is fitted the correct way round, i.e. so that the wider diameter of the ring is uppermost, otherwise it cannot prevent drips. All bottles should be maintained in good condition; any showing chips or cracks, especially around the neck, should be discarded. Bottle volumes ranging from 100 ml to 1 litre are most often used, although it is important that volumes of media and diluents that are to be sterilised by auto-claving are kept to a minimum so that they receive an adequate heat process (see Chapter 7 and 'Sterilisation of media' later in this chapter).

Small quantities e.g. 10 ml of broth media and diluents, required ready for direct inoculation or for use in preparing serial dilutions of the test sample respectively, are ideally dispensed into 1 ounce 'Universal' bottles (for an example, see Plate 10) or 1 ounce MacCartney bottles that have a smaller cap size; alternatively, capped test tubes may be used but these are not hermetically sealed and may allow evaporation of the contents during autoclaving or storage so great care is required to control and monitor final liquid volumes.

Apart from the laboratory autoclave (see below), essential items of capital equip-ment include balances, pH meters and colony counters, although larger laboratories may also possess one or more laminar air-flow cabinets, safety cabinets, automatic plate pourers and other equipment (see Table 6.3).

'Top pan' balances are the most useful type of balance for weighing out media and test samples (Plate 17). Depending on the types and quantities of media that need to be prepared, it may be necessary to invest in two or more balances with different weighing ranges and sensitivities; for quantities up to a few kilograms an accuracy of only ± 0.1 g, i.e. '1 decimal place' is required, but smaller quantities may need to be weighed out more accurately, for which a '2 decimal place' balance is required. The accuracy of balances should be checked daily by weighing an item of known weight and the result should be recorded (see Table 6.3). Analytical balances that have an accuracy of 3 or 4 decimal places, as used in chemistry laboratories, may be required for weighing supplements for selective-diagnostic media; however most supplements are now supplied in pre-dispensed quantities and need only to be re-hydrated, before adding to the basal medium, thus eliminating the need for an analytical balance.

For health and safety reasons, it is important to avoid the inhalation of dust when weighing out powdered materials and suitable safety precautions must be taken, e.g. wearing a face mask that provides an adequate level of filtration, or weighing out the material under an extraction hood; if the latter approach is used, then it is important to ensure that the flow of air created or any vibration from the extraction fan does not affect the accuracy or performance of the balance.

The pH value of a microbiological growth medium or diluent can significantly affect the viability or growth potential of microorganisms suspended in it or placed on it and the pH values of all in-house prepared media and diluents must therefore be checked (see Chapter 7), usually before autoclaving, although occasional post-autoclaving checks are also advisable. pH value is measured using a pH meter (Plate 18), which is an electronic instrument with an electrode connected to it by a cable and an in-built temperature compensation system. Either the electrode is immersed in the solution to be measured or, a special contact electrode is placed flat on the surface of solidified agar. An electrical change occurs at the contact surface of the electrode (the material of which has a high electrical conductivity) and a solution and this varies with the pH value of the solution (see Chapter 2). It is this electrical change that is detected via the electrode of a pH meter and this is converted to a digital readout of pH value. It is essential that the pH meter is calibrated using solutions of known pH value before use (see Chapter 6). It is important that laboratory staff understand the operation of a laboratory's specific pH meter and how it should be maintained and pH meters must always be calibrated before use each day; generally, this is done by measuring the pH of chemical buffer solutions of known pH values, usually pH 4 and pH 7, and adjusting the instrument according to the readings obtained.

pH probes are delicate and must be handled with care. Sometimes, it is necessary to measure the pH value of a food material but it is important that the pH probe used for media and diluents is never used for this purpose. High levels of fat and protein in a food can coat the surface of a pH probe and significantly affect the reading obtained, so a separate probe should be kept if the pH values of foods is to be measured regularly; a probe used for this purpose should be subject to a reliable cleaning regime and may need to be 'de-proteinised' or 'de-fatted' periodically using reagents specifically designed for the purpose.

Balances and pH meters are essential in the operations of a food microbiology laboratory and their accuracy and precision must be maintained by regular calibration, checking and servicing (see Chapter 7).

Staff in food microbiology laboratories often try to count colonies that have developed on agar plates (see Chapter 10, '10.2.1 Colony counts') without the aid of a colony counter, yet the colony counter is one of the most simple but important items of equipment that a laboratory may purchase. The colony counter has two important functions: it illuminates and magnifies the colonies so that small colonies can be seen more easily and particulate matter of non-microbial origin, e.g. food particles, can be distinguished from colonies; in addition, some colony counters incorporate a counting device (tally counter) that is used to record the number of colonies counted without the technician having to remember it. An example of an illuminated colony counter, with magnifying glass, is shown in Plate 19.

Air management forms an essential part of the control of microbiological contamination in food microbiology laboratories, especially in larger laboratories and in those handling pathogens, and several types of cabinet that control air-flow and quality are available for different purposes (Plate 20; see Table 6.3). When manipulating microbiological media using aseptic techniques, e.g. adding selective agents, or filter-sterilising solutions, a laminar air-flow cabinet is commonly used (Plate 20(a)) to protect the materials being handled from air-borne contamination in the environment. In this type of cabinet, the materials being handled are protected by a horizontal flow of sterile air, from the back of the cabinet towards the worker and, before use, all internal surfaces of the cabinet should be disinfected using a non-corrosive disinfectant.

In the UK, the use of an appropriate class of microbiological safety cabinet is advocated by the Advisory Committee on Dangerous Pathogens in accordance with the level of containment required (see Chapter 6). Class I microbiological safety cabinets are most commonly found in food microbiology laboratories. Class II and Class III cabinets are not generally found in routine food microbiology laboratories; however, each of these classes of cabinet provides different, but high, levels of protection / containment for the worker and the work and they are used depending upon the infectivity of, i.e. risk from, the organisms being handled or anticipated in the materials being examined. Class I safety cabinets (Plate 20(b)) draw air into the cabinet, i.e. away from the worker; these are used when handling food-borne pathogens, especially when undertaking confirmation tests that may involve broths or suspensions containing very high numbers of potentially pathogenic organisms. They should also be used when undertaking microbiological challenge tests (see Chapter 5) to minimise the potential for contamination of any routine sample being handled.

Class II and III safety cabinets provide high levels of containment that protect both the worker and the work, but Class III safety cabinets that provide 'total containment', being totally sealed with all test materials being handled via gloves sealed into the front of the cabinet, are only required for use when handling highly infectious organisms usually categorised in Hazard Group 4 (Advisory Committee on Dangerous Pathogens, 2001) and these are not the subject of work in food laboratories that undertake routine pathogen testing.

Although the types of cabinet used for food microbiology provide protection for either the worker or the materials being handled, they only protect against air-borne contamination, and contamination spread by direct contact is not prevented; in this respect, hands can become contaminated easily; this is especially important when undertaking confirmation tests (see Chapter 11) during which a lack of hygienic practice can spread pathogens very quickly around the laboratory, particularly to door and cupboard handles, telephones, calculators and computer keyboards. Therefore, **laminar air-flow and safety cabinets cannot replace good aseptic technique** and good hygiene practices and must only be considered to provide some added safety and security during work.

9.3 Microbiological media

Microbiological growth media occur in both liquid and gel forms. Their origins go back to the nineteenth century when bacteria were grown on media made from ingredients that were readily available, mainly materials from abattoirs or hospitals, e.g. blood, beef carcasses and offal.

Many modern media have similar basic chemical and biochemical constituents, that provide protein-derived nutrients in the form of peptides / amino acids for nitrogen and carbohydrates for energy. An important constituent of many microbiological growth media is hydrolysed protein known as peptone but, in addition, they may contain other ingredients such as sugars, yeast or malt extract, salt, trace elements, vitamins and chemical buffers to maintain a stable pH value.

Over the years, microbiological growth media have become increasingly complex and many are now specially adapted for specific organisms; for example, they may contain milk protein hydrolysates instead of meat peptones. Vegetable (soya) derived peptones are also used. Such media also contain a range of other constituents for specific purposes. For example, they may contain specific, growth-enhancing ingredients that encourage the target organism to grow while the growth of other organisms that may be present in the sample is restricted because they cannot utilise the added ingredients (elective media); they may contain ingredients that change colour or appearance as a result of a particular microbial metabolic process so that the target organism causes a distinctive reaction (diagnostic media), or they may contain inhibitors that prevent the growth of organisms that compete with the target organism (selective media). These features are often combined as in selective-diagnostic agars.

For media required in gel (solid) form, a gelling agent, usually agar, is added. *The Oxoid Manual* (Bridson, 1998) provides some useful information about the general development and formulation of culture media as well as guidance concerning the use of media. Table 9.1 indicates the purpose and qualities / characteristics of different constituents of culture media.

9.3.1 *Diluents*

Many microbiological tests are designed to provide quantitative results, i.e. they are designed to allow the enumeration (count) of a particular microorganism in a test sample. To do this requires accurate dilution of the test sample (see Chapter 10, 'Preparing a dilution series of the test sample') which, in turn, requires the use of a suitable diluent. In addition, because most foods are solids they need to be prepared initially as a suspension in a diluent for convenient handling prior to any further dilution.

Ideally, a diluent should protect microorganisms, especially those that may have been damaged or stressed by the food-processing operations or the intrinsic or extrinsic properties of the food (see Chapter 2) so that they remain viable but neither multiply nor die. To achieve protection, a solution of physiological saline (salt) that is isotonic with the microorganisms suspended in it is used, i.e. the solution has a similar osmotic pressure to the cytoplasm of an 'average' food-borne microorganism. Historically, water was used, later a saline solution containing sodium chloride at approximately 0.85% w/v was used and this was superseded by $\frac{1}{4}$ strength Ringer solution containing a mixture of mineral salts in addition to sodium chloride. To help maintain the viability of food-borne microorganisms further, this formulation was enhanced by the addition of 0.1% w/v peptone (see Table 9.1). Today, perhaps the most widely used diluent formulation is known as Maximum (or Maximal) Recovery Diluent (MRD) (or peptone-saline diluent); this contains 0.85% w/v saline and 0.1% peptone.

MRD can be used as a diluent for most food-borne microorganisms but, for some specific applications, different diluent formulations may be required. In particular, these relate to mycological tests for which 0.1% peptone is the preferred diluent, also, a detergent such as Tween 80 may need to be added to reduce the surface tension of the diluent so that hydrophobic mould spores (spores that repel water) can be evenly suspended in the diluent, or a diluent containing either 40, 50 or 60% w/w glucose is needed to maintain a high osmotic pressure when isolating xerophilic moulds and yeasts from low water activity foods.

9.3.2 *Liquid growth media*

Liquid growth media, called broths, are usually used to grow microorganisms so that their numbers are amplified to a level that is useful as an inoculum for other forms of test, e.g. antibody-based assays. However, for many decades, broth media provided the main means for conducting batteries of tests to determine the biochemical profile of microorganisms for their identification (see Chapter 11) and are still used for

Table 9.1 Some examples of the constituents of microbiological culture media, adapted from Bridson, 1998.

Constituent	Purpose / qualities	Contained in
Major ingredients		
Proteins / peptides / amino-acids	nutrients / source of amino-nitrogen	almost all bacteriological media
(1) meat infusion / meat extract	low water-soluble peptide content	
(2) peptones (acid- and enzyme-hydrolysates)		
a) hydrolysates of meat proteins	high water-soluble peptide content	
b) hydrolysates of other animal proteins, e.g. casein	high tryptophan content	
c) hydrolysates of vegetable proteins, e.g. soya	high carbohydrate content	
Carbohydrates:	energy source / carbon source	most media
(1) glucose	provides a general source of energy for most organisms	general purpose and diagnostic media
(2) other (specific) carbohydrates	provide specific energy requirements, or are used to promote specific diagnostic reactions	media for specific organisms and diagnostic / confirmatory media
Essential metals and minerals:	essential for microbial metabolism	most media
e.g. sodium potassium chlorine phosphorus sulphur calcium magnesium iron	typically present in 'mg/l' to 'g/l' quantities	
Horse blood / sheep blood:	source of growth factors	general purpose media diagnostic media
defibrinated oxalated haemolysed / laked	necessary for diagnostic reactions	
Agar	gelling agent	all solid and semi-solid media

Contd.

Table 9.1 *Contd.*

Constituent	Purpose / qualities	Contained in
Minor ingredients		
Essential metals and minerals:	essential for microbial metabolism	most media
zinc		
manganese		
bromine		
boron	typically present	
copper	in 'µg/l to mg/l'	
cobalt	quantities	
molybdenum		
vanadium		
strontium		
Complex anions:	buffers	most media
phosphates		
acetates		
citrates		
specific amino acids		
Dyes:	indicators of change in pH value	diagnostic media
phenol red		
bromocresol purple		
Miscellaneous chemicals:	selective agents	selective media
antimicrobials		
bile salts		
dyes		
selenite		
tetrathionate		
tellurite		
azide		
Elective agents	required for growth of specific microorganisms	elective media
lactose		
ethanol		
vitamins		
thioglycollate	reduces E_h in media for the	
cysteine	growth of anaerobes	
peptic digest (Fildes extract)		

carrying out 'most probable number' tests (see Chapter 10). They can be made elective or selective and also diagnostic. Broths are usually dispensed into tubes or bottles of varying capacity between about 5 ml and 1 litre.

Broths are often used for the primary culture of organisms from foods, e.g. in detection tests for *Salmonella* or *Listeria*; however, all of the microorganisms in a broth culture are mixed up together and sometimes the target organism does not grow to high enough numbers to allow its presence to be visualised or otherwise detected. Thus, microbiological growth media prepared in gel form are used so that microorganisms can be immobilised and grown separately from each other and thus, be more readily observed. Specific organisms growing on a selective agar medium can be detected by their specific reactions on the medium e.g. *Salmonella* and *Listeria*.

9.3.3   Gel (solid) growth media

Agar can be added to liquid media to make them into a firm gel. Agar is a complex galactan (polymer of galactose) obtained from some seaweeds and is usually added to media at a concentration of around 1–2% w/v. Sometimes, lower concentrations are used to form a 'sloppy gel' known as semi-solid agar (see Chapter 11), which is particularly useful when studying motile organisms that can swim through the viscous matrix.

Agar is usually sold as a powder and has the unusual property that it only dissolves in water (or melts if already dissolved) when heated to above 84°C but, upon cooling, it remains liquid until the temperature falls to less than about 42°C, when a firm gel is formed. Organisms can thus be suspended in cooled molten agar which is allowed to set and then is incubated until visible growth has occurred, as in 'pour plates' commonly used for enumerating (counting) microorganisms, e.g. total colony counts (TCC). Alternatively, agar can be dispensed into Petri dishes and allowed to set before the microorganisms are spread over its surface as in 'spread plates'.

Although agar is not an inert material and it can contribute nutrients and toxic agents to culture media, depending on the production method (Bridson, 1998), most microorganisms of interest in routine food microbiology do not use agar as a nutritional source; thus, it remains firm after the organisms have grown. If a gelling agent such as gelatin is used, many more microorganisms can, and do, break down the gelatin by hydrolysis, and the gelatin slowly liquefies as the organisms grow. This characteristic is actually used as one of a range of diagnostic tests for the identification of some organisms, e.g. *Clostridium* species and Enterobacteriaceae.

9.3.4   General purpose growth media

General purpose media are used to grow as wide a range of microorganisms as possible from the particular test sample under examination, and so do not contain selective agents. However, no single medium is ever able to grow all microorganisms because different species have different growth requirements, e.g. different specific nutrient requirements, optimal growth temperatures and oxygen levels etc. Thus, different general purpose media have been developed to grow different microbial groups such as bacterial aerobes and anaerobes, yeasts and moulds, Total Enterobacteriaceae; but even these media do not grow all types of the targeted but still broad groups of organisms and the laboratory's selection of media for use must take into account (among other considerations) the sample types to be examined and the type of information required (Institute of Food Science and Technology (UK), 1999).

Probably the most widely used application of general purpose media is for making general bacterial colony counts. These counts are known as (most correctly) 'total colony counts' (TCC) but also as (technically incorrectly) 'total viable counts' (TVC), or 'aerobic plate counts' (APC), and other names. Such counts are used to estimate the overall numbers of microorganisms present in a test sample. However, the medium used can only allow the detection of those organisms capable of growth on it under the conditions of temperature and gas atmosphere used for incubation, so the description of the result from an enumeration test should be supported by information about both the medium and growth conditions used. Thus, for example, a more correct reporting of a total colony count carried out by the pour plate technique using plate count agar (PCA) incubated at 30°C in aerobic conditions is 'total aerobic mesophilic colony count; PCA pour plate at 30°C'. However, the microbiological shorthand of TVC or APC or TCC is still widely used and acceptable if supported by the laboratory's microbiological methods documentation which should give detailed information about the techniques, media and incubation conditions to be used for all the sample types examined and tests carried out on these in the laboratory.

9.3.5 Enrichment media

Enrichment media are usually broths and are used to increase the numbers and / or proportion of a target organism from a test sample in the presence of competitive microflora.

Elective media contain a specific nutrient(s) that the target organism can utilise but many others cannot so the target organism grows while other microbial growth is discouraged e.g. ethanol is an elective agent used for growing acetic acid bacteria and lactose is an elective agent for growing spoilage yeasts from dairy environments.

Selective media contain substances that selectively inhibit the growth of many competitor microorganisms present in the test material but allow the target organism to grow. For example, many bacteria are inhibited by sodium biselenite but most salmonellae are able to grow in the presence of this chemical, so sodium biselenite forms the basis of an important selective enrichment medium for *Salmonella* spp. that enables these organisms to grow more prolifically i.e. be 'enriched', when in mixed populations; similarly, potassium tellurite is used as a selective agent in media for growing *Staphylococcus aureus*. Selective agents, however, are rarely perfect: too low a concentration allows too many non-target organisms to grow whilst too high a concentration often inhibits the target organism itself especially if it has been damaged.

9.3.6 Pre-enrichment media

In food microbiology, selective media are much more widely used than elective media. However, organisms that have been exposed to a food production process are often sub-lethally injured or damaged and if placed directly into a selective medium, they would die or, at least, fail to grow. Microbiological methods may therefore include an initial resuscitation step to allow the target organisms to recover before being exposed to the stress of a selective agent. This is done by inoculating the test sample into a **resuscitation medium**.

Perhaps the best-known resuscitation medium is buffered peptone water (BPW), a dilute general purpose broth used widely in the isolation of *Salmonella* from many foods. Because this medium is used prior to the selective enrichment stage, it is often referred to as a **pre-enrichment medium** or, more correctly, as a **non-selective enrichment medium** since the numbers of salmonellae as well as competitor organisms usually increase in this medium during incubation in the absence of any selective agents, before being sub-cultured into a selective enrichment broth.

Some different pre-enrichment medium formulations are required when isolating *Salmonella* from specific types of food and these are discussed in Chapter 10 (see Table 10.1).

9.3.7 *Diagnostic media*

Microorganisms metabolise a wide variety of chemicals. The products of their metabolism can be useful to distinguish one family, genus or species of organism from another. Diagnostic media are formulated so that specific products of the metabolism of particular chemicals can be detected easily, usually visually, and used to differentiate readily between cultures of different organisms and aid in their identification. A widely used test is that of acid production from sugars, e.g. glucose, lactose, sucrose, and pH indicators are used to detect acid production by means of a colour change produced as the pH decreases. Such media may be broths, e.g. MacConkey broth used to detect coliforms, but are frequently agars, e.g. violet red bile glucose agar (VRBGA) used to detect Enterobacteriaceae.

Diagnostic media very often also contain selective agents to inhibit competitors that may give similar diagnostic reactions to the target organism, and hence media formulations can become quite complex. For example, xylose lysine desoxycholate agar (XLD) contains inhibitory agents to some competitor organisms to *Salmonella* and a number of substances that allow the differentiation of five diagnostic reactions: xylose fermentation, lactose fermentation, sucrose fermentation, lysine decarboxylation and production of hydrogen sulphide; together these diagnostic reactions can be used to distinguish *Salmonella* from *Shigella*, *Providencia* and *Edwardsiella*. On XLD agar, *Salmonella* appears as black-centred colonies with a transparent perimeter on cherry-red agar (Plate 21).

Many of the diagnostic media in use today were developed in the first half of the twentieth century and are still very effectively used in food microbiology, but new formulations are continually being developed for example:

(1) 'MUG' (4-methylumbelliferyl-β-D-glucuronide) is a diagnostic reagent that can be added to media for the detection of *E. coli*. This organism possesses the enzyme glucuronidase (see Chapter 11) that cleaves (splits) MUG to produce 4-methyl-umbelliferone, a compound that produces a green-blue fluorescence when exposed to ultra-violet light. MUG can be added to most selective broths and agars used to detect *E. coli*; after incubation the medium is observed during exposure to ultra-violet light when the presence of *E. coli* is seen as fluorescence, either of the broth or of colonies growing on agar.

(2) Another innovation in the formulation of microbiological media is the development of chromogenic agars. Whereas conventional diagnostic reactions on selective agars are usually based upon a colour reaction in the agar surrounding a particular colony, organisms growing on chromogenic agars produce a coloured reaction product actually within the cells of the colony that has grown and which does not diffuse into the agar; this can improve the discrimination between different types of organism growing on the same agar plate because the colour reactions of different organisms on some conventional diagnostic agars may be masked by the diffusion of colour(s) in the agar. These new media contain a chromogenic substrate, e.g. 5-bromo-4-chloro-3-indolyl-β-D-glucuronate (BCIG), that is specifically taken up and acted upon by enzymes within the microbial cells of organisms growing on the agar.

The chromogenic molecule is split by enzyme activity to yield an insoluble chromo-phore (coloured substance) that builds up within the cell to colour the colony.

Because of similarities of biochemical characteristics between related species of microorganisms and sometimes the diversity of these characteristics in strains within a single species, diagnostic media are rarely perfect. Typical (positive) diagnostic reactions are, therefore, considered only as a 'presumptive' identification of the target organism. For pathogen tests in particular, it is important that false-positive results are avoided and more detailed identification procedures have to be applied to confirm the initial, presumptive result, as discussed in Chapter 11.

9.3.8   *Preparation of microbiological media*

All microbiological media must be prepared by accurate measurement of weights and volumes, thorough mixing and dissolution of the dehydrated medium using heating as appropriate, neither under-heating nor over-heating, and finally sterilising by an appropriate heat process or other means. The water used for bacteriological media must be of high quality (see Chapter 6); for mould media, however, in particular when the identification of mould isolates is required, this can be a disadvantage. This is because moulds require trace elements (especially copper and zinc) for the production of pigments and to ensure good growth, that allows the structures and colours upon which their identification is based to be clearly visible. Traditionally, trace elements were present as impurities in the peptones and other media constituents including water; however, nowadays, microbiological media contain few impurities and, when media for growing moulds are prepared using very pure water, this can lead to poor growth of moulds and poor colonial and microscopic morphologies, and also, a noticeable lack of pigments. To overcome this problem, **a trace element solution should be added to all mould growth media**, especially if poor growth and / or pigment production is observed consistently. The composition of the trace element solution is as follows:

$ZnSO_4.7H_2O$	1.0 g
$CuSO_4.5H_2O$	0.5 g
water	100 ml

This solution of zinc and copper sulphates, which does not require sterilisation because it is toxic to microorganisms, should be added to agars to be used for mould growth, before autoclaving, at the rate of 1 ml of trace element solution per litre of agar.

For further details of the sterilisation of media and preparation ready for use see Chapter 7 and '9.4.6 Sterilisation and disposal' in this chapter.

9.4   Basic techniques of food microbiology

9.4.1   *Aseptic technique*

All procedures employed in a food microbiology laboratory must incorporate the fundamental principle and operation of aseptic technique, i.e. technique that aids the prevention of contamination. Aseptic techniques are the means of protecting the sample from contamination with organisms present in the laboratory, on equipment and on staff, **and** of protecting the laboratory worker from contamination with organisms present in the sample or in and on cultured media. This means that everything that comes into contact with a factory or test sample during laboratory procedures must be sterile and also that the samples must be manipulated in such a way that the microorganisms they contain are not allowed to contaminate the laboratory or personnel. Aseptic technique is therefore not a single procedure; it is a fundamental way of doing things that should be applied to all laboratory operations that involve handling food and food-related samples or live organisms. In practice, different aseptic techniques may be employed depending on the nature of the task and the risk of contamination.

The risk of contamination depends largely on the type of material being handled and whether or not it has been incubated, which can result in the presence of a large number of organisms. If a test sample or test sample dilution becomes contaminated with organisms from the laboratory or staff before it has been inoculated into or onto media and incubated, then the contaminant may outgrow the target organism and the result obtained is likely to be invalid (this is especially critical if the contaminant is a pathogen and the factory sample was not contaminated with it when it entered the laboratory); in this case, the prime objective is to protect the sample material. Conversely, after a test sample or its dilution has been inoculated into / onto media and incubated, the number of microorganisms in or on the incubated media can often be many million times greater than before incubation and the risk of infecting the laboratory and staff is greatly increased (this is especially critical if the sample contained human pathogens); in this case, the objective is primarily to protect the staff and working environment. However, in both cases, the main objective is to prevent cross-contamination by maintaining rigorous aseptic practices, personal hygiene practices and by careful disposal of contaminated materials.

The basic practical operations for sample processing that take place in most food microbiology laboratories comprise:

- preparing a factory sample for examination, i.e. taking a test sample of a specific weight or volume and, for many tests, making a primary dilution and sometimes appropriate further dilutions, then
- adding the dispensed and / or diluted test sample to growth media, either liquid (broths) or solid (agar plates or slopes),
- sub-culturing, where required, between incubation steps, to other growth media and, finally,
- assessing and recording the results (see Figures 6.4 and 7.2).

Each of these operations requires a slightly different 'style' of aseptic technique but in all cases the basic principles remain the same.

There are many potential sources from which a test sample may be contaminated, including staff, laboratory fittings, equipment, un-sterile reagents and the air. **All** sources must be controlled:

- by using sterile growth media and sample contact materials and equipment,
- by working quickly, but cleanly and carefully,

- by preventing the sample or sterile equipment from contacting any un-sterile materials or surfaces, and
- by minimising air-borne contamination.

It is generally considered good practice to 'flame-sterilise' the necks of culture media tubes and bottles after removing the lids or closures to pour the media (molten agar poured into plates) or to inoculate them. However, this technique, which is often poorly understood and practised, adds very little additional security to well conducted alternative aseptic practices that, for many operations, can provide a highly satisfactory approach to controlling the risk of cross-contamination. Cross-contamination may be minimised effectively by working quickly and efficiently in a clean, quiet area (free from high activity and air turbulence) and by ensuring that the media containers are never held or left in a vertical position when open.

On some occasions, extra special care needs to be taken to protect either the samples or the worker, and manipulations should then be carried out in an appropriate laboratory cabinet, either a laminar air-flow cabinet that protects the sample from the worker or a safety cabinet that also protects the worker from the sample (see Plate 20). More detailed descriptions of the ways in which aseptic practices protect the sample and the operator are given in Table 9.2.

9.4.2 *Pouring an agar plate*

Once agar has been prepared, sterilised and 'tempered' (cooled to 45–47°C and kept, molten, at that temperature until required for use), it is poured, usually into Petri dishes. This is done, either to prepare agar plates that are to be used for surface inoculation, or for 'pour plates' (see Chapter 10, '10.2.1 Colony counts') into which the dilution of the test sample has been placed prior to pouring the agar.

For both purposes, pouring of molten agar must always be done with care, yet speedily, to avoid extraneous contamination. If a water bath is used for agar tempering, the outside of the container of agar must be wiped dry to avoid depositing / dripping un-sterile water from the water bath in the Petri dishes, which are laid out on the workbench. After removing the lid of the bottle of molten agar which is held at an angle, the lids of the Petri dishes are lifted (usually by tilting), one by one, and a suitable quantity, often 12–15 ml (judged by practice and experience) for a conventional 90 mm diameter dish, is poured cleanly into the centre of the base of the dish, avoiding spillages and splashes, especially over the rim of the dish; once a plate has been 'poured', the lid is replaced.

Plates that are to be used for surface inoculation are left undisturbed until the agar has solidified *fully*; they may then be collected, stacked and, if not required for immediate use, may be stored refrigerated in plastic 'sleeves' to prevent their drying out. Before surface inoculation, the surface of all pre-poured plates must be dried to prevent the inoculated organisms from spreading, as this may result in confluent growth and unreliable results. This may be achieved by various means and the laboratory should have a Standard Operating Procedure for this operation.

Plates that have been prepared as 'pour plates' require mixing of the inoculum with

Table 9.2   Some common aseptic practices and working procedures.

Source of contamination	General procedures	Specific aseptic procedures
A. Laboratory staff	*Laboratory staff should wear dedicated laboratory clothing and conduct themselves in a manner that minimises contamination and cross-contamination of themselves, the laboratory environment and the work in progress*	
Clothing	Laboratory coats should cover personal clothing completely Laboratory coats should always be kept properly fastened Laboratory coats should be changed frequently	Test materials should be held away from the face and body Laboratory clothing should not be allowed to come into contact with samples, cultures or incubated test materials
Hands, fingers	Hands should be washed before and after a work 'session' and if contamination is suspected	Laboratory staff should not touch the test material or growth media Laboratory staff should not touch the openings of sterile containers Hands should not become contaminated when handling cultures or incubated test materials
Breath	A laminar air-flow cabinet should be used for 'high care' work, e.g. sterility tests and for sampling clean materials	Laboratory staff should not talk or cough directly over exposed test material or media
Hair	Long hair should be tied well back or kept covered	The test materials should be held away from the face and body
B. Laboratory and fittings	*The laboratory should be designed for easy cleaning and be maintained in a clean condition*	
Work surfaces	Work surfaces should be disinfected before and after each work 'session' Work surfaces should be dry when in use	Laboratory staff should always work cleanly Spillages and contamination should be cleaned up immediately using disinfectant Contaminated sampling utensils / blender bags / bottle caps / pipettes / loops / spreaders should not be allowed to touch any equipment or work surfaces
Incubators and refrigerators	The interior surfaces and handles should be cleaned and disinfected regularly	Spillages or contamination should be cleaned up immediately using disinfectant
Sinks and other laboratory fitments	Sinks should be cleaned and disinfected regularly Door handles, tap handles and cupboard door handles, telephones and computer keyboards should be cleaned and disinfected regularly	After handling cultures, especially cultures of pathogens, hands should be washed thoroughly before opening doors or touching other laboratory fitments or equipment and work

Contd.

Table 9.2 *Contd.*

Source of contamination	General procedures	Specific aseptic procedures
Air	Windows should always be kept closed Doors should be kept closed while work is in progress Staff movements should be kept to a minimum whilst work is in progress A laminar air-flow cabinet should be used for handling 'high care' samples, e.g. sterility tests A Class I safety cabinet should be used for handling 'hazardous' samples, e.g. positive control cultures Air-conditioning equipment should be cleaned regularly Air-conditioning equipment should not generate a positive air pressure or contribute to the dispersal of aerosols Equipment that generates steam should be kept away from sample and culture processing areas	Laboratory staff should work quickly but carefully Sterile containers / Petri dishes should not be kept open for longer than necessary Containers / Petri dishes that contain cultures / incubated test materials should not be kept open for longer than necessary Opened containers should be protected from contamination e.g. by holding bottles / tubes at an angle, by holding the Petri dish lid at an angle over the base.
Safety cabinet (Class I)	The cabinet and its environment should be kept clean and dry The cabinet should be fumigated periodically and before undertaking any maintenance work	The cabinet interior should be disinfected before and after a work 'session' Cultures or incubated test materials should not be allowed to contaminate the interior of the cabinet
C. Equipment	*The method of use of equipment should minimise the risk of it becoming contaminated*	
Discard containers for pipettes and small disposable items such as pipette tips, loops, spreaders and microscope slides	Discard containers should be kept filled with an appropriate disinfectant at an appropriate concentration Discard containers should be emptied at least daily, then washed and fresh disinfectant added	Used pipettes, loops, etc. should be ***fully immersed*** in disinfectant The creation of splashes and aerosols should be avoided
Pipetters and tips / pipettes / spreaders / loops / culture bottles / blender bags, etc.	Disposable / sterile, single-use items should be used in preference in reusable tools that need to be cleaned and re-sterilised between each use. When sub-culturing, especially from liquid cultures and test materials, the creation of aerosols should be avoided	Sterile equipment / utensils should not be exposed to the air unnecessarily before use Sterile items should not be allowed to touch any unsterile equipment or surfaces Open culture tubes and bottles should be kept angled to minimise extraneous contamination Used items should be sterilised or placed in a discard container immediately after use

Contd.

Table 9.2 *Contd.*

Source of contamination	General procedures	Specific aseptic procedures
Bottles of molten agar	When using a water bath to temper molten agar, the necks of bottles should be kept above the water level at all times but the level of agar should be below the water level Any bottles of tempering agar that have capsised should be discarded Tempering baths should be kept clean and the water should be changed regularly If a hot cabinet is used to temper molten agar, it should be kept clean	If a water bath is used to temper molten agar, when the bottles are removed from the water bath, the outsides of bottles should be thoroughly dried before pouring the agar
D. Reagents / test materials	*Reagents and test materials should be handled in a manner that minimises the risk of their contamination*	
Diluents / growth media	Growth media should always be stored under appropriate conditions, e.g. in the dark at 2–4°C Diluents are normally stored at room temperature in the dark Part-used diluents and growth media should be discarded and not stored for use on another occasion	Growth media and diluents that have been sterilised by autoclaving should be used only if the autoclave process and QC results are satisfactory
Reagents / supplements for growth media	Reagents and supplements should always be stored under appropriate conditions, e.g. in the dark at 2–4°C All reagents supplied sterile should be handled aseptically at all times	When filter-sterilising reagents, fingers should be kept well away from the 'sterile side' of the filter; this procedure should preferably be undertaken in a laminar air flow cabinet The integrity of equipment used for filter-sterilisation should be confirmed after the procedure has been completed, e.g. by inspecting the filter for damage
Factory sample packaging	Where possible, the outside surfaces of factory sample containers should be disinfected before opening for sub-sampling	Where practical, the sub-sample should be taken from an area of the factory sample that is distant from the immediate area where it is opened

Contd.

Table 9.2 *Contd.*

Source of contamination	General procedures	Specific aseptic procedures

'High risk' waste materials, especially:
– highly contaminated or dusty samples
– colonies on agar plates and slopes
– incubated broth cultures
– dilutions prepared from cultures of known pathogens, e.g. positive control cultures
– Blender bags that have been used to contain test samples under incubation
	Incubated plates should be discarded by placing gently into discard bags or containers. To avoid the creation of aerosols, they should never be dropped into a container from a distance.
	   A dedicated area should be kept for short-term storage of properly contained waste materials and it should be cleaned and disinfected regularly
	   Waste containers should not be allowed to become overfull
	   Highly contaminated samples and positive control cultures should be handled in a separate room, or at the end of the working day when less highly contaminated samples have already been processed
	   Dust extraction equipment should be used when handling dusty samples and hands should be washed when work has been completed
	   If used for incubating samples, e.g. *Salmonella* pre-enrichment, blender bags should be sealed securely before incubation and should not be allowed to leak in the incubator
		When incubated plates / broths have been assessed and work with them, e.g. sub-culturing, has been completed, they should be discarded immediately into an autoclavable, leak-proof container or incinerator container
		   Incubated test samples should never be discarded into a sink or waste disposal unit without prior sterilisation

'Low-risk' waste materials, such as: – diluted test samples used to inoculate growth media – excess food samples	Food waste must never be taken out of the laboratory for consumption but disposed of safely in secure, dedicated waste containers	Diluted test samples should preferably be autoclaved before disposal. In any case, diluted test samples should be disposed of in a hygienic manner, avoiding aerosol formation.

the molten agar immediately after the agar is poured and before it cools to the temperature at which it solidifies; this is to ensure even distribution of the inoculum in the agar so that the colonies that develop during subsequent incubation are evenly dispersed. This is a critical procedure that requires a careful, yet thorough technique (see Chapter 10, '10.2.1 Colony counts'); mixing too gently is likely to result in poor distribution of the inoculum, especially for low dilutions of viscous products, e.g. chocolate, too vigorous mixing is likely to result in the molten agar spilling out of the dish and may lead to contamination. After mixing, 'pour plates' are left undisturbed until the agar has solidified *fully*, they are then transferred to an incubator.

9.4.3   *Streaking an agar plate*

The single cells of microorganisms such as bacteria and yeasts are invisible to the naked eye; however, when they are inoculated onto a 'solid' nutrient medium surface, such as agar in a plate, and incubated, the single cells divide to become eventually the many millions of cells that form a visible colony. Often, different species of micro-organism are found side by side as 'mixed cultures' in broths and, even on selective agar media. In order to study one species individually the target cells in individual colonies or broths need to be separated and removed from the unwanted ones and purified. This is done using a technique known as streaking. The objective of streaking the surface of an agar is to separate out the cells of organisms from a culture (a colony on a plating medium or organisms growing in a broth medium) so that, when the streaked agar plate is incubated, each individual species of organism grows as isolated colonies, each of which has been derived from a single cell, or from a clump of a few identical cells, known as a colony-forming unit (cfu).

A diagram showing a common and 'good practice' approach to streaking a plate is shown in Figure 7.1. The object of the technique is to spread out the cells of the organisms by distributing the culture over the agar surface in stages (Figure 7.1(a) [1]–[6]), so that the numbers of cells are successively reduced, as follows:

- An inoculum is obtained, using a sterile loop, either from a broth culture, taking a loopful of broth, or from a colony, by just touching the lower edge of the loop onto the colony to be purified.
- The loop is wiped over a section of the agar to form a 'well' (Figure 7.1(a) [1]) that usually produces confluent growth as it may contain many millions of organisms, especially if the inoculum was taken from a colony.
- Because many cells remain on the loop, the loop is then flame sterilised and cooled (or discarded and replaced if disposable loops are being used) and a few streaks are made from the well across the next section of agar (Figure 7.1(a) [2]).
- The procedure is repeated another two or three times, usually without sterilising the loop again for broth cultures, but re-sterilising the loop for cultures taken from agar (Figure 7.1(a) [3] to [5]).
- A final zig-zag is made from the last section [5] into the centre of the Petri dish (Figure 7.1(a) [6]).

The results of good and poor streaking techniques are shown in Figure 9.1. To be consistently successful in obtaining isolated colonies from broth or agar cultures and achieving pure cultures, the streaking procedure has to be practised. Success depends on many factors including the amount of inoculum initially picked up on the loop as well as the source of the inoculum. A single bacterial colony may well contain around 10^{10}–10^{11} cells, compared with less than 10^6–10^8 cells/ml of a liquid culture. Since a loopful of a liquid culture may amount to only 0.01 ml, the number of organisms taken from a liquid culture is often some orders of magnitude lower than that of an inoculum taken from a colony. Consequently, the number of streak stages and the number of times the loop is sterilised or replaced need to be adapted to suit the circumstances. Confluent growth resulting from poor streaking practice is due mainly to removing too much growth (inoculum) from a colony, not using sterile loops for each streak stage or streaking a wet plate.

Figure 9.1   The results of good and bad streaking.

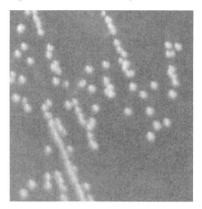

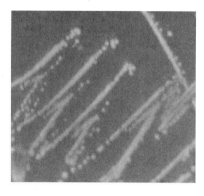

(a) The results of a well-streaked plate showing well-separated colonies obtained at the 6th level (see Figure 7.1(a)).

(b) The results of a poorly streaked plate showing confluent growth at the 6th level (see Figure 7.1(a)) and no well-separated colonies.

In addition, particular aspects of aseptic practice need to be observed to prevent contamination of the laboratory environment from aerosols during the sub-culturing procedure from broths or colonies and during streaking. For the streaking procedure, the plate is placed inverted, with the agar-containing base uppermost, on the bench; the base, containing the agar, is then picked up and held at an angle sufficient to perform the streaking operation whilst also presenting a barrier to help minimise aerial contamination of the agar surface (Figure 9.2a); alternatively, the base, containing the agar, may be left on the bench and the lid of the plate tilted sufficiently to allow the streaking operation to be performed, thus providing some protection from aerial contamination (Figure 9.2b). During the streaking operation, care must be exercised to ensure that the loops used to streak the culture do not touch the edge of the plate (see Figure 7.1(a)), otherwise undesirable, confluent growth may occur.

9.4.4   *Slopes and stab techniques*

For some purposes, microorganisms are grown on agar 'slopes' (sometimes also known as 'slants'). After initial preparation (dispersing and dissolution), the agar medium is dispensed, usually in 7–10 ml quantities, into screw-capped containers, e.g. Universal bottles, or into test tubes that are then capped, and sterilised by autoclaving. The containers are removed from the autoclave whilst still molten and placed an angle to allow the agar to solidify, so that it forms a 'butt' and a 'slope' (Figure 9.3).

Slopes are a convenient means of storing reference cultures of aerobes, in which case the butt is small and the slope long. A culture is streaked onto the slope and incubated so that confluent growth occurs on the surface; this is one of the few occasions in microbiology when confluent growth is desirable as the object of the

Figure 9.2   Diagram of alternative ways to hold a plate whilst streaking.

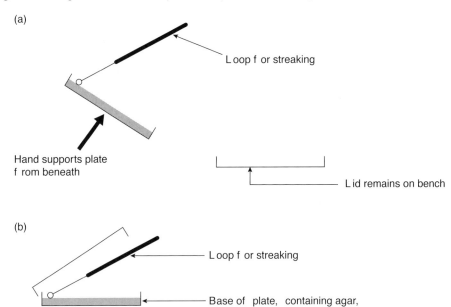

(a)

Loop f or streaking

Hand supports plate
f rom beneath

Lid remains on bench

(b)

Loop f or streaking

Base of plate, containing agar,
remains on bench

(a) *Key points:*
• The base of the plate, containing the agar, is rotated after each streak so that it is positioned correctly for
  the next streak. To minimise contamination of the agar and lid, the streak is completed carefully, but as
  swiftly as possible.
• The lid remains on the bench during streaking and, to help minimise contamination of the agar, the base
  can be replaced into the lid immediately each streak has been completed and while rotating the plate.

(b) *Key points:*
• The lid of the plate is held at an angle sufficient to allow access to the agar surface for streaking, but also to
  minimise air-borne contamination of the agar during streaking.
• The lid is replaced onto the base between each streak to allow the plate to be turned to the correct position
  for the next streak.

technique is to produce as much 'biomass' as possible from a culture that has already
been purified. When required for use, sub-cultures are taken, using aseptic technique,
by collecting a small quantity of growth from the surface of the slope with a loop.

  Some diagnostic media are also prepared as slopes (see Chapter 11), in which case a
deep butt is required, in addition to a large / long slope. This is because the diagnostic
reactions caused by the organisms require different levels of oxygen present in different parts of the agar and the oxygen level present at the bottom of a deep butt is
substantially lower than that at the surface of a slope. These media are generally
inoculated by 'stabbing the butt and streaking the slope'. An inoculum is collected
from a pure culture using a straight wire or needle that is then plunged or 'stabbed'
straight and deep into the butt so that the inoculum is transferred all the way to the
bottom, then the needle is withdrawn and the slope is streaked with the needle or a
separate loop. The culture is then incubated and observed for diagnostic reactions,
which are usually different in the aerobic and microaerobic / anaerobic parts of the

Figure 9.3   Slopes and stab techniques.

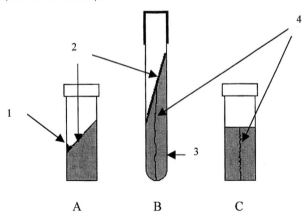

Key points:
(A) A slope used for storing reference cultures.
- After the agar has been sterilised and allowed to set at an angle, a small amount of condensation collects at point (1); if this has evaporated, the slope is too dry and the agar will not support good microbial growth.
- The slope (2) is streaked with the inoculum, using a loop, and the slope is incubated. After incubation, cultures that have grown on slopes should be stored refrigerated until required for use. Slopes may be used to produce 'working cultures' from laboratory 'reference stocks' for use as control cultures. Once opened, working cultures should be used for a limited period, e.g. 1 week, to provide inocula for use in media quality control etc.; this minimises the risk of contamination. Control cultures should not be sub-cultured repeatedly from one slope to the next as they may mutate, i.e. change their characteristics (see Figure 7.4) or become contaminated.
(B) A slope and a butt technique used with some diagnostic media for identifying microorganisms.
- The agar is sterilised and allowed to set at an angle so that it forms a long slope (2) and deep butt (3). The agar is inoculated using a wire (needle) by stabbing the butt (4) and streaking the slope (2). The culture is then incubated, after which the reactions of the organism are read and recorded.
(C) The semi-solid ('sloppy') agar technique used to determine motility, especially for anaerobic organisms.
- The semi-solid agar is inoculated by careful stabbing using a wire (needle) (4) and then incubated. Motile organisms can 'swim' away from the stab and produce diffuse growth seen as turbidity (cloudiness) throughout the medium. Non-motile organisms remain close to the stab and produce heavy growth along the line of the stab.

agar. Normally, there is no need to further sub-culture from this type of culture after the diagnostic reactions have been assessed and recorded.

9.4.5   *Detection of gas production in broth cultures*

Many organisms produce gas from specific substrates and the results from tests that allow an organism's ability to produce gas from different substrates to be assessed can make a useful contribution to the detection and identification of microorganisms. Gas production is most easily detected by means of a Durham tube (Durham, 1898). This is a small tube, sealed at one end, that is submerged with the sealed end uppermost in a broth medium before autoclaving. The prepared broths are then heat-sterilised, during which operation the air in the Durham tube is expelled and replaced by the medium as it cools. The broth is then inoculated with a culture or dilution of a test sample and incubated; gas that is produced by the organism during incubation collects in the Durham tube and forms a bubble that is recognised easily.

9.4.6　*Sterilisation and disposal*

Probably the single most important operation in the microbiology laboratory is sterilisation, not only of media and equipment to be used for sample and culture handling, but also of waste materials. Sterilisation is usually accomplished by means of an autoclave, which employs similar principles to those of the domestic pressure cooker. A sealed chamber is filled with saturated, wet steam under pressure that enables temperatures higher than the boiling point of water to be reached. This is necessary because many types of bacterial spores are only inactivated at temperatures substantially higher than 100°C. Typical laboratory autoclaves are shown in Plate 22.

Microbiological media, equipment and waste materials all have different thermal properties and some media contain heat sensitive ingredients so different autoclave processes are necessary in order to achieve effective sterilisation or decontamination, as well as to avoid overheating of media. Modern autoclaves are fitted with complex electronic process controls and many process time and temperature combinations are possible. It is therefore important that the specific autoclave manufacturer's instructions are always read, understood and applied if the autoclave is to be used consistently, correctly and efficiently.

Sterilisation of media

All microbiological growth media, other than those that contain highly inhibitory selective agents, must be sterilised before use so that organisms present in the ingredients or as extraneous contaminants from the laboratory environment or staff do not grow and compete with the target organisms in the samples to be tested and that the media are intended to culture. All autoclave processes used in microbiology laboratories employ saturated (wet) steam to displace the air in the chamber and heat the autoclave contents, because this is a very efficient means of heat transfer.

Many of the constituents of microbiological growth media are heat sensitive and can easily be overheated; the autoclave process must therefore provide a compromise between adequate destruction of microorganisms and retention of the nutritional and selective properties of the media. The widely used process that provides this 'happy medium' for one litre volumes of most media is the time–temperature combination of 121.1°C for 15 minutes (or, for some media, a lower process temperature of 115°C is used for 10, 15 or 20 minutes); i.e. after the autoclave chamber has heated up to 121°C, it is held at that temperature for 15 minutes before cooling down to room temperature. Autoclaves should always be loaded in such a way as to allow free circulation of the steam to all parts of the load.

Large volumes of media, e.g. 2 litres or more, in large containers cannot be adequately processed using this time–temperature profile (121.1°C / 15 minutes) as the heat takes some time to penetrate throughout the contents, especially for agar media, and single large volumes of media may therefore be inadequately sterilised if this process is applied. In some circumstances, this can be partially overcome by extending the 'holding time' but this presents a considerable risk of overheating the media near the container surfaces whilst the media in the centre of the container may not be sterilised. Autoclave manufacturers have attempted to help overcome this problem by providing a process control thermocouple probe that is inserted into a 'dummy' container or that can be placed in a container of medium, i.e. a load probe. This probe is linked to the timer switch which operates when the temperature of the medium reaches the pre-set process temperature, e.g. 121°C. However, this approach may not be entirely satisfactory for some media types as media can receive a more severe heat process in load probe controlled autoclaves than in autoclaves controlled by a chamber probe (see Plate 16) and this may damage some more heat-sensitive constituents of media. A more consistently reliable solution is to limit the volume of each medium to be autoclaved to a maximum of 1 litre and, preferably, a maximum of 500 ml.

Media manufacturers provide instructions concerning media preparation and the heat treatments to be applied, and these must be followed and, where necessary, advice sought concerning appropriate sterilisation processes to be used in particular circumstances e.g. load versus chamber controlled. In any case, staff should carry out studies of the different autoclave types and processes used for the different load types to establish an understanding of the heat processes actually delivered to the loads. Some effects of overheating media are indicated in Table 7.6.

After the autoclave process has been completed, the autoclave must be opened and unloaded. Autoclaves have safety mechanisms that prevent the door from being opened when the autoclave contents are above 80°C and these should not be over-ridden. Many laboratories process autoclave loads, including media, at the end of the working day, leaving the autoclave to switch off automatically overnight, so that the contents are safely at ambient temperature when work resumes the next morning.

If an autoclave is not unloaded immediately its process cycle is complete, any agar in the load may have solidified, or, agars may be allowed to solidify after autoclaving and kept in storage until required for use. In either case, it must then be re-melted before it can be used. Agar should be re-melted only once, because repeated heating can damage heat-sensitive constituents and the medium may not perform satisfac-torily when it is inoculated and incubated, i.e. microorganisms may fail to grow (or fail to grow well), deterioration of selective agents may mean that too wide a range of organisms grows, or the required diagnostic reactions may be incorrect. If agar is prepared for stock purposes well in advance of the time it is required for use, then it must be stored in appropriate conditions e.g. that prevent drying out, and used within its assigned shelf life.

Agar is re-melted by heating it to boiling point, usually in a boiling water bath; it is important that the caps of the bottles are slightly loosened before placing them in a water bath so that pressure does not build up inside the bottles as they heat up and that the water level is maintained above the level of agar in the container. The bottles should be checked at regular intervals during re-melting and should not be removed until the agar has *completely* re-dissolved; gentle swirling of the agar usually reveals whether or not the agar is completely molten.

Sometimes, other means of re-melting agar are used: e.g. in autoclaves that permit 'free-steaming' of the load, i.e. the load is heated unpressurised and therefore the temperature does not exceed 100°C. Occasionally, if molten agar is required urgently, laboratories re-melt agar in a microwave oven. This is a hazardous procedure as the heating is uncontrolled, and if the agar boils, bottles may spill their contents and this can be dangerous. The agar may also be over-heated and damaged; microwave heating is not therefore, generally recommended.

Freshly melted agar must be allowed to cool to around 80°C before being placed in a water bath (with circulated water) for tempering; this is a safety precaution so that hot, glass bottles do not shatter when placed in much cooler surroundings. The lids, that were loosened before re-melting, must be tightened before tempering so that unsterile water does not enter the bottles and contaminate the contents. It is especially important to ensure that bottles of agar remain upright when in a tempering bath and, if they capsize, the agar must be discarded. It is important that bottles of agar are tempered for a minimum length of time to ensure that the contents have cooled to the correct temperature before being used to pour plates (hot agar kills microorganisms); the time required varies with the volume of agar in a bottle and the number of bottles placed in the water bath and is likely to differ from one laboratory to the next. The laboratory should therefore, have its own written procedure, with a 'minimum holding time' for each medium determined by experiment. Tempered molten media should be used within 3 hours and any unused agar should be discarded.

Sterilisation of equipment

Equipment is sterilised by autoclaving or by dry heat in an oven, e.g. at 160°C for 2–3 hours, and the choice of method largely depends on the type of equipment being sterilised. If an autoclave process is used, the equipment is likely to be wet when removed from the autoclave; paper and cotton wool, in particular, can become quite 'soggy'. This increases the chance of recontamination greatly because many bacteria are motile and can 'swim' very efficiently. Therefore, dry heat is often used for glassware e.g. pipettes and for other items that are, or contain absorbent materials. Items to be sterilised by dry heat must, of course, be capable of withstanding the high temperature and care must be taken to ensure that plastic materials do not melt. Metal equipment, e.g. filtration units and sampling tools, however, are more usually sterilised by autoclaving, often at 121.1°C for 30 minutes, which ensures adequate heat penetration to all internal surfaces. The laboratory's Standard Operating Procedure should detail the appropriate type of process to be applied to each type of material and equipment.

Disposal of waste materials

Very high numbers of microorganisms are usually present in / on incubated broths and agar plates, as well as identification test kits / systems, such as biochemical test strips, and these organisms must be destroyed after completion of the test; all such contaminated waste must therefore be sterilised. This is generally accomplished by autoclaving the waste before it leaves the laboratory. However, this can be a time-consuming process and an alternative is to seal the waste into specially designed containers for disposal by a specialist licensed incineration contractor service. In the UK, laboratory managers employing incineration services have a 'Duty of Care' to assure themselves that the service is operating legally and safely.

Waste materials can be difficult to autoclave satisfactorily as they often contain pockets of air that cannot be displaced by steam during the autoclave cycle. Consequently, the contents may not always achieve a temperature high enough to destroy pathogenic organisms. It is therefore essential that the bags / containers of materials to be autoclaved are left open to allow the steam to circulate as thoroughly as possible and that a process of **at least** 121.1°C for 30 minutes is used (ISO 7218, 1996, Part 0).

Items that have not been cultured / incubated, e.g. remnants of food samples or used diluents, normally contain much lower numbers of organisms that may well be innocuous. The volumes of waste generated by even a small laboratory can quickly exceed the capacity of the laboratory autoclave so laboratories may dispose of these materials without sterilisation if they are considered to be non-hazardous, e.g. foodstuffs that would normally have been destined for human consumption taken from a production line. However, it is preferable that all microbiological waste materials should be sterilised before disposal and this should be ideally carried out in a dedicated autoclave that is not used for media preparation or destroyed by a suitable incineration contractor service.

Validation and verification of autoclave processes

Because the types of materials that need to be autoclaved in a laboratory vary widely, it is essential that each type of load / process, i.e. media, equipment and waste, is checked to ensure that the processes to be used routinely are adequate for each type of load (but not excessive for media). This is done in two ways:

(1)   When a new autoclave is commissioned for use, typical 'dummy' loads should be processed, with thermocouples placed in different parts of the chamber and load so that the temperature profile across the chamber and load can be derived. In the case of media, this should include loads to represent the different volumes of media likely to be processed by the laboratory, e.g. 10 ml, 250 ml and 500 ml. If an unsatisfactory profile is recorded, then the process and / or load must be re-designed until a satisfactory profile is achieved. This is known as process *vali-dation* and, usually, needs to be done only on first commissioning of the auto-clave; however, if the autoclave is moved or has a major engineering repair / refit, then process validation may need to be repeated (Anon, 1999 (2)).

(2)   The efficacy of the autoclave process must also be *verified* each time a load is processed. This can be done in different ways. The autoclave may be fitted with a continuous chart recorder linked to a 'roving' thermocouple probe; this device does not control the process but independently displays the time–temperature profile achieved at the point where it is placed. It is important that the probe is placed in a suitable position e.g. a point determined during process validation as having the slowest heat penetration i.e. the 'coldest spot'. In the case of media, the probe may be placed in the chamber between containers of media or in a dummy container of water; in the case of waste, the probe is usually placed between containers of waste because, for safety reasons, it is not recommended that a probe is inserted into the contaminated waste containers. In addition, independent time–temperature chemical monitors, e.g. tubes or paper strips with chemical indicators that change colour at specific temperatures, may be placed in the autoclave in or on suitable containers located at suitable points. These

indicators should be checked and the result recorded every time the autoclave is unloaded. So-called 'autoclave tape' should only be used to distinguish between processed and unprocessed loads and is not suitable for use in verifying processes because the darkened stripes appear at temperatures lower than those required for sterilisation processes. Biological indicators containing heat resistant spores are sometimes used to verify the efficacy of a sterilisation process but the heat processed spores then have to be inoculated into media and incubated before a 'no growth' (satisfactory) result can be recorded.

The laboratory's Standard Operating Procedures should indicate the autoclave process verification methods to be used for different loads and the recording and reaction (corrective) procedure to be followed in the event that unsatisfactory results are obtained.

9.5 Incubation conditions

As for all living creatures, different microorganisms have different optimal conditions for growth and they can easily fail to grow, or give an incorrect diagnostic reaction if the incubation conditions are inappropriate, leading to an incorrect (false-negative) result. Culture requirements for specific tests are discussed in detail in Chapters 10 and 11 but it is pertinent here to outline the fundamental requirements for optimal microbial growth. In addition to the use of an appropriate growth medium, the other conditions requiring control in microbiological tests are the composition of the gas atmosphere used for incubation, the incubation temperature and the incubation time (Table 9.3).

9.5.1 *Atmosphere composition*

The atmosphere in which cultures are incubated, and particularly the presence or absence of oxygen and carbon dioxide, can dramatically affect the ability of organisms to grow. As discussed in Chapter 2, microorganisms capable of growing in atmospheric air are known as **aerobes** and those capable of growth in the absence of air (oxygen) are known as **anaerobes**. Some aerobes can grow only if oxygen is available and are known as **obligate aerobes** whilst those anaerobes that can grow only if oxygen is absent are known as **obligate anaerobes**. Many normally aerobic organisms, however, can grow almost equally well in the absence of oxygen and are known as **facultative anaerobes**. A few microorganisms grow best in atmospheres containing lower oxygen levels and higher carbon dioxide levels than are present in air and these are said to be **micro-aerobes**.

The diversity of responses of microorganisms to the presence and levels of oxygen and carbon dioxide means that no single atmosphere composition is optimal for all organisms and it is important to provide an atmosphere that is suitable for the particular organisms sought. Obligate aerobes grow best on the surface of agar plates and slopes and their growth may be restricted or completely inhibited if they are buried within a depth of agar, as in the pour plate technique. Similarly, poor growth may occur in broth cultures unless these are shaken throughout the incubation period. Facultative anaerobes are capable of growing either on the surface or in the depths of agar; however, they employ different metabolic processes in aerobic and anaerobic

Table 9.3 Some examples of different types of microbiological growth media and their incubation conditions.

Microbiological growth medium			Usual incubation conditions		
Name	Used for:	Type	Temperature	Time	Atmosphere
General purpose media:					
Plate count agar (PCA)	Aerobes	Non-selective	30°C or 37°C	24–72 h	Air
Tryptone soya broth / agar (TSB / TSA)	Aerobes / anaerobes	Non-selective	30°C or 37°C	24–72 h	Air or anaerobic
Horse blood agar (HBA)	Aerobes / anaerobes	Non-selective, semi-diagnostic	30°C or 37°C	24–72 h	Air or anaerobic
Reinforced clostridial broth / agar (RCM / RCA)	Anaerobes	Non-selective	30°C or 37°C	48–72 h	Anaerobic
Media for growing Enteric bacteria					
MacConkey broth and agar	Coliforms / *E. coli*	Selective diagnostic	30°C or 37°C	24 h	Air
Violet red bile lactose agar (VRBLA, VRBA)	Coliforms	Selective diagnostic	30°C or 37°C	24 h	Air (agar overlay)
Violet red bile glucose agar (VRBGA)	Enterobacteriaceae	Selective diagnostic	37°C	24 h	Air (agar overlay)
Brilliant green bile broth (BGBB)	Coliforms / *E. coli*	Selective	30°C or 37°C or 44°C (*E. coli*)	48 h	Air
Media for isolating Salmonella:					
Buffered peptone water (BPW)	Pre-enrichment	Non-selective	37°C	16–20 h	Air
Selenite-cystine broth (SC)	Enrichment	Selective	37°C	18–24 h	Air
Rappaport-Vassiliadis medium (RV)	Enrichment	Selective	42°C	24–48 h	Air
Brilliant green agar (BGA)	Isolation	Selective diagnostic	37°C	18–24 h	Air
Xylose lysine desoxycholate agar (XLD)	Isolation	Selective diagnostic	37°C	18–24 h	Air
Media for isolating Listeria:					
Listeria enrichment broth (LEB)	Enrichment	Selective	30°C or 37°C	48 h	Air
Fraser broth (FB)	Enrichment	Selective	37°C	24–28 h	Air
Oxford agar (OX)	Isolation	Selective diagnostic	30°C or 37°C	48 h	Air
Polymixin acriflavin lithium chloride ceftazidime aesculin mannitol agar (PALCAM)	Isolation	Selective diagnostic	30°C or 37°C	48 h	Air or micro-aerobic

Media for isolating other Gram-positive pathogens:

Baird-Parker agar (BP)	*Staphylococcus aureus*	Selective diagnostic	37°C	48 h	Air
Mannitol egg-yolk polymyxin agar (MYP)	*Bacillus cereus*	Selective diagnostic	30°C	24–48 h	Air
Oleandomycin polymixin sulphadiazine perfringens agar (OPSP)	*Clostridium perfringens*	Selective diagnostic	37°C	18–24 h	Anaerobic
Tryptose sulphite cycloserine agar (TSC) (egg yolk-free)	Sulphite-reducing clostridia	Selective diagnostic	30°C	48–72 h	Anaerobic + agar overlay

Media for growing spoilage bacteria:

Cetrimide fucidin cephaloridine agar (CFC)	*Pseudomonas* spp.	Selective	25°C or 30°C	48 h	Air
de Man Rogosa Sharpe agar (MRS)	'Lactic acid bacteria'*	Semi-selective	25°C or 30°C or 37°C	48–72 h	Air or microaerobic or anaerobic
de Man Rogosa Sharpe agar + sorbic acid (MRS-S)†	'Lactic acid bacteria'*	Semi-selective			

Media for growing yeasts and moulds:

Dichloran rose bengal chloramphenicol agar (DRBCA)	General purpose	Selective	25°C	5 days	Air
Dichloran 18% glycerol agar (DG18)	General purpose**	Selective	25°C	5 days	Air
Rose-bengal chloramphenicol agar (RBCA)	General purpose	Selective	25°C	5 days	Air
Oxytetracycline glucose yeast extract agar (OGYE)	General purpose***	Selective	25°C	5 days	Air
Czapek yeast agar (CYA)	Identification	Non-selective	25°C	7 days	Air
Malt extract agar (MEA) (Blakeslee's formulation)	Identification	Non-selective	25°C	7 days	Air

h = hours

* The lactic acid bacteria are a heterogeneous group of organisms and the choice of incubation temperature and atmosphere conditions depends upon the sample type under examination and the species sought.

** Also isolates moderately xerophilic yeasts and moulds.

*** Most useful for yeasts, plates may be overgrown by rapidly spreading mould species.

† The presence of sorbic acid inhibits the growth of yeasts.

atmospheres, which may affect the results of biochemical reactions observed in some diagnostic media. For example, in enumeration tests for coliforms in foods using violet red bile lactose agar (VRBLA), a pour plate technique is used. When the agar layer has solidified and entrapped any organisms present in the test sample inoculum, the oxygen level available to the organisms is reduced by overlaying the agar with a layer of the sterile agar in order to enhance the development of the characteristic, deep red coliform colonies (see Plate 25(a)).

Anaerobes are usually grown on agar plates in a sealed jar, known as an anaerobe jar (see Plate 11). An oxygen-free atmosphere is created in the jar, usually be means of commercially available gas producing kits, before incubation. Micro-aerobes are often also grown on agar plates in an anaerobe jar but a gas composition is used that contains reduced oxygen and elevated carbon dioxide. However, some micro-aerobes can be grown using a 'deep pour plate', i.e. using approximately 25 ml of agar instead of the more usual 12–15 ml and this may or may not be overlaid with more agar before incubating the plates in air. Traditional equipment for growing micro-aerobes that is not widely used nowadays is the 'candle jar'. This is similar to an anaerobe jar that is filled with agar plates but a lighted candle is placed inside at the top of the jar before it is sealed; as the candle burns it uses the oxygen inside the jar and produces carbon dioxide which extinguishes the candle before the oxygen has been completely used up. Thus, the correct gas composition is produced. Although this technique has not been standardised as normally required today, it is, nevertheless, a very effective method for modifying the atmosphere within a jar.

9.5.2 Incubation temperature

Many different incubation temperatures are used in food microbiology practical work, ranging from 55°C for thermophiles, through 37°C for most enteric pathogens, 25–30°C for general counts and some specific groups of organisms, 22°C or 25°C for yeasts and moulds to 5–20°C for psychrotrophic organisms. In some tests, different incubation temperatures are required at different stages; e.g. media for growing coliforms may be incubated at 30°C or 37°C, but the subsequent confirmation test media for *E. coli* may be incubated at 44°C.

For pure cultures, the incubation temperature used is generally the optimum growth temperature of the organism, i.e. the temperature at which the target organism grows most quickly on the medium and under the atmosphere conditions used. In mixed culture, however, the optimum incubation temperature for the target organism may also encourage the growth of competitors and better recovery is sometimes obtained by incubating the medium at a higher or lower temperature than the optimum growth temperature of the target organism. Rappaport-Vassiliadis medium (a selective enrichment broth used in tests for the detection of *Salmonella*), for example, is incubated at 42°C, which is higher than the optimum growth temperature for *Salmonella* but still allows its growth while suppressing the growth of other Gram-negative bacterial species.

Unfortunately, there is, as yet, no international convention regarding incubation temperatures to be used in tests for the detection or enumeration of different organisms; different countries and even different food industry sectors use their own preferred temperatures (and, indeed, methods). In the USA, for example, TCC and pathogen tests are often incubated at 35°C, whereas many European countries use 30°C for TCC and 37°C for most pathogens. Similarly, meat industry laboratories commonly incubate media for coliform tests at 37°C since the organisms sought are associated with warm carcasses, whereas dairy industry laboratories undertake the same test using the same medium but incubate it at 30°C because 'dairy coliforms' are more commonly associated with cool, wet environments.

Thus, because of the range of incubation temperatures that may be required to be used in tests for different organisms in different foods or food-related samples, a laboratory may require several incubators and, for incubation temperatures lower than 30°C, specially designed, refrigerated incubators are required. Because of the importance of microbiological tests in international trade, there is currently a desire to harmonise test methods in general and incubation temperatures in particular. It could take some years, however, before such harmonisation is completed.

9.5.3   *Incubation time*

Microbiology differs considerably from chemistry in its unique requirement for the prolonged incubation of inoculated media necessary to allow microorganisms to grow. Microorganisms differ considerably in their rates of cell division, and incubation times must be varied accordingly; the overall 'elapse time' for microbiological tests can therefore be highly variable. Members of the Enterobacteriaceae are among the most rapidly growing food-borne microorganisms and are capable of producing visible colonies within 18 hours at their optimum growth temperature; conversely, some mycobacteria are probably the most slow-growing and incubation times for these organisms can be up to one year before the test is considered negative. However, most tests in routine food microbiology can normally be completed within 1–5 days unless confirmation tests need to be undertaken.

A single incubation period is used for some simple tests in food microbiology, e.g. for total aerobic colony counts, a plate count agar (PCA) pour plate is commonly incubated in air at 30°C for 48 or 72 hours ± 4 hours, whilst more complex techniques often require one or more broth sub-cultures before plating onto an agar to allow colonies to develop, e.g. for the detection of *Salmonella* or *Listeria* spp. The overall test time therefore depends on the number of sub-cultures required and the incubation times at each stage between them. It also depends on whether any presumptive positive reactions (or 'suspect' colonies) have been observed (see Chapter 11). If no colonies, or no typical colonies, develop, then a test can often be declared negative at an early stage; however, if 'suspect' colonies require confirmatory testing, the test time can be extended by a variable amount which can cause difficulties in cases where food cannot be sold or processed further until a negative result is reported, as is the case in positive release systems. If false-negative results are to be avoided, it is essential that the required 'full incubation time' is allowed at each stage of a microbiological test.

9.6   Microbiological techniques

The microbiological examination of food samples begins with sample collection, i.e. a procedure for taking a 'factory sample' from a production line or warehouse, which should also include an appropriate means of handling and storing the factory sample during transport to the microbiology laboratory. In the laboratory, the factory sample is often sub-sampled before the actual microbiological test begins. (For some explanations of the different types of sample see Chapter 7 and Figure 7.2.)

The different types of test that may be applied in the laboratory depend on the types and numbers of organisms expected; these may be simple tests involving a single incubation step or complex methods that include several sub-cultures followed by confirmation tests for 'suspect' colonies. Almost all laboratories employ 'routine test methods' which, of necessity, need to be as cost-effective as possible; however, occasionally laboratories are required to use 'reference methods' that are accepted nationally and / or internationally. These might be necessary in cases of a dispute between customer and supplier, for international trade purposes or in the event of court proceedings. Reference methods can be quite involved and time-consuming and are not always suitable for day-to-day use. These methods are usually prepared by committees of experts, e.g. the International Organisation for Standardisation (ISO), the International Dairy Federation (IDF) and the Association of Official Analytical Chemists (AOAC), and are also published by these bodies. Most routine laboratory methods are simplified versions of reference methods, often employing the same media and basic test conditions as in the reference method, and the laboratory's specific test methods should be documented in a methods manual indicating the sample types for which they are applicable in the laboratory.

All microbiological test methods, whether reference or 'industry accepted' methods, should be verified for their efficacy with the food and food-related sample types to be examined in the laboratory. If a laboratory develops its own method or adapts a reference or industry accepted method, then the adaptation must be properly validated in the laboratory using relevant sample types before it is used routinely. Method validation is an expensive and extensive process that should only be undertaken by or under the strict supervision of an experienced food microbiologist because the procedure, when involving pathogens, can be hazardous and because of the risk of cross-contaminating 'genuine' samples. The process usually includes testing large numbers of samples that contain varying but known levels of the target organism (added deliberately / spiked, as part of the method validation protocol) and comparing the test results obtained from the relevant reference / 'current' laboratory method used alongside the 'new' method. Even when validated, **all** methods need to be regularly verified, which involves periodically testing real samples 'spiked' with the target organism to confirm that the test method is working as expected; this is often referred to as a method efficacy test (see Chapter 7 and Figure 7.3).

The work in a food microbiology laboratory, however, does not only involve the examination of food samples; environmental monitoring is also widely undertaken in which swab samples may be collected from product contact surfaces or the hands of production staff, together with swabs or debris from floors, drains, vacuum cleaner contents, air samples and other non-product contact areas (see Chapter 5). The results from such tests can provide a 'map' of contamination 'hotspots' that may need to be given particular attention in cleaning operations.

9.7 Obtaining and handling samples

Microbiological testing and the results obtained are useless if the factory sample has been taken incorrectly or has been mis-handled, as the results will not reflect the microbiological status of the material at the time it was sampled. It is therefore important that factory samples are only taken by properly trained staff, whether they work in the microbiology laboratory, in quality control or in production areas.

The food microbiologist also needs to be aware of the statistics involved in taking factory samples, so that the samples taken are as representative as possible of the batch or lot to be examined. This is important because microorganisms are rarely evenly distributed throughout a large batch of product and it is necessary that the factory sample(s) taken for examination has the highest likelihood of being representative of the material sampled and of containing the organisms sought if they are present.

Ensuring that the factory sample(s) taken for examination is, as far as possible, representative of the batch from which it was taken may mean taking one or more random samples, or a structured sampling scheme (plan) may be used where, e.g. a series of samples is taken at the start, middle and end of a production run. For finished products, whole packs are usually taken as factory samples, e.g. whole cheese, cartons of cream, yoghurt and other dairy products, bottles / containers of milk or fruit drinks, cans of soup, meat, vegetable or fish-based products, packs of sandwiches, recipe dishes / ready-meals, delicatessen products.

For bulk raw materials and intermediate production samples, individual ingredients or mixes are usually sampled by aseptically removing quantities sufficient for testing and placing them in a suitably labelled sterile container for transfer to the laboratory. Some commodities however, can be particularly problematic with regard to determining an effective means of sampling, e.g. animal carcasses. For such materials, it may be necessary to carry out an investigation over a period of time and over several batches, taking several small meat samples or swabs across different carcasses within each batch to establish the most appropriate sampling method and those areas that best represent the microbiological status of the material; these are usually, but not always, the most regularly highly contaminated areas found on the carcases and these may then be used in a routine sampling plan. In addition, some raw materials, e.g. powders, granules and liquids, are received in large loads of several tonnes and samples taken from such large consignments of material should, in general, be as large as possible in order to be representative and also to ensure that there is enough material to undertake several such tests, if required. Factory sample sizes are therefore often in the range 100 g to 1 kg and several samples may be taken from different parts of large consignments e.g. of raw materials, for bulking together in the laboratory for examination.

Sampling of water supplies needs particularly careful thought. When an environmental health officer / public health inspector takes a water sample, the objective is usually to determine the microbiological quality of the water in the pipework, but taps and nozzles are exposed to the environment and can become quite heavily contaminated; the normal sampling procedure for these is therefore to clean and flame-sterilise the tap outlet and then let the water run for some time before collecting a water sample. However, when food factory staff actually use the water supply they just turn the tap on and use the water immediately delivered; a sample of water to be examined in the factory laboratory may therefore provide more useful information if it is taken as it leaves the tap under normal operating conditions and, in this case, water is run directly into a sterile sampling container without first cleaning and sterilising the tap / nozzle. Water supplies are generally made potable by treating with chlorine, and a small amount of a neutralising agent, e.g. sodium thiosulphate, is normally added to the water sample container to reduce the chlorine stress or damage to microorganisms during transport to the laboratory. Neutralising agents may also be required when taking swab samples of disinfected surfaces and the appropriate neutraliser must be used for the particular disinfectants used in the factory (Table 9.4). The laboratory should have standard documented procedures detailing all the factory sample types, their locations and the numbers of samples to be taken, frequency of sampling and the method of sampling to be employed.

Factory samples must always be taken and handled aseptically. This means that sampling tools / utensils and containers must always be sterile before use and the method of sampling must not expose the sample to extraneous contamination. Samples must always be stored and transported to the laboratory in appropriate conditions. Since many foods are perishable, this often means storing the factory samples under proper refrigeration or in cool boxes containing pre-frozen ice packs. For such samples it may be useful to measure the temperature upon arrival at the laboratory to check the temperature has been maintained within an acceptable range, especially if there has been a delay in transit. Frozen samples should always be kept frozen but foods that would normally be stored chilled should only be frozen in exceptional circumstances, as freezing can damage microorganisms and sometimes kill them completely, leading to a false microbiological profile of the sample. Even for frozen samples, it is essential to prevent unnecessary freeze–thaw cycles as these also can alter the microbiological profile of the material.

Ideally, all factory samples should be received by the laboratory and tested within an hour or so of being taken but this is often impractical, especially when the samples are being sent to a contract testing service laboratory. Wherever possible, however, factory samples should be examined on the day that they were taken and certainly no later than the following day, unless the product is ambient-stable or a particular examination schedule is required, e.g. for a shelf-life study.

Table 9.4 Agents for neutralising disinfectant residues when swabbing factory environments and equipment.

Neutralising agent	Final concentration (%w/v)	Disinfectant types neutralised
Sodium thiosulphate	1.0	Halogen-based compounds, e.g. chlorine (hypochlorite); iodine (iodophor)
Tween 80	3.0–5.0	Organic acids and esters Quaternary ammonium compounds Biguanides Amphoteric compounds

Note: So-called 'universal neutralising solutions' are sometimes used; these usually contain a mixture of sodium thiosulphate, lecithin, Tween 80 and, sometimes, sodium bisulphite in a general diluent, e.g. maximum recovery diluent.

9.7.1 *Laboratory handling of factory samples*

It is important to check that the factory sample packaging / container and its general condition is satisfactory on receipt into the laboratory as damaged packaging can lead to extraneous microbiological contamination and false results.

Liquids often need to be stirred thoroughly before sub-sampling or taking a test sample. For solids, a decision may need to be made whether to take the test sample from the surface or the interior of a product, such as a whole cheese where the consumer may not eat the rind. Composite foods such as trifles or layered pasta dishes can also pose problems as they have different compositional characteristics in different parts of the product; the fruit jelly in a trifle is likely to be acidic which favours the development of yeasts and moulds, whilst the cream layer has a near neutral pH in which Gram-negative psychrotrophs may predominate. Thus, the microbiologist may need to decide whether the test sampling needs to reflect the overall contaminating microflora or whether only those areas of the product at risk from specific spoilage organisms should be examined.

Samples such as powders and granular materials may need to be mixed well before sampling. For general colony counts, it is normal practice to use just a few grams e.g. 10 g or 20 g as a test sample quantity, but for tests for the detection of pathogens (presence / absence tests), a 25 g test sample is often examined; however, multiple test samples are sometimes 'bulked' (composited), e.g. 4×25 g or 12×25 g, making a 100 or 300 g final test sample respectively. In these cases, the limiting factors for the size of test sample used are usually the size of the containers required, incubator space, and the disposal facility and its capacity.

The product or raw material specifications provide a guide as to the quantity of material to be used as a test sample in the examination to be carried out and these can often be found in the laboratory's documentation for microbiological methods and interpretation of results.

9.7.2 *Statistical sampling*

Whatever practical test sample size (quantity) is used, examination of just a single test sample from a consignment cannot provide a representative view of the microbiological quality of the whole batch, especially when examining the batch for pathogens that may be present in very low numbers and, for this reason the concept of Hazard Analysis Critical Control Point (HACCP) was developed and is promulgated (see Chapter 5). The problem is that microorganisms are rarely distributed evenly within a batch of food, even if liquid, and the chance of finding a pathogen if it is present in, say, a 10 tonne consignment by examining 25 g or even 100 g of the material is usually negligible. Even in the case of colony counts where much higher numbers of microorganisms are usually being sought, the levels present can vary significantly from one part of the batch to another.

To improve the probability (chances) of obtaining more accurate information about the microbiological load of and particular organisms present within a consignment of food, statistically based sampling plans are sometimes used. Use of these plans guides the microbiologist as to how many test samples need to be examined and how test results should be interpreted to provide a given probability of correctly or incorrectly accepting (or rejecting) the batch. Different plans, i.e. different numbers of test samples and interpretation criteria, are used for different levels of concern about the safety or quality of a particular food; the greater the concern e.g. food destined for consumption by babies, the more test samples need to be examined and the more stringent is the interpretation of the results.

The most widely used sampling plans are based on those developed by the International Commission on Microbiological Specifications for Foods (ICMSF) (1986). The principles discussed by the ICMSF are generally for application for international trade purposes and aimed at port-of-entry sampling and examination of foods, circumstances in which very little may be known concerning the food production processes. Food manufacturers, however, should have full knowledge of the sources and quality of raw materials, food processing technology and other factors contributing to the final product's microbiological characteristics. With this knowledge, plans for product sampling and examination, and interpretation of results in food businesses are applied in a more practical way (Institute of Food Science and Technology (UK), 1999).

9.8 Safety in the food microbiology laboratory

Laboratories contain hazards of various types, namely biological hazards, chemical hazards and physical hazards, and microbiology laboratories contain all three. It is stating the obvious to say that microbiology laboratories contain microorganisms, some of which are human pathogens but a wide range of chemicals is often also held, e.g. acids, alkalis and salts, disinfectants, solvents and reagents used in diagnostic reactions, dyes and stains, antibiotics and growth media. Microbiology laboratories also have mechanical and electrical devices, from pH meters to autoclaves. Safe working practices must therefore be in place to ensure that accidents are prevented.

The working environment is subject to various Regulations and Codes of Practice and to EU Directives that have been incorporated into the UK national legislation and that of other European member states. The Health and Safety at Work Act (Anon, 1974) is an 'Enabling Act' that forms the basis of several, more specific Regulations, and associated, approved Codes of Practice. All of these place emphasis on the individual responsibility of all staff to ensure safety in the working environment. The most directly relevant Regulation for food microbiology laboratory staff is the Control of Substances Hazardous to Health (COSHH) (Anon, 1994), which covers biological as well as chemical hazards. However, other relevant Regulations cover Personal Protective Equipment at Work and Display Screen Equipment, i.e. the use of computer monitor screens.

More specific for microbiology are the guidance documents produced by the UK Advisory Committee on Dangerous Pathogens (ACDP, 1995 and 2001) on the categorisation of biological agents according to hazard and categories of containment, which is discussed in more detail in Chapter 6. Here, it is sufficient to state that, with a few exceptions, food-borne microorganisms are categorised in Hazard Groups 1 and 2 whilst Hazard Groups 3 and 4 contain more infectious agents most of which are generally not considered to be food-borne, with the exception of vero-cytotoxigenic *E. coli* (Hazard Group 3).

Hazards must be assessed in terms of the risk to staff, i.e. the probability of exposure to a hazard causing an adverse health effect and the severity of the effect, and this must be assessed for each task undertaken. If the risk of exposure, including exposure to microorganisms, is such that the worker is likely to be harmed in any way, then precautions to limit that risk must be taken. For further information see Chapter 6. Effective training of staff and the ready availability of relevant safety information that is read and understood by staff is essential for a microbiology laboratory to be operated consistently safely.

9.9   Further reading

Bridson, E.Y. (1998) *The Oxoid Manual*, 8th edn. Oxoid Ltd, Basingstoke, UK.

Corry, J.E.L., Curtis, G.D.W. and Baird, R.M. (2003) *Handbook of Culture Media for Food Microbiology. Progress in Industrial Microbiology Volume 37*. Elsevier Science, B.V., Amsterdam, The Netherlands.

Downes, F.P. & Ito, K. (eds) (2001) *Compendium of Methods for the Microbiological Examination of Foods*. Compiled by the APHA Technical Committee on Microbiological Methods for Foods, American Public Health Association, Washington D.C., USA.

Food and Drug Administration (1995) and Revision A (1998) *Bacteriological Analytical Manual*, 8th edn. Association of Official Analytical Chemists International, Arlington, Virginia, USA.

Hocking, A.D., Arnold, G., Jenson, I., Newton, K. & Sutherland, P. (eds) (1997) *Foodborne Microorganisms of Public Health Significance*, 5th edn. Australian Institute of Food Science and Technology, NSW Branch, Food Microbiology Group, Sydney, Australia.

Holbrook, R. (2000) Detection of Microorganisms in Foods – Principles of Culture Methods. In *The Microbiological Safety and Quality of Food Volume II*. (eds B.M. Lund, T.C. Baird-Parker, G.W. Gould). Aspen Publishers Inc, Maryland, USA. pp. 1761–1790.

Institute of Food Science and Technology (UK) (1999) *Development and Use of Microbiological Criteria for Foods*. IFST (UK), London, UK.

Yousef, A.E. & Carlstrom, C. (2003) *Food Microbiology: A Laboratory Manual*. Wiley-Interscience, John Wiley & Sons, Inc., New Jersey, USA.

10 Conventional Microbiological Methods II: Microbiological Examination of Samples

10.1 Introduction

As discussed in Chapter 9, two important questions have to be answered when a factory sample arrives in a food microbiology laboratory: these are 'What organisms should I be looking for in this sample?' and 'How many of these organisms am I expecting to find?' A range of test methods is available to the food microbiologist and the answers to these questions determine the general test method that should be used, as well as whether and how the method needs to be 'adapted' for the particular sample. Previous chapters have presented some answers to the first of these questions whilst in this chapter, aspects of and approaches to the second question are discussed.

Microbiological tests can either be quantitative or qualitative. Quantitative tests are known as enumeration (counting) tests and these allow an estimate to be made of the number of a particular type of microorganism present in a certain quantity of sample. Two types of technique are used for enumeration, colony counts and most probable number (MPN) tests; colony count results for food samples are usually expressed as the number of 'colony-forming units' per gram of sample (see 10.2.1 'colony counts' for an explanation of 'colony-forming units'), whilst the results of MPN tests are usually expressed as the 'most probable number' of organisms per 1 g, 10 g or 100 g / ml (these terms are explained later in this chapter). Thus, MPN tests are generally more sensitive than colony counts, although they are also more cumbersome to undertake.

Qualitative tests are known as detection, or presence / absence, tests. These tests are used to estimate the presence (or absence) of a particular organism in a certain quantity of sample, usually 25 g, irrespective of the number of organisms present. Absence in this context means 'not detected by the method used'. Thus, these tests are generally used for the detection of pathogenic organisms as they are quite sensitive. Sometimes, samples for presence / absence tests are 'bulked up' (composited) so that several small test samples are tested together as one large test sample in a single test, e.g. 4×25 g samples may be composited to form a single test sample weighing 100 g.

10.1.1 *The need for confirmation*

Because of the diversity of microorganisms and the similarities between related species, microbiological media rarely grow / detect just those organisms that the food microbiologist is interested in. Most microbiological tests are therefore done in stages. The first stage consists of the steps required to count or detect the target organism; if,

Plates 9, 10, 13, 15 and 19 are located in the colour plate section at p. 100.
Plates 21 and 23–31 are located in the colour plate section at p. 228.

at the end of this stage, no colonies, or no typical colonies, or no growth or positive reactions (in the case of broth media) are observed, then the test is terminated and the result is reported. For colony counts, this result is often reported as 'less than x colony-forming units/gram of test sample' (where x is the lowest sensitivity of the test) or, for detection tests, 'not detected/y gram of test sample' (where y is the quantity of test sample examined).

If, however, 'suspect' colonies or growth are observed at the end of the first stage, further confirmation tests may need to be undertaken in order to establish whether the organisms that have grown are the target organisms or some other species. This chapter discusses the first stage of testing whilst confirmation tests are discussed in Chapter 11.

10.2 Enumeration techniques

10.2.1 *Colony counts*

Growing microorganisms on a solid agar surface or in an agar gel allows them to develop into colonies that eventually become visible to the naked eye. Counting the number of colonies that develop on or in agar provides a means of estimating the number of organisms present in the original sample.

A suspension of the food is made by blending a known quantity of test sample with a known quantity of diluent (usually 10 g of test sample and 90 ml diluent, Plate 23; see colour plate sections for all plates) and, from this initial suspension, further dilutions are prepared as required. The appropriate agar medium (chosen according to the organisms sought) is then inoculated with known quantities of each required test sample dilution. There are two ways of doing this: either by adding the inoculum (usually 1 ml) to an empty sterile Petri dish and pouring on molten agar, mixing the inoculum into the agar and allowing the agar to set (known as the 'pour plate' technique) or by spreading the inoculum (usually 0.1 ml to 0.5 ml) evenly over the surface of an agar gel that has been pre-poured into a Petri dish and allowed to set (known as the 'spread plate' technique (see Figure 7.1(b))); a semi-automated version of the spread plate technique employs a 'spiral plater' to deposit a spiral inoculum onto the agar; this requires fewer dilutions to be prepared (see '10.6.1 Automation of repetitive procedures'). For both techniques, separate Petri dishes are prepared for each dilution to be plated.

After inoculation, the prepared dishes are incubated under the conditions of temperature and atmosphere and for the time appropriate for the organisms sought, after which the number of colonies that have developed, or typical colonies in the case of selective-diagnostic agar, is counted. If typical colonies of the target organism are observed on selective-diagnostic agar, then further confirmation tests may be required to confirm the identity of the colonies before the 'colony-forming units per gram of test sample' can be calculated.

The results of colony count tests are normally expressed as the number of 'colony-forming units' (cfu) per gram or ml of test sample. This is because microorganisms often stick together when growing in foods or any other environment or may be present as microcolonies (microscopic colonies containing a few hundred cells) and the dispersion of a food material in diluent when the test sample is prepared does not guarantee that small clumps of cells are separated into all of their individual cells.

When a single cell and a small clump of cells grow on or in agar, each produces a single colony but it is not possible to determine whether the colony was derived from the single cell or the small clump. Therefore, the test estimates the number of 'colony-forming units' in a given quantity of test sample, not the number of individual cells of organisms. Despite the inaccuracy inherent in colony counts caused by the methods used, colony counts have served the food industry well over many decades and have provided a simple means for making some estimate of the numbers of microorganisms encountered in most types of food; this has, in turn, been valuable in trend analyses.

Test sensitivity and types of colony counts

The colony count technique allows a fairly wide range of numbers of organisms in a food to be counted. At the lower end of the counting range, colony counts can be used to enumerate 1–100 cfu/g of test sample, the lower limit most commonly used being 10 cfu/g for a solid food or 1 cfu/g or ml for a liquid, e.g. milk. At the higher end, there is no limit, as the number of dilutions that can be prepared is theoretically endless, but it is rare to find organisms in foods in numbers greater than 10^9 cfu/g.

As explained in Chapter 9, in the food industry, general aerobic bacterial colony counts on food samples are the type of test probably most widely undertaken; a variety of general-purpose media can be used, the most widely used being plate count agar (PCA). Such counts are known as 'total colony counts' (TCC), but are also referred to as 'total viable counts' (TVC), 'aerobic colony counts' (ACC) or 'aerobic plate counts' (APC). It is important to note that none of these descriptions is entirely accurate, e.g. the test does not detect the total number of organisms present in the test sample, only those capable of growing on the specified medium under the selected conditions and time of incubation, and microbiologists do not count plates, they count colonies. Perhaps, the most accurate description is e.g. total aerobic mesophilic colony count (PCA incubated at 30°C for 48 hours).

Although the aerobic mesophilic TCC is the most widely used colony count method, general purpose (non-selective) media can also be used to undertake other types of colony count for more restricted groups of organisms by varying the incubation temperature or incubation atmosphere, or by treating the test sample dilution before plating (a heat-treatment is often used) so that the range of viable organisms in the test sample dilution mixture is restricted to just a few specific types. Indeed, these techniques may be used in combination to restrict the range of organisms capable of growth during incubation even further.

Some examples of incubation conditions or test sample treatment for different types of colony count that use general purpose media are as follows:

- Psychrotrophic colony counts: incubation in air at 7°C for 7–10 days.
- Mesophilic colony counts,* e.g. TCC: incubation in air at 30°C for 2 or 3 days.
- Thermophilic colony counts:* incubation in air at 55°C for 2 days.
- Mesophilic spore-formers:* heat-treatment of a test sample dilution, e.g. 80°C for 10 minutes, then incubation of prepared plates in air at 30°C for 3 days.

*Anaerobic versions of these types of count may also be undertaken by incubation of the plates in an atmosphere of suitable composition, e.g. 95% hydrogen, 5% carbon dioxide.

When plates are incubated at high temperatures, e.g. at 55°C for the enumeration of *Bacillus stearothermophilus*, evaporation of water from the agar must be prevented, so that the correct moisture level is maintained in the agar to facilitate growth of the organism. This is usually accomplished by enclosing the plates in a plastic bag, although it is important to ensure that aerobes still have sufficient oxygen available for growth. The type of plastic and the method of sealing the bags can affect the results, so individual laboratories need to establish their own method of preventing moisture losses from agar, possibly after some initial experimentation. Alternatively, containers of water are sometimes placed in the incubator to maintain a high humidity that inhibits evaporation of water from the agar in the plates. In these cases, it is important to ensure the water and its container do not harbour microorganisms.

Although colony counts using general purpose media are widely undertaken, they do not allow the specific detection / enumeration of particular organisms or groups of organism, e.g. *Escherichia coli*, *Bacillus cereus*. To do this, selective-diagnostic media are usually used; these restrict the range of organisms capable of growth on the medium to the target organism (and usually a few related species) and enable the target organism to produce typical diagnostic reaction(s) that allow it to be recognised on the agar.

How to do a colony count

The colony counting method involves a well-defined sequence of steps, the main decision that needs to be made being how many dilutions of the test sample to prepare. The procedure, in summary, is as follows.

Preparing a dilution series of the test sample

The aim of preparing a dilution series is to make accurate, quantitative, usually 10-fold, dilutions of the test sample (see Plate 24). Dilution of high numbers and closely crowded microorganisms in a test sample reduces the number of organisms at each level of dilution and means that, when the dilutions of the test sample are inoculated onto or into agar, individual cells or clumps of cells can grow into colonies that are separated from each other and hence, are easily countable.

(1) A 'top-pan' balance with an accuracy of ± 0.1 g is checked and tared (adjusted) to zero. A clean container e.g. 500 ml beaker, is placed on the balance and a sterile plastic bag is placed in the container (for support), then the mouth of the bag is opened aseptically. The balance is re-tared to zero.

(2) As quickly as possible (but maintaining aseptic techniques), a portion of the factory sample (either solid or liquid) is weighed out into the sterile plastic bag. This entire test sample must be placed in the bottom of the bag and not allowed to touch or stick to the upper parts of the bag's internal surfaces otherwise an inaccurate test sample dilution may result. Sterile diluent, e.g. maximum recovery diluent (MRD), is added to make a 10-fold, or 10^{-1}, dilution. Usually, to improve the accuracy of this primary dilution a 10 g test sample is weighed and 90 ml diluent is added by weight, i.e. 90 g.

Note: Gravimetric diluters (see '10.6.1 Automation of repetitive procedures') may be used to aid primary sample dilution, in which case, the manufacturer's instructions must be followed.

(3)   For solid samples, the weighed test sample is physically dispersed in the diluent to dislodge and distribute the organisms that are present homogeneously throughout the suspension, usually using a laboratory blender (see Plate 9). Blending should continue for at least 30 seconds, or more if the test sample is of a type that is difficult to disperse. A laboratory Standard Operating Procedure should specify the blending times to be used for each sample type. For liquids, such physical dispersion is normally unnecessary and gentle hand-shaking is often sufficient to disperse them in the diluent.

(4)   The 10^{-1} primary dilution is further diluted 10-fold to create a 10^{-2} dilution (1 in 100) of the original test sample, by transferring 1 ml of the primary dilution into 9 ml of sterile diluent in a Universal bottle or capped test tube, using a sterile pipette (see Plates 15 and 24). The 10^{-2} dilution is then mixed thoroughly, often using a rotating (vortex) mixer (see Plate 10).

(5)   The dilution series is then extended, as required, depending upon the number of organisms anticipated, i.e. 1 ml of the 10^{-2} dilution is transferred to a container of 9 ml sterile diluent and mixed to create a 10^{-3} dilution (1 in 1000) etc. (Plate 24). Because this procedure works **down** the dilution series to more dilute suspensions of the test sample and organisms, a new, sterile pipette / tip must be used to prepare each new dilution, otherwise excess microorganisms are carried from one dilution to the next in food material and droplets that adhere to the outside of the pipette / tip and this makes the dilution series and hence the final results inaccurate.

(6)   The dilutions are then used to prepare agar plates using either the pour plate or the spread plate technique. If a spiral plater is used (see '10.6.1 Automation of repetitive procedures'), it is not normally necessary to dilute the sample beyond the 10^{-1} or 10^{-2} stage. If not transferred to growth medium immediately, the dilutions must be re-mixed immediately before an inoculum volume is removed to ensure that the organisms are as evenly distributed as possible; however, test sample dilutions must always be plated within 20 minutes of their preparation to minimise the growth or death of the microorganisms in the suspension.

Note: It is usual to plate consecutive test sample dilutions, e.g. 10^{-1}, 10^{-2}, 10^{-3}, etc., according to the number of organisms anticipated in the test sample; however, in some laboratories, in an effort to minimise the number of Petri dishes required, alternate dilutions are plated, but this makes the test less accurate because, after incubation, the plates may contain either too many colonies to count or too few. Thus, this practice is to be discouraged. Some laboratories prepare two plates (duplicate) of each test sample dilution and if only very low numbers i.e. 10s of cfu/g of organisms are anticipated only two plates of the primary dilution, usually 10^{-1}, may be used. The laboratory's Standard Operating Procedures should specify the dilution levels and number of replicates to be plated for each sample type.

Making pour plates

The aim of the pour plate technique is to disperse and immobilise single cells or small clumps (groups) of cells in a layer of agar so that, when incubated, each cell or clump of cells (colony-forming unit) grows into a single, isolated colony.

(1) Empty sterile Petri dishes are labelled with a unique sample identifying code, a unique medium identifying code (see Chapter 8 and Figure 8.3) and the test sample dilution level to be added to the dish. The base of the dish, i.e. the part that contains the agar, should always be labelled in case the lids become mixed up. Ideally, duplicate plates should be used for each dilution to be plated but, in routine laboratories, single plates per dilution are often prepared.

(2) An aliquot (usually 1 ml) of the most dilute suspension required for the test is transferred, using a sterile pipette, to each Petri dish corresponding to that dilution.

(3) An aliquot of the next dilution (more concentrated suspension) is then dispensed into each of its corresponding dishes. Because this procedure works **up** the dilution series to more concentrated dilutions of the sample and organisms, there is no practical compromise of accuracy due to cross-contamination from the more dilute suspensions and the same pipette may be used throughout. Further aliquots are then dispensed from each consecutive and more concentrated dilution, as required, into the corresponding Petri dishes.

(4) As quickly as possible, and before the inoculum has time to start drying out (see Plate 25b), a fresh bottle of the appropriate molten agar, tempered to $46 \pm 1°C$, is removed from the tempering cabinet or water bath and, if necessary, the outside is dried thoroughly; then, approximately 15 ml of the agar is poured carefully (to avoid splashing) into each dish. It is essential that the agar is no hotter than 47°C or some of the microorganisms inoculated into the Petri dish may be killed, leading to an inaccurate result.

(5) Before the agar has solidified, the inoculum and agar are mixed together quickly, but gently and thoroughly, by sliding the dishes in a circular motion 5 times clockwise then 5 times anti-clockwise, so that the agar and inoculum gently swirl together, followed by 5 up-and-down movements and 5 side-to-side movements. This must be done without splashing the agar over the inside of the lid or over the rim of the base of the Petri dish. Experienced staff may stack the plates 2 or 3 high to accomplish this operation more rapidly.

(6) The inoculated, poured and mixed plates are then spread out and the agar is allowed to set thoroughly. For some tests, an agar overlay is added after the first

layer has set completely: following the procedure outlined in stage (4), about 4 ml of molten agar is poured onto the surface of the set agar layer and the plate is gently tilted so that the base agar surface is covered entirely and evenly, and then this second agar layer is allowed to set.

Note: Automated equipment is available that dispenses agar into plates to which the inoculum has already been added (the equipment may be programmed for pouring different agars in sequence) and mixes the inoculum / agar mixture. Use of any equipment to assist plate preparation must be carefully controlled and moni-tored to ensure that aseptic conditions are maintained and the correct agars and volumes are dispensed. As with all equipment in the laboratory, the manufacturer's instructions must be followed with respect to maintenance, cleaning and operating conditions.

(7)   When **completely** set, the plates are inverted. If this is attempted before the agar has set completely, the centre section of agar may fall into the lid rendering the plate useless.

(8)   Finally, the plates are incubated under the appropriate conditions for the necessary time.

Making spread plates

The aim of the spread plate technique is to disperse and immobilise single cells or small clumps of cells on the surface of a layer of agar so that, when incubated, each cell or clump of cells (colony-forming unit) gives rise to a single, isolated colony.

(1)   Sterile agar (approximately 15 ml) is poured into empty Petri dishes labelled with the medium's unique identifying code, and allowed to set. This may be done some days in advance of being required for a test sample examination, provided the plates are stored under refrigeration and an appropriate shelf life is applied, supported by suitable quality control tests.

(2)   When required for use, the surfaces of the agar of the pre-poured agar plates are dried, e.g. by placing the plates in a warm air incubator or in a laminar air flow cabinet. This procedure removes excess water from the agar surface and facil-itates the absorption of the inoculum onto the medium. It is important that the agar is not over-dried and the exact time and temperature for this procedure depends on many factors and varies from laboratory to laboratory; each laboratory should have a specified, validated procedure for drying the surface of plates.

(3)   The plates of pre-poured agar are labelled with a unique sample identifying code and the test sample dilution level to be inoculated. The base, i.e. the part that contains the agar, should always be labelled in case the lids become mixed up. Ideally, duplicate plates should be used for each dilution to be plated but in routine laboratories single plates per dilution are often prepared.

(4)   An aliquot, usually 0.1 ml, 0.2 ml or 0.5 ml, of the most dilute suspension is transferred, using a sterile pipette, to each Petri dish corresponding to that dilution. The inoculum is deposited onto the agar surface in the centre of the dish. Care must be taken to avoid damaging the agar.

Note: The inoculum size for spread plates is normally in the range 0.1 ml to 0.5 ml. The volume added should not be more than 0.5 ml because the medium may not absorb the inoculum adequately and the resulting wet agar surfaces can lead to confluent microbial growth, giving inaccurate results.

(5)   An aliquot of the next dilution (more concentrated suspension) is then dispensed into each of its corresponding dishes. Because this procedure works **up** the dilution series to more concentrated dilutions of the sample and organisms, there is no practical compromise of accuracy due to cross-contamination from the more dilute suspensions and the same pipette may be used throughout. Further aliquots are then dispensed from each consecutive dilution, as required, into the corresponding Petri dishes.

Note: In theory it is possible to reduce the number of dilutions required to be made for test samples containing high numbers of microorganisms by plating a small quantity, such as 10 μl (0.01 ml), of the more concentrated (lower) dilutions. This practice should be discouraged as it makes the test results less accurate as such small volumes of diluted food materials are difficult to dispense accurately and spread efficiently.

(6)   Before it dries into the agar, the inoculum from the most dilute suspension is quickly but carefully spread evenly over the agar surface using a sterile spreader; care must be taken to avoid touching the edges of the Petri dish and spreading the inoculum up to the edges (see Figure 7.1b). If duplicate plates are prepared from the dilution, both plates are spread before spreading the inoculum for the next (more concentrated) dilution. Care must be taken to avoid damaging the agar during inoculum spreading.

Note: The spreader must not be allowed to touch the edge of the Petri dish and the inoculum should be kept 3–4 mm away from the edge at all points, otherwise undesirable confluent growth may occur around the perimeter of the agar which may lead to an inaccurate count. The spreader should be touched firmly but gently on the surface of the agar so that the full length of the spreading portion remains evenly in contact with the agar whilst spreading (see Figure 7.1(b)); if this is not done, the inoculum may be spread unevenly, which can result in patchy growth.

(7)   Consecutive dilutions are spread until all plates have been surface inoculated. Because this procedure works **up** the dilution series, it is not essential to use a new sterile spreader for each inoculum dilution although it is essential to ensure that aseptic practice is maintained to prevent extraneous contamination.

Note: If a single spreader is used for several plates then, should an agar plate be contaminated with an invisible microcolony, contamination will be transferred to and spread over all subsequent plates; for this reason, it is the policy in many laboratories to require the use of a separate spreader for each plate or the duplicate plates for each test sample dilution. Providing the laboratory has satisfactory control of media and plate preparation procedures, quality and storage conditions, this type of contamination and resultant problem should be an unlikely event.

(8)   A spiral plater may be used as an alternative to the manual spreading procedure described in steps (4) to (7) above (see '10.6.1 Automation of repetitive procedures').

(9)   The inoculated and spread plates are spread out to allow the inoculum to dry into the agar and are then inverted (if required) and incubated under appropriate conditions for an appropriate time.

Counting colonies and calculating the results

The aim of colony counting is to obtain an accurate count of the numbers of colonies that have developed on / in the medium in the plates during incubation. On general purpose agars, all colonies are counted; on diagnostic agars, only those colonies that show specific colony characteristics and diagnostic reactions are counted, and, sometimes, further confirmation tests are required before the colony count of the target organism per gram of the test sample can be calculated.

(1)   At the end of incubation, the plates are removed from the incubator and inspected for the presence of visible colonies (Figure 10.1). Plates containing 'countable numbers' of colonies are selected for counting. For general-purpose agars, this usually means between 30 and 300 colonies per plate, or 15–150 colonies for selective and diagnostic media. The laboratory's methods docu-

Figure 10.1   Picture of visible colonies. Yeast colonies growing in a pour plate of wort agar.

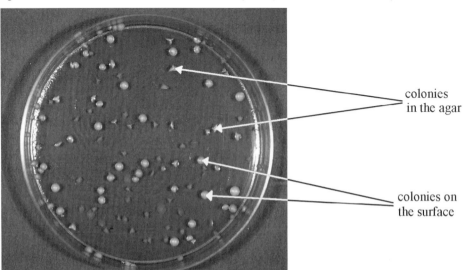

colonies in the agar

colonies on the surface

• Note the even distribution of colonies on the surface and throughout the depth of the agar.

mentation should indicate how plates should be selected for counting. Different counting procedures are applied to plates prepared by means of a spiral plater and the manufacturer's instructions must be followed to ensure the correct procedure is followed.

(2) Colonies are usually counted 'manually' with the aid of an illuminated, magnifying colony counter with an inbuilt tally counter (see Plate 19) and a marker pen. In the case of selective-diagnostic agars, only colonies showing appropriate diagnostic reactions are counted, unless there is a specific reason for counting colonies showing atypical reactions. For each level of dilution counted, the number of colonies per plate is recorded, together with the associated sample dilution level and volume of inoculum. If the test has been performed correctly, the duplicate plates at any given dilution level should contain approximately equal numbers of colonies whilst the numbers of colonies from plates of successive dilutions of the test sample should show 10-fold differences (Figure 10.2).

Note: After dilution and blending, some test samples, e.g. cooked egg yolk, flour, produce highly dispersed and very small particles in a pour plate that can easily be confused with microbial colonies; it is essential that every care is taken to ensure that only microbial colonies are counted and not food particles from the test sample. Additional magnification may be required to help distinguish between particles and colonies, the former usually being of irregular shape and the latter being more regular and smooth in shape.

(3) If the numbers of colonies on all of the plates prepared for a test sample show a satisfactory numerical pattern as described in stage (2), then the number of colony-forming units per gram of test sample can be calculated from the raw data; this can be done in several different ways, depending on the number of countable plates per sample.

Example 1: Colonies counted on a single plate at a single dilution

Count/g = number of colonies counted $\times$ reciprocal of dilution $\times$ reciprocal of the dilution volume plated

(A) Pour plate technique
180 colonies counted on a plate that was inoculated with 1 ml of a 10^{-2} dilution.

$$\text{Count/g} = 180 \times \frac{1}{10^{-2}} \times \frac{1}{1}$$

$$= 180 \times 10^2 \times 1 = 18\,000 \quad \text{or} \quad 1.8 \times 10^4\,\text{cfu/g}$$

(B) Spread plate technique
180 colonies counted on a plate inoculated with 0.1 ml of a 10^{-1} dilution.

$$\text{Count/g} = 180 \times \frac{1}{10^{-1}} \times \frac{1}{0.1}$$

$$= 180 \times 10^1 \times 10 = 18\,000 \quad \text{or} \quad 1.8 \times 10^4\,\text{cfu/g}$$

Figure 10.2 Illustration of an accurate dilution series for colony counts.

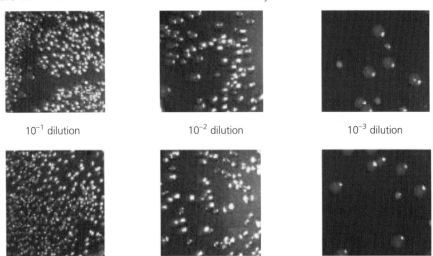

10^{-1} dilution 10^{-2} dilution 10^{-3} dilution

(a) Duplicate spread plates, i.e. surface plating.
Notes:
(1) All colonies have developed on the agar surface. The presence of colonies within agar inoculated by surface plating indicates contamination or poorly prepared plates and such plates should be discarded.
(2) At each dilution, the duplicate plates contain approximately similar numbers of colonies.
(3) The plates at each successive dilution contain approximately 10 times fewer colonies than those of the previous dilution, i.e. the plates for the 10^{-1} dilution contain approximately 10 times more colonies than those for the 10^{-2} dilution, which, in turn, contain approximately 10 times more colonies than the plates for the 10^{-3} dilution.

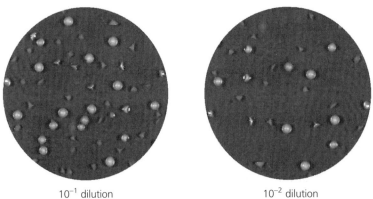

10^{-1} dilution 10^{-2} dilution

(b) Single pour plates.
Notes:
(1) Some colonies are located within the depth of agar whilst others have developed on the surface. Surface colonies often have a different appearance from colonies that are buried in the agar and may be considerably larger.
(2) The plate for the 10^{-2} dilution contains approximately 10 times fewer colonies than the plate for the 10^{-1} dilution.

Example 2: Colonies counted on duplicate plates at a single dilution

Count/g = (total number of colonies counted ÷ number of plates selected)* × reciprocal of dilution × reciprocal of the volume plated

*(total number of colonies counted ÷ number of plates selected) provides the arithmetic mean number of colonies per plate.

(A) Pour plate technique
165 and 195 colonies counted on each of two plates, each inoculated with 1 ml of a 10^{-2} dilution.

$$\text{Count/g} = \frac{(165 + 195)}{2} \times \frac{1}{10^{-2}} \times \frac{1}{1}$$

$$= 180 \times 10^2 \times 1 = 18\,000 \quad \text{or} \quad 1.8 \times 10^4\,\text{cfu/g}$$

(B) Spread plate technique
165 and 195 colonies counted on two plates, each inoculated with 0.1 ml of a 10^{-1} dilution.

$$\text{Count/g} = \frac{(165 + 195)}{2} \times \frac{1}{10^{-1}} \times \frac{1}{0.1}$$

$$= 180 \times 10^1 \times 10 = 18\,000 \quad \text{or} \quad 1.8 \times 10^4\,\text{cfu/g}$$

Example 3: Colonies counted on duplicate plates at each of two dilutions

If colonies are counted at more than one dilution, then the 'weighted mean' calculation is used.

$$\text{Count/g} = \frac{C}{V \times (n_1 + (0.1 \times n_2)) \times d}$$

where C = the total number of colonies counted on all plates
  V = the volume of inoculum applied to each plate
  n_1 = the number of plates counted from the lower dilution*
  n_2 = the number of plates counted from the higher dilution
  d = the lower dilution* from which counts were obtained.
* The lower dilution means the least dilute sample counted, i.e. the more concentrated, e.g. the 10^{-2} dilution rather than the 10^{-3} dilution, for the following pour plate example (A), or the 10^{-1} dilution rather than the 10^{-2} dilution, for the following spread plate example (B).

(A) Pour plate technique
168 and 196 colonies counted on two plates at a 10^{-2} dilution, and 17 and 19 colonies counted on two plates at a 10^{-3} dilution; each plate inoculated with 1 ml.

$$\text{Count/g} = \frac{168 + 196 + 17 + 19}{1 \times (2 + (0.1 \times 2)) \times 10^{-2}}$$

$$= \left(\frac{400}{1 \times 2.2}\right) \times 10^2 = 18\,182 \quad \text{or} \quad 1.8 \times 10^4\,\text{cfu/g}$$

(B) Spread plate technique
168 and 196 colonies counted on two plates at a 10^{-1} dilution, and 17 and 19 colonies counted on two plates at a 10^{-2} dilution; each plate inoculated with 0.1 ml.

$$\text{Count/g} = \frac{168 + 196 + 17 + 19}{0.1 \times (2 + (0.1 \times 2)) \times 10^{-1}}$$

$$= \left(\frac{400}{0.1 \times 2.2}\right) \times 10^{1} = 18\,182 \quad \text{or} \quad 1.8 \times 10^{4}\,\text{cfu/g}$$

(4)   Counts are usually expressed in exponential form to 2 significant figures or in $\log_{10}$ form to 2 decimal places. Thus, 18 182 cfu/g is expressed as 1.8×10^{4} cfu/g or $\log_{10}$ 4.26 cfu/g.

Spreading colonies

Some microorganisms (notably *Bacillus* spp. and *Proteus* spp.) produce large colonies that can spread or 'swarm' over a high proportion of the agar surface. This can occur both in pour plates and on spread plates, although it is perhaps more common with spread plates. On pour plates with an overlay, spreading colonies are unlikely to be observed if the first agar layer has been properly overlaid. Overgrowth by spreading colonies can lead to oxygen being excluded from organisms beneath the spreading colony and these may fail to develop visible colonies; even if colonies do develop, it is often difficult to see them and so count them accurately underneath the spreading colony.

   This problem may be resolved in a number of ways. If only occasional plates are affected and other plates prepared from the same test sample contain countable colonies, then the affected plates may be disregarded and the remaining plates used for counting and calculating the result. If the spreading colony affects only a proportion of a plate, then an unaffected sector of the plate, e.g. half or quarter, may be counted and the relevant additional multiplication factor applied when calculating the results, i.e. $\times 2$ if half the plate was counted and $\times 4$ if a quarter of the plate was counted. If a persistent problem is encountered with particular sample types, e.g. some dried goods, then, to inhibit spreading colonies, a pour plate may be overlaid with sterile agar or, alternatively, the strength of agar in the medium formulation may be increased by adding additional, purified bacteriological agar. In either case, the modification to the method used should be validated.

10.2.2 *Most probable number (MPN) techniques*

The Most Probable Number or MPN technique (also known as the Multiple Tube Method) employs a statistical approach for estimating the numbers of microorganisms in a test sample by assessing the presence or absence of the target organisms at different dilution levels of the test sample. It is mostly used for liquid samples, especially water, but can also be used for solid foods. The MPN technique is generally regarded as being inaccurate because it employs a statistical basis for calculating the numbers of organisms present in a test sample. However, it is more sensitive than the colony count (agar plating) technique when only low numbers of organisms are present and is therefore useful for application to samples which contain only very low numbers of microorganisms i.e. $< 10/\text{g}$ or ml; for samples with numbers of organisms expected to be greater than about 100 cfu/g, colony counting provides greater accuracy.

To obtain satisfactory results from an MPN test, a dilution series is prepared from the test sample that must extend sufficiently to ensure that some higher dilutions used for inoculation do not contain the target organism. In addition, at least 3 sequential dilutions must be inoculated into broth media, either 3 or 5 replicate tubes of broth, i.e. multiple tubes, being inoculated for each test sample dilution level used. After incubation, the presence or absence of the target organism in each tube is determined; depending on the target organism, e.g. for a coliform MPN test, this may involve direct inspection of media for growth, acid or gas production, but for some other organisms, e.g. sulphite-reducing clostridia, further confirmation tests involving plating onto agar are required before the confirmed MPN can be calculated.

By this means, the number of tubes positive and negative for growth of the target organism(s) at each test sample dilution is determined. This creates a 'profile', e.g. 3–1–0, that is translated into the most probable number of organisms / test sample quantity by the use of statistical tables (International Commission on Microbiological Specifications for Foods, 1978; International Standard ISO 7218, 1996).

Sensitivity of MPN tests

Often, there is confusion over the sensitivity level of MPN tests and how to calculate the results. This is because MPN tests were originally devised as a sensitive means of detecting low numbers of microorganisms, e.g. coliforms, in water supplies for which the results are expressed as the most probable number of organisms / **100 ml** of water, based upon a test method in which large quantities of test sample were (and still are) examined, e.g. 10 ml or larger.

The statistical tables used to derive the results of MPN tests can be adapted for use with foods by (simply) accounting for the different quantities of test sample examined. In the case of liquid test samples, this is not too complicated as the undiluted test sample can be examined, e.g. 1 ml or 10 ml; it should, however, be noted that, for the purposes of this test, 1 ml of an undiluted liquid test sample is regarded as 10 times more 'dilute', i.e. contains $10 \times$ fewer organisms, than 10 ml of the same undiluted test sample and, in mathematical terms, represents the next 10-fold level of dilution.

Thus, for liquid samples, the minimum level of detection for the so-called 3-tube MPN method is 3 microorganisms / 100 ml (0.3 / 10 ml or 0.03 / ml), when the lowest 'dilution' examined per tube is 10 ml of the undiluted test sample. If the lowest 'dilution' examined per tube is 1 ml of the undiluted test sample, then the test sensitivity is 30 organisms / 100 ml, or 3 organisms / 10 ml, i.e. 0.3 organisms / ml.

Solid foods present a further complication because they have to be dispersed in a diluent to form a 10^{-1} dilution before they can be inoculated into microbiological growth media and an initial inoculum level of either 1 ml of the 10^{-1} dilution (0.1 g of test sample) or 10 ml of the 10^{-1} dilution (1 g of test sample) may be used. Therefore, it is important to ensure that the calculation of results for solid test samples takes into account both the dilution level and the quantity (volume) of each dilution inoculated so that the decimal point for the result is in the correct place.

Thus, for solid foods, the sensitivity of a 3-tube MPN method is 3 organisms / 10 g, i.e. 0.3 organisms / g, when the highest quantity i.e. first / lowest dilution, examined per tube is 10 ml of the 10^{-1} dilution of the test sample, i.e. 1 g of test sample per tube. When the highest amount examined per tube is 1 ml of the 10^{-1} dilution of the test sample, i.e. 0.1 g of test sample per tube, the test sensitivity is only 3 organisms / g (International Standard 7218, 1996).

The above values relate to the 3-tube MPN test; however, several other versions of the test exist that require different quantities of test sample to be examined and employ different numbers of tubes at each dilution level. Examples include a 5–5–5 tube test, i.e. 5 tubes for each of 3 dilutions, and a 6-tube test that employs 6 × 18 ml (or 18 g) of one test sample dilution level only; other combinations may also be found. Each of these combinations has its own sensitivity and associated statistical foundation, so it is important that the appropriate MPN tables are used with each test format.

How to do an MPN test

Although the statistical principles of MPN tests are well-defined, in practical terms, the precise details of any particular method used must be suited to the organism sought; for example, a MPN test for coliforms may be a simple broth method that requires few or no confirmation tests, whereas the same MPN format applied to *Salmonella* requires each tube to be sub-cultured to further broths and then plated, followed by confirmation tests on 'suspect' colonies, before the MPN result can be calculated.

However, MPN tests are nearly always broth methods, employing selective broth media, and the general principles are as follows:

Preparation of dilutions and enrichment

(1) A dilution series is prepared from the test sample using the same procedure as for colony counts. The dilution series must extend to at least 3 dilutions and far enough so that the target organism is 'diluted out', i.e. the target organism must not be present in high numbers or preferably is absent in the highest (final) dilution.

(2) An appropriate broth medium is inoculated by adding aliquots of each test sample dilution to multiple containers (called 'tubes' because the media are often prepared in test tubes) of the medium, usually a selective medium. Normally, 1 ml aliquots of a dilution are added to 10 ml volumes of broth. Sometimes, however, 10 ml quantities of the test sample dilution may be added to 10 ml volumes of double-strength medium to provide a lower level of test sample dilution (containing a higher quantity of sample) thus making the test more sensitive; in this case, an MPN tube format might be:

3×10 ml of a 10^{-1} dilution
3×1 ml of a 10^{-1} dilution
3×1 ml of a 10^{-2} dilution.

Three tubes per dilution are prepared for the 3-tube MPN test whilst 5 tubes per dilution are inoculated for the 5-tube MPN test. At least 3 sequential dilution levels are inoculated into broths, which must include the dilution at which the target organism is expected to have been diluted out. Thus, a **minimum** of 9 tubes

(3-tube test) or 15 tubes (5-tube test) per test sample must be inoculated. The principles of a 3-tube MPN format are shown in Plate 26.

(3) The inoculated selective broths are incubated as appropriate for the target organism and, after incubation, the tubes are assessed for growth and / or reactions giving a pattern of positive and negative results. For simple tests such as coliform tests, the tubes may only require inspection for the presence of characteristic reactions, e.g. growth causing turbidity of the medium and acid and / or gas production; for more complex tests such as for the enumeration of *Salmonella*, sub-culture and identification of presumptive positive isolates is likely to be needed before presumptive positive reactions can be confirmed.

Recording the results and calculating the MPN of organisms in the test sample

(1) The pattern or 'profile' of positive and negative results is recorded, e.g.:

10^{-1} dilution   3/3 positive
10^{-2} dilution   1/3 positive
10^{-3} dilution   0/3 positive

which gives a profile of 3–1–0.

(2) The profile is converted to a Most Probable Number by referring to published MPN tables. In the example described above for a 3-tube MPN test, and assuming that the highest quantity inoculated per tube was 1 ml of a 10^{-1} dilution, i.e. 0.1 g of test sample per tube, then the profile, 3–1–0, corresponds to an MPN of 43 organisms / g (International Standard 7218, 1996). As stated above, if the general method is adapted by using different dilution series or different volumes of inoculum, then an appropriate multiplication factor needs to be calculated and applied to derive the final result.

10.3   Detection tests

Detection tests are designed to determine the presence or absence of a specific microorganism, often a particular pathogen, in a given quantity of test sample rather than to count the number of organisms present. They normally employ enrichment-plating techniques, i.e. the test sample is incubated in a broth, perhaps followed by sub-cultures into further broths, before finally being plated onto selective / diagnostic agar(s). For these tests, confirmation of suspect colonies that develop on the agar is essential.

Detection tests are most commonly used to detect pathogens such as *Salmonella* or *Listeria monocytogenes* in food and food-related samples. A typical conventional method for the detection of *Salmonella* is given in Figure 10.3 and the procedure for carrying out a *Salmonella* test is described in Section 10.3.2.

10.3.1   *Sensitivity of detection tests*

Tests for the detection of pathogens are usually much more sensitive than colony count tests, this is because enrichment procedures are used that allow very low numbers of the organisms to be detected.

Figure 10.3   Conventional method for the detection of *Salmonella* spp. in foods, adapted from Bell and Kyriakides, 2002.

Stage of the method	Incubation conditions
Pre-enrichment Inoculate pre-enrichment medium appropriate to the sample type (1 part test sample + 9 parts medium)	Day 0
↓ ↓	Incubate 37°C / 16–20 hours
Selective enrichment Sub-culture from the pre-enrichment broth to 2 selective enrichment broths at the appropriate ratio (1 + 9 or 1 + 100 depending on the selective medium used)	Day 1
↓ ↓	Incubate at the appropriate temperature for the appropriate time, e.g.: Selenite-Cystine (SC) medium, 37°C / 24 hours + 24 hours and Rappaport-Vassiliadis (RV) broth, 42°C / 24 hours + (if necessary) 24 hours
Selective plating Streak a loopful of culture from each of the selective enrichment media onto each of 2 selective agars, e.g. XLD agar, Brilliant Green agar (modified) or Hektoen Enteric agar	Days 2 and 3
↓ ↓	Incubate 37°C / 20–24 hours and a further 18–24 hours if necessary
Inspect plates for the presence of characteristic colonies and any primary biochemical reactions	Days 3 / 4 or 4 / 5
Confirmation of suspect colonies Purify suspect colonies by streaking on nutrient agar	
↓	Incubate 37°C / 18–24 hours
Serology using 'O' & 'H' antisera Inoculate media or test strips to obtain biochemical profile	Days 4–6
↓ ↓	Incubate according to the manufacturer's instructions, usually 37°C / 18–24 hours
Read reactions	Days 5–7

Plate 21   An example of diagnostic reactions on selective-diagnostic agar. The medium shown is xylose lysine desoxycholate agar (XLD agar) that is used for the isolation of *Salmonella* spp.; the diagnostic reactions develop after incubation at 37°C for 24 hours.

(a) *Salmonella* Typhimurium.

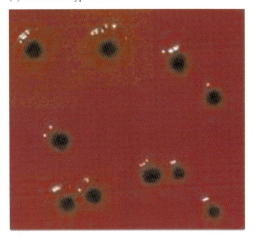

(Courtesy of bioMérieux© SA. Photographer Noël Bouchut.)

Note: After 24 hours incubation on XLD agar, most *Salmonella* serotypes produce transparent colonies with a black centre due to the production of hydrogen sulphide, whilst the surrounding medium changes from its original orange-red colour to a deep cherry-red, i.e. the pH indicator is showing that the medium has become alkaline.

Salmonella cannot produce acid from the lactose or sucrose in the medium and produce only a small quantity of acid from the xylose present, turning the medium immediately surrounding the colony yellow after about 18 hours incubation. Upon further incubation, *Salmonella* decarboxylates lysine, which produces highly alkaline end products that neutralise the small quantity of acid formed from xylose and, after incubation for 24 hours, the medium becomes red. The change from an initial, slight acid reaction to a highly alkaline reaction is known as an 'alkaline reversion'. The production of hydrogen sulphide, indicated by blackening of the colony, is used to differentiate *Salmonella* from organisms such as *Shigella* that also produce reddening of the medium.

(b) *Salmonella* Typhimurium, *Escherichia coli* and *Enterobacter cloacae*.

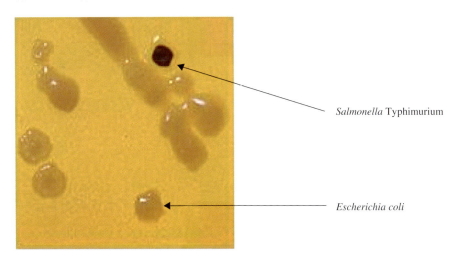

Salmonella Typhimurium

Escherichia coli

(Courtesy of bioMérieux© SA. Photographer Noël Bouchut.)

Note: Organisms, e.g. *E. coli* and *Enterobacter*, which produce acid from the lactose and/or sucrose present in XLD agar, develop a strong yellow colour in the medium. Often, the acid production is so strong that the yellow colour spreads throughout the agar. In mixed culture, therefore, *Salmonella* may appear to be growing on a yellow medium and may only be recognised on XLD agar by its ability to produce a black-centred colony due to the production of hydrogen sulphide. Colonies without a black centre and growing on a strongly yellow background medium are usually disregarded.

Plate 22 Laboratory autoclaves. (Courtesy of PriorClave Ltd.)

(a) A front-loading, 40-litre capacity, bench-top autoclave.

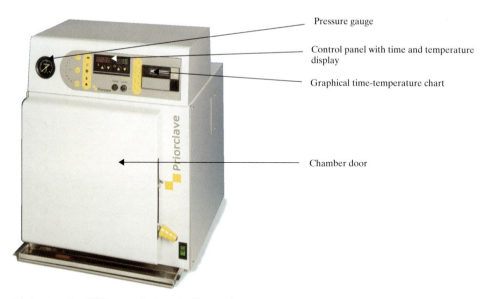

Pressure gauge

Control panel with time and temperature display

Graphical time-temperature chart

Chamber door

(b) A top-loading, 100-litre capacity, floor-standing autoclave.

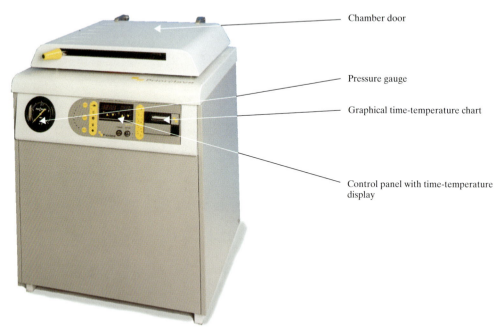

Chamber door

Pressure gauge

Graphical time-temperature chart

Control panel with time-temperature display

Key points:
- Autoclaves must be used in accordance with the manufacturer's instructions.
- Autoclaves must be of adequate size for the volumes of materials required to be autoclaved.
- Materials to be autoclaved must be loaded in such a way as to allow free circulation of steam to all parts of the load.
- All processes must be appropriate for the types of load, e.g. media, waste, equipment; different media may have different process requirements for sterilisation. Incompatible loads, e.g. media and waste, must not be processed together.
- All processes must be monitored by means of the integral equipment supplied, i.e. temperature chart recorder, and independent in-load monitors, e.g. liquid- or paper-based colour indicators, should also be used to assist verification of each process (see Table 6.3).

Plate 23 Illustration of the effect of correct blending of a food sample.

    (a) Before blending   (b) After blending.

Courtesy of Norpath Laboratories Ltd.

Note:
• The test sample is deposited cleanly into the bottom of the blender bag with no material adhering to the upper internal surfaces of the bag.
• The action of blending, e.g. in a 'Stomacher', squeezes and shears the test sample material to release microorganisms into the diluent.

Plate 24 Preparing a dilution series.

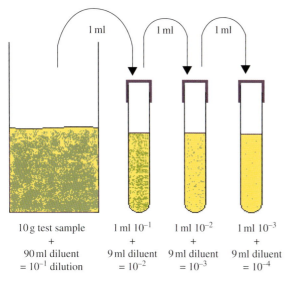

10 g test sample   1 ml 10^{-1} 1 ml 10^{-2} 1 ml 10^{-3}
+   +   +   +
90 ml diluent   9 ml diluent 9 ml diluent 9 ml diluent
= 10^{-1} dilution   = 10^{-2} = 10^{-3} = 10^{-4}

Key points:
• Microorganisms adhere to the internal and external surfaces of a pipette when it is used to transfer a suspension from one test sample diluent container to another. Therefore:
    (1) a new, sterile pipette / pipette tip must be used for each transfer as each successive dilution contains fewer organisms than the previous one, i.e. transfers are made **down** the dilution series (higher dilution = lower concentration of test material).
    (2) the pipette must not be immersed in the diluent when expelling the contents; this prevents organisms from being added from the outside surface of the pipette.
• After adding the inoculum to the next diluent volume, thoroughly mix the suspension, e.g. using a vortex mixer, especially before pipetting 1 ml to the next tube.

Plate 25 Colonies of coliforms growing in violet red bile lactose agar (VRBLA) prepared as a pour plate with overlay.

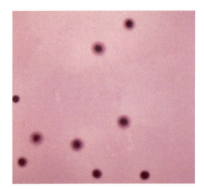

(a) Characteristic coliform colonies showing a deep red-purple colour, an entire edge, diameter usually greater than 0.5 mm and good separation of colonies within the agar.

(b) Confluent growth within the agar occurs occasionally e.g. when an inoculum is allowed to start drying out before agar is added, that makes interpretation of the colony morphologies and hence counting of the number of coliform colonies extremely difficult. Such plates should be disregarded and the test repeated; if this difficulty occurs persistently, the cause should be investigated.

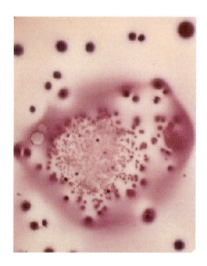

Plate 26 Example of a most probable number (MPN) test for coliforms using MacConkey broth incubated at 37°C for 48 hours.

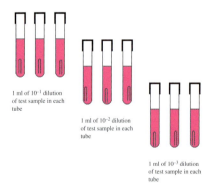

1 ml of 10^{-1} dilution of test sample in each tube

1 ml of 10^{-2} dilution of test sample in each tube

1 ml of 10^{-3} dilution of test sample in each tube

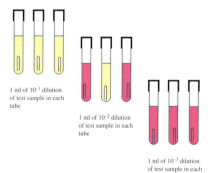

1 ml of 10^{-1} dilution of test sample in each tube

1 ml of 10^{-2} dilution of test sample in each tube

1 ml of 10^{-3} dilution of test sample in each tube

(a) Inoculated tubes before incubation.
Note the absence of gas in the Durham tubes and the even purple colour throughout all tubes.

(b) Tubes after incubation.
Note the presence of acid in the culture medium denoted by a change in colour from purple to yellow; evidence of gas production can be seen by the presence of a bubble in the Durham tube. A **presumptive** result is indicated by the presence of acid and gas in the same tube. Further tests may be required to produce a **confirmed** result. The MPN profile for the example shown is recorded as 3–1–0 and gives an MPN of 43 organisms/g (International Standard 7218, 1996).

Plate 27 Illustration of swab dilutions.

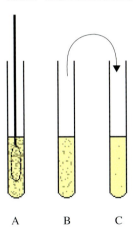

A   B   C

Notes:
A = The swab is placed in 10 ml neutralising broth or maximum recovery diluent and the organisms collected from the swabbed surface by the swab are dispersed into the broth by shaking manually or using a vortex mixer.

B = The swab is removed from the tube and any retained liquid squeezed out back into the tube by rolling the swab gently around the inside of the tube. This is the **primary suspension**. If 1 ml of the primary suspension is plated, then the count is calculated by multiplying the colony count obtained by 10 to obtain an estimate of the total number of colony-forming units in the total volume of the primary suspension, i.e. the count per swab.

C = 1 ml of the primary suspension from B is added to 9 ml of diluent, e.g. maximum recovery diluent, to produce a 10^{-1} suspension. If 1 ml of the 10^{-1} suspension is plated, then the count is calculated by multiplying the colony count obtained by 10 to obtain an estimate of the total number of colony-forming units per ml of the primary suspension. This number is multiplied again by 10 to provide the count per swab.

Although for calculation purposes, the multipliers ×10 and ×100 used to obtain the results for the first two dilutions i.e. the primary suspension and the 10^{-1} dilution respectively are the same as those used to obtain results for the first two dilutions of a food test sample i.e. 10^{-1} and 10^{-2} (see Plate 24), the result obtained for the swab relates to the *Microbial load on the whole swab* whereas for the food test sample, the result calculated is only *per g* of the test sample which is *not* the entire / whole test sample (often 10 g).

Plate 28 A binocular microscope (sometimes referred to as a plate microscope).

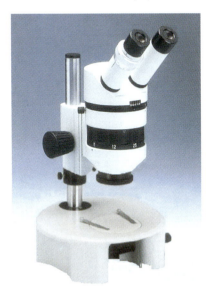

Courtesy of Pyser-SGI Ltd.

This is used for observing colonies under low magnification, e.g. ×60, to help distinguish them from particles of non-microbial origin.

Key points:
• The microscope is usually used with an external light source which illuminates the surface of the plate being viewed. It is essential that the stage and all optical parts are maintained in a clean condition at all times.

Plate 29   An example of colony reactions on a selective-diagnostic agar.

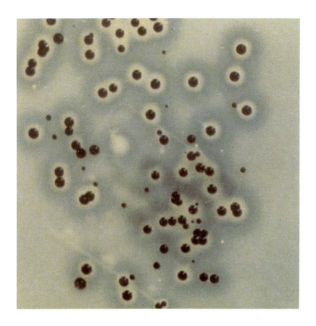

Colonies of *Staphylococcus aureus* on Baird-Parker agar, showing typical, characteristic reactions, that should be interpreted as presumptive *Staphylococcus aureus* (or coagulase-producing staphylococci) when assessing incubated plates of this medium.

Note the characteristics of colonies of typical strains of *Staphylococcus aureus* are as follows:
(1) black / grey-black, shiny and convex, 1.0 mm to 1.5 mm in diameter after incubation for 24 hours, and up to 3.0 mm after incubation for 48 hours;
(2) surrounded by a narrow white, entire margin, within a zone of clearing 2–5 mm wide.

Key points:
• Colonies of presumptive *Staphylococcus aureus* should be confirmed by the tube coagulase test (see Chapter 11).
• Results are usually recorded as confirmed coagulase-producing staphylococci.

Plate 30 Gravimetric diluter

This equipment provides a means for obtaining accurate standardised dilutions of solid and liquid samples in sterile bags prior to homogenisation. The equipment is microprocessor controlled and has a load cell and either one or two peristaltic pumps for dispensing liquids. Labels and reports can be generated to assist documentation of sample preparation.

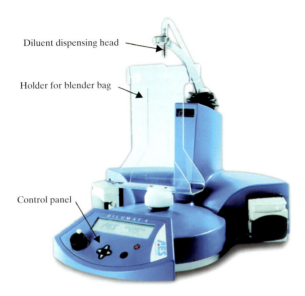

Diluent dispensing head

Holder for blender bag

Control panel

(The Dilumat MK4, courtesy of Don Whitley Scientific Ltd).

Method of operation:
(1) Programme the equipment to provide the dilution level required.
(2) Check that the correct diluent(s) / pre-enrichment media and appropriate sterile tubing are in place.
(3) Ensure the balance has been checked and tared (adjusted) to zero.
(4) Place an empty sterile blender bag in the bag holder.
(5) Re-tare the balance.
(6) Add an appropriate weight of sample cleanly to the bag.
(7) Initiate diluent / medium addition according to the manufacturer's instructions for the specific equipment.
(8) When the diluent / medium volume addition has been completed, raise the dispensing arm.
(9) Remove the sample bag.

Key points:
• Ensure the correct dilution ratio has been selected.
• Check there is sufficient diluent / medium in the reservoir.
• Ensure new sets of sterile tubing are in place between sample types or between different media used in accordance with the laboratory's Standard Operating Procedures.
• Ensure the diluent dispensing tip does not become contaminated.

Plate 31   Spiral plating technique.

(a) Spiral plater.

This equipment is used for examining food, pharmaceuticals, cosmetics and water for the presence and levels of bacteria. The heart of the spiral plater is a precision-built, microprocessor-controlled dispenser that distributes a liquid sample onto the surface of a rotating agar plate. Depending on the count range expected, the dispensing system can inoculate up to 400 μl of the liquid test sample either uniformly across the plate or as a continuously decreasing volume of sample via a tube on a dispensing arm which moves from near the centre of the plate towards the outside edge, depositing the sample in an Archimedes spiral. This results in a sample concentration range of up to 1000 : 1 on a single plate and ensures that the volume of sample on any particular segment of the plate is known and constant. After incubation, colonies appear along the spiral track made by the deposited sample with the spacing between colonies along the track increasing from the centre to the edge. Concentration is determined by counting only the well-spaced colonies and dividing this number by the volume of sample contained in the area that has been counted. Counting can be done manually by using a specially designed viewing grid. Using a 9 cm Petri dish, colonies on a standard 50 μl spiral inoculum will be countable if the sample contains between 400 and 4×10^5 colony forming units per ml. The lower limit of count is reduced by depositing larger sample volumes on a single plate. The upper count limit can be extended by serial dilution of the sample.

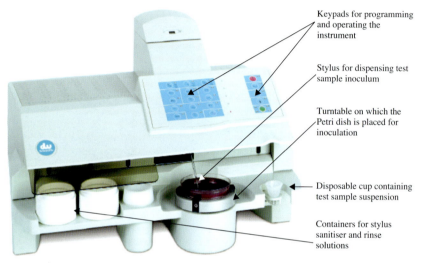

Keypads for programming and operating the instrument

Stylus for dispensing test sample inoculum

Turntable on which the Petri dish is placed for inoculation

Disposable cup containing test sample suspension

Containers for stylus sanitiser and rinse solutions

(Model WASP 2, courtesy of Don Whitley Scientific Ltd.).

Method of operation:
(1) Disinfect the spiral plater stylus by drawing up sodium hypochlorite solution.
(2) Rinse the stylus with sterile deionised water.
In the WASP 2 instrument, steps 1 and 2 are carried out in an automated cleaning cycle.
(3) Locate the disposable cup containing the test sample dilution in position so that the stylus can take up the required volume.
(4) Locate the agar plate on the spiral plater turntable.
(5) Start automatic sample deposition and wait until the cycle of operation is complete.
(6) Allow the spiral track to dry before inverting the agar plate.
(7) Incubate the plate using an appropriate temperature / atmosphere / time combination.
(8) Enumerate colonies in appropriate spiral plate sectors.

Key points:
• Always use appropriately dried, level and entirely smooth surface agar plates.
• Check for correct sample deposition by daily spiral plating of water-soluble ink.
• Rinse and disinfect the stylus between samples.
• Allow spiral plates to dry before incubation.

(b) Examples of plates that have been inoculated with test sample dilutions of 10^{-1}, 10^{-2} and 10^{-3} using a spiral plater.

Plate 32 Examples of common biochemical reactions used in confirmation tests for members of the Enterobacteriaceae.

(a) The indole test.

Positive   Negative

Note:
The indole test indicates whether or not a microorganism can produce indole from tryptophan; this is an important test used in the identification of *Escherichia coli*. The organism is grown in tryptone water (tryptone is a peptone that contains a high concentration of tryptophan) and, after incubation, 0.5 ml of Kovács' reagent (Kovács' reagent is a solution of *p*-dimethylaminobenzaldehyde in amyl alcohol with added concentrated hydrochloric acid) is added and the mixture is shaken. The tubes are allowed to stand and the results observed after 1 minute. The reagent settles into a layer on top of the aqueous medium and its colour changes from yellow to deep red if indole is present; this indicates a positive reaction. If no colour change is observed, the test reaction is recorded as negative.

(b) The Voges-Proskauer (VP) test.

Positive   Negative

Note:
The VP test indicates an organism's ability to produce acetylmethylcarbinol (acetoin). After the organism has been incubated in a simple glucose-phosphate buffer medium, acetoin production is demonstrated by adding 'VP reagents' (α-naphthol followed by a potassium hydroxide / creatine solution). The culture is shaken vigorously and allowed to stand; in the presence of air and potassium hydroxide, acetoin is spontaneously oxidised to diacetyl that reacts with the guanidine group of creatine to form a red complex. Thus, the test is positive if a strong red colour develops in the medium; the time required for colour development can vary from 30 seconds to 1 hour depending on the specific method used. The VP test is important for distinguishing between members of the Enterobacteriaceae.

Plate 33 An example of the application of the API® 20E test strip for the identification of *Escherichia coli*.

(a) API® 20E test strip after inoculation with a pure culture of *E. coli* and incubation at 37°C for 24 hours.

Courtesy of bioMérieux® SA: photographer Noël Bouchut.

(b) Interpretation of the results from the incubated API® 20E test strip (the reactions to the tests and the results are 'scored' from left to right of the test strip).

API® 20E reaction codes	Test reactions	Result*	API® score	API® profile
ONPG	β-galactosidase	+	1	
ADH	Arginine dihydrolase	−	0	5
LDC	Lysine decarboxylase	+	4	
ODC	Ornithine decarboxylase	+	1	
CIT	Citrate utilisation	−	0	1
H₂S	Hydrogen sulphide	−	0	
URE	Urease	−	0	
TDA	Tryptophan deaminase	−	0	4
IND	Indole production	+	4	
VP	Voger-Proskauer (acetoin production)	−	0	
GEL	Gelatin liquefaction	−	0	4
GLU	Glucose utilisation	+	4	
MAN	Mannitol utilisation	+	1	
INO	Inositol utilisation	−	0	5
SOR	Sorbitol utilisation	+	4	
RHA	Rhamnose utilisation	+	1	
SAC	Sucrose utilisation	−	0	5
MEL	Melibiose utilisation	+	4	
AMY	Amygdalin utilisation	−	0	
ARA	Arabinose utilisation	+	2	2
OX	Oxidase reaction	−	0	

*+ = positive
− = negative

The API® 20E profile reads 5144552 which identifies the culture as *Escherichia coli*.

Note:
• The oxidase test is not included on the API® 20E test strip and must be carried out as a separate test.
• A test for the production of nitrite from nitrate can also be carried out if necessary using reagents added to the Glucose cupule.

Plate 34   An example of the application of the API® 20E test strip for the identification of *Salmonella*.

(a) API® 20E test strip after inoculation with a pure culture of *Salmonella* and incubation at 37°C for 24 hours.

Courtesy of Westward Laboratories Ltd.

(b) The interpretation of the results from the incubated API® 20E test strip (the reactions to the test and the results are 'scored' from left to right of the test strip).

API® 20E reaction codes	Test reactions	Result*	API® score	API® profile
ONPG	β-galactosidase	–	0	6
ADH	Arginine dihydrolase	+	2	
LDC	Lysine decarboxylase	+	4	
ODC	Ornithine decarboxylase	+	1	7
CIT	Citrate utilisation	+	2	
H₂S	Hydrogen sulphide	+	4	
URE	Urease	–	0	0
TDA	Tryptophan deaminase	–	0	
IND	Indole production	–	0	
VP	Voger-Proskauer (acetoin production)	–	0	4
GEL	Gelatin liquefaction	–	0	
GLU	Glucose utilisation	+	4	
MAN	Mannitol utilisation	+	1	7
INO	Inositol utilisation	+	2	
SOR	Sorbitol utilisation	+	4	
RHA	Rhamnose utilisation	+	1	5
SAC	Sucrose utilisation	–	0	
MEL	Melibiose utilisation	+	4	
AMY	Amygdalin utilisation	+	1	3
ARA	Arabinose utilisation	+	2	
OX	Oxidase reaction	–	0	

*+ = positive
 – = negative

The API® 20E profile reads 6704753 which identifies the culture as *Salmonella* sp.

Notes:
- The oxidase test is not included on the API® 20E test strip and must be carried out as a separate test.
- A test for the production of nitrite from nitrate can also be carried out if necessary using reagents added to the Glucose cupule.

Plate 35 An automated, benchtop luminometer. This is used to measure the light produced in an ATP assay.

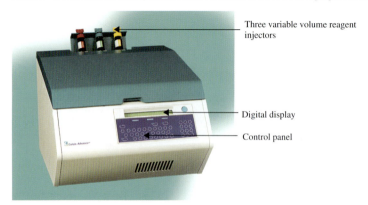

Three variable volume reagent injectors

Digital display

Control panel

(Celsis Advance™ courtesy of Celsis Ltd.)

Note:
The Celsis Advance™ luminometer has a capacity of 150 samples and a read time of between 0.1 to 99 seconds.

Plate 36 A hygiene monitoring swab ('*snap*shot'/ Ultrasnap™) that uses the ATP-bioluminescence reaction.

Liquid reagents contained in sealed capsule

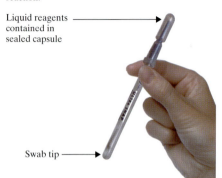

Note:
In this unit, the reagents are sealed into a capsule that forms part of the swab unit; the reagents are released after swabbing by snapping the device and breaking the seal. This allows the reagents to react with any ATP on the swab and the light produced is measured in a luminometer.

Swab tip

Courtesy of Hygiena International Ltd.

Plate 37 A hand-held luminometer ('SystemSUREII™').

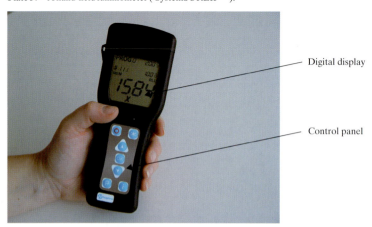

Digital display

Control panel

Courtesy of Hygiena International Ltd and Celsis Ltd.

Notes:
• This unit is readily portable and therefore may be used in food production areas, e.g. for reading the results of the '*snap*shot' hygiene monitoring swab (see Plate 36).
• The light output from the swab is displayed numerically as Relative Light Units (RLUs).
• Hygiene monitoring tests require initial calibration according to the nature of the product contact surface and hygiene standards required so that 'Action' and 'Alert' values can be set.

Plate 38 Diagrammatic representation of a 'sandwich' enzyme linked immunosorbent assay (ELISA), adapted from Jones (2000).

Step 1
The test well is coated with an antibody to *Salmonella*

Step 5
Detection antibody, with enzyme, is added to the well

Step 2
The test sample, containing *Salmonella*, is added to the well

Step 6
Detection antibody binds to the antigen, to make a 'sandwich'

Step 3
Antibody captures *Salmonella* antigen

Step 7
Unbound detection antibody is washed away

Step 4
The non-binding components are washed away

Step 8
The enzyme substrate is added to the well and is acted upon by the enzyme to produce a coloured end-product

KEY

Y   Antibody

◇   Antigen, e.g. *Salmonella*

Ӻ   Detection antibody, with enzyme

Plate 39   An enzyme-linked immunosorbent assay (ELISA) for *Salmonella*.

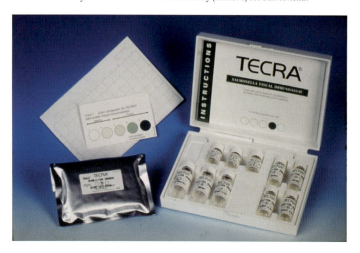

Courtesy of Tecra Diagnostics UK.

Key points:
• Reagents should be reconstituted in clean containers using distilled or deionised water only.
• Reconstituted reagents should be labelled with their expiry date and should not be used after that date.
• Unused reagents should be stored at 2–8°C.
• Reconstituted reagents should be allowed to reach room temperature before use.
• Reagents from different batches should not be mixed / used in the same test.
• The correct incubation times and temperatures should be used.
• Temperature differences between different wells in a plate should be minimised.
• During incubation, plates should be covered to prevent evaporation.

Plate 40 Pipetting test samples and controls into the wells of a micro-titre place for an enzyme-linked immunosorbent assay (ELISA).

Courtesy of Tecra Diagnostics UK.

Key points:
• The precision and accuracy of the pipetter should be checked before use.
• A fresh tip must be used for each sample.
• The tip should not touch the plate / top of wells.
• Separate pipetters should be used for dispensing the conjugate and substrate.
• The reagents must not be allowed to come into contact with metal.

Plate 41 An automated micro-titre plate washer for an enzyme-linked immunosorbent assay (ELISA) micro-titre plate.

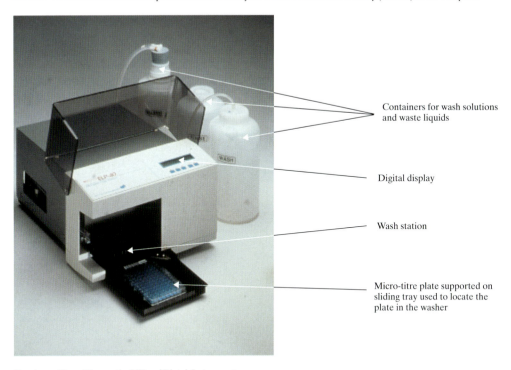

Containers for wash solutions and waste liquids

Digital display

Wash station

Micro-titre plate supported on sliding tray used to locate the plate in the washer

Courtesy of Tecra Diagnostics UK and Biotek Instruments.

Key points:
• The wash dispenser should be checked before use to ensure that it is functioning correctly.
• All equipment should be regularly serviced and periodically checked to ensure that it is operating correctly.
• Equipment should be cleaned regularly, especially after a spillage.
• Basic maintenance should be carried out according to the manufacturer's instructions.

Plate 42 Positive (coloured) and negative (clear) test reactions in an enzyme-linked immunosorbent assay (ELISA) micro-titre plate.

Courtesy of Tecra Diagnostics UK.

Key point:
• When reading plates visually, the colour intensity of the wells should always be compared with the correct colour comparison card.

Plate 43   A micro-titre plate reader for an enzyme-linked immunosorbent assay (ELISA)

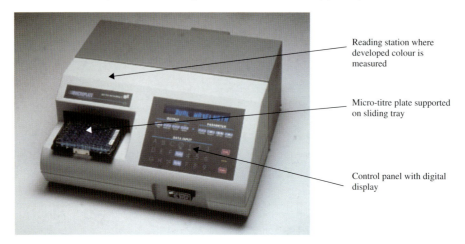

Reading station where developed colour is measured

Micro-titre plate supported on sliding tray

Control panel with digital display

Courtesy of Tecra Diagnostics UK and Biotek Instruments.

Key points:
• The correct wavelength filter should always be used.
• Sufficient time should be allowed for the instrument to warm-up before use.
• The kit manufacturer's 'blanking procedure' should always be followed.
• Plate reader results can be affected by drops of liquid under the wells, scratches or fingerprints under the wells and dust particles: these should always be avoided.

Plate 44   A 'dipstick' format ('Path-Stik') for an immuno-chromatographic assay.

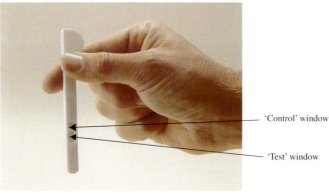

'Control' window

'Test' window

Courtesy of Celsis Ltd.

Notes:
• 'Path-Stik' tests are available for examining test samples for the presence of *Salmonella* and *E. coli* O157.
• The test sample is first enriched in conventional test media and the 'Path-Stik' is then dipped into the culture for a few seconds.
• A few minutes later, the test result is read from the reaction widows: the upper window is a 'control' that produces a visible line for all tests; a line appears in the lower (test) window when the target organism has been detected, i.e. 'positive'.

A 25 g test sample is usually examined but for foods of high concern, such as baby foods, the application of statistical sampling plans may increase this to as many as 60 × 25 g, i.e. 1.5 kg. In this case, the sample would probably be divided into, say, 5 × 300 g test samples, so that 5 separate tests would actually be undertaken, although a single result could be derived, i.e. 'detected' or 'not detected' in 1.5 kg.

10.3.2 *How to do a detection test for* Salmonella

Resuscitation and selective enrichment

(1) A 'top-pan' balance with an accuracy of ±0.1 g is checked and tared (adjusted) to zero. A clean container is placed on the balance and a sterile plastic bag is placed in the container (for support) and then the mouth of the bag is opened aseptically. The balance is re-tared to zero.

(2) As quickly as possible (but maintaining aseptic techniques), a portion of the factory sample (either solid or liquid) is weighed out cleanly into the sterile plastic bag to provide the test sample; usually a test sample weighing 25 g or a composite test sample of several 25 g quantities is examined. An appropriate resuscitation (pre-enrichment) broth is added at 9 times the weight of the test sample, i.e. 225 g (for 25 g test sample) or the appropriate multiple. The most usual pre-enrichment broth used for *Salmonella* is buffered peptone water but more specifically-designed broths are used for certain commodities, e.g. milk powders, chocolate and products containing garlic or onion (Table 10.1).

Table 10.1 Alternative pre-enrichment systems used for some food types, Bell and Kyriakides, 2002.

Food type	Pre-enrichment	Effect
Foods with a high fat content, e.g. cheese	Buffered peptone water with 0.22% w/v Tergitol 7	Aids fat dispersion
Highly acidic or alkaline products	Adjust pH value of pre-enrichment broth to 6.6–7.0 before incubation	Neutralises acid or alkali
Chocolate and confectionery products	Reconstituted skim milk powder (10% w/v) with brilliant green dye (final concentration of 0.002% w/v)	Reduces inhibition of *Salmonella*
Garlic and onion	Buffered peptone water with potassium sulphite (0.5% final concentration)	Reduces inhibition of *Salmonella*
Products that may contain inhibitory substances or products that may be osmotically active, e.g. some herbs and spices (oregano, cinnamon, cloves), honey	1 in 100 dilution of test sample in buffered peptone water, e.g. 25 g + 2475 ml	Reduces inhibition of *Salmonella*

Note: Gravimetric diluters (see '10.6.1 Automation of repetitive procedures') may be used to aid primary sample dilution, in which case, the manufacturer's instructions must be followed.

(3)   For solid samples, the weighed test sample is physically dispersed in the broth to dislodge and distribute the organisms that are present homogeneously throughout the suspension, usually using a dedicated laboratory blender (see Plate 9). Blending should continue for at least 30 seconds, or more if the test sample is of a type that is difficult to disperse. A laboratory standard operating procedure should specify the blending time to be used for different sample types. For liquids, such physical dispersion is normally unnecessary and gentle hand-shaking is often sufficient to disperse them in the diluent. After blending, the bag is sealed.

(4)   The test sample / pre-enrichment broth mixture is incubated, usually for 16–20 hours, at an appropriate temperature, usually 37°C.

(5)   Following incubation, the pre-enrichment broth mixture is sub-cultured into each of two selective enrichment broths of different formulations, e.g. selenite-cystine broth (SC) and Rappaport-Vassiliadis (RV) medium, by transferring aliquots of the pre-enrichment mixture using a sterile pipette. Normally, a quantity of either 1 ml or 10 ml is transferred e.g. to 100 ml RV and 90 ml SC respectively but it is important that the correct volume of broth is sub-cultured to ensure that the selective enrichment broth is not over-diluted by the inoculum; this is particularly important for the RV medium. Information concerning the use of specific selective enrichment media is readily available from media manufacturers.

(6)   The selective enrichment broths are incubated, usually for 24 hours and some-times, an additional 24 hours depending on the media and method employed, at a temperature appropriate for the medium used; usually this is 37°C for selenite-cystine broth and 42°C for RV medium.

Selective, diagnostic plating

(1)   Following incubation of the selective enrichment broths, each broth is sub-cultured after 24 and 48 hours (where an additional 24 hours is required), by streaking onto a plate each of at least two selective / diagnostic media of different formulations, so as to produce isolated colonies (see Figure 7.1(a) and Chapter 9). The plates are incubated, usually at 37°C for 24 hours and sometimes for an additional 24 hours depending on the media used. Modified brilliant green agar (BGA) and xylose lysine desoxycholate medium (XLD) are two widely used selective plating media in tests for the detection / isolation of *Salmonella*.

(2)   After incubation, the plates are inspected for colonies characteristic of the target organism which may include an assessment of colony morphology and bio-chemical reactions (see Plate 21); their presence indicates a **presumptive** positive

result. All presumptive results require further confirmatory tests to be made before the result can be **confirmed**. Depending on the commodity being examined, 10% or more of test samples may be expected to yield presumptive positive results that require confirmation.

Confirmation and reporting

(1) Colonies designated as 'suspect', and therefore presumptively positive for *Salmonella*, are further sub-cultured onto other plating media by streaking to obtain a pure culture of isolated colonies; the medium used may be non-selective or selective and the particular method protocol specified in the laboratory's documented procedures must be followed. The pure culture is then subjected to a series of biochemical, and possibly immunological, tests to determine its identity. These are discussed in detail in Chapter 11.

(2) It is important that negative results are not reported to the customer as 'absent' because the target organism may have been present but not detected by the procedures employed; negative results should therefore always be expressed as 'not detected'. Conversely, positive results are usually expressed as 'detected' rather than 'present' in the test sample quantity taken.

10.4 Environmental monitoring

The work of most food microbiology laboratories includes the examination of factory samples other than food materials taken from a production line and these samples are mainly associated with hygiene monitoring of the production environment. Examination of manufacturing equipment cleanliness, and assessment of the levels of airborne microorganisms and organisms in / on many non-food contact environmental surfaces, e.g. floors, drains, overheads, require specialised sampling techniques to be used. There are few standard methods for environmental monitoring, although the general techniques are well-established. Therefore, only general descriptions are given in the following sections and a degree of initiative may be needed to adapt them to specific circumstances; further information is given in Chapter 5 ('5.5 Hygiene monitoring and microbiology').

10.4.1 *Swabs*

Swabs and sponges are used to recover microorganisms from surfaces, whether they are food contact surfaces such as on production equipment or non-food contact surfaces in and around the production building itself. Generally, swabs and sponges are moistened lightly to facilitate the swabbing process (unless the surface being swabbed is already wet / moist). After swabbing, swabs are replaced in their containers which are labelled and returned to the laboratory as quickly as possible for immediate examination. This minimises the possibility of changes occurring in the microbial populations collected due to growth on the moist swab or death if the swab dries out. Sometimes a swab may be placed in a 'transport medium' immediately after use at the point of swabbing, to help keep the organisms viable whilst in transit to the laboratory. Neutralising broth media should be used to moisten the swab for use and may also be used for making the primary dilution of the swab material in the laboratory; this helps to maintain the viability of microorganisms on the swab by neutralising any residual sanitiser / disinfectant on the surfaces that were swabbed (see Table 9.4).

If several tests are required to be carried out on swabs of the same item / area, especially if a mixture of colony counts and detection tests is included, it is preferable that separate swabs are taken for each test as it is difficult (although not impossible) to test a single swab for two or more target organisms.

In the laboratory, the organisms on the swab are first dispersed in a suitable liquid medium, either neutralising broth or more usually maximum recovery diluent (MRD) for colony counts or a pre- or selective-enrichment broth for detection tests. This initial suspension is then processed by the normal method used for the target organism.

For detection tests, the results are either 'detected' or 'not detected' per swab. However, the calculation of colony counts must take account of the effect of preparation (dilution) of the swab in the laboratory prior to media inoculation. Often, the count obtained per incubated plate is reported to the customer, but because there is no industry standard for the volume of diluent that a laboratory should use to disperse a swab, this value may not be very helpful as the customer cannot relate the result to the swab sample taken. This is especially important if a 'less than' count was obtained. For example, if a small cotton (medical) swab was dispersed in 10 ml diluent, 1 ml was plated and no colonies were observed on the agar medium, then a count of < 1 cfu / plate recorded actually means < 10 cfu / swab; this difference in reporting might make a difference to the customer who is responsible for production hygiene (it is incorrect to record a count of zero which is why < 1 cfu / plate is used). However, the latter method of recording is the most appropriate and technically correct and the number of cfu per swab should be calculated by taking into account the volume of diluent in which the swab was dispersed. Additionally, from this approach, the number of cfu per area swabbed can be calculated if this is known.

Different laboratories may use different terminology to describe the suspension of microorganisms produced when a swab is dispersed in diluent; the primary suspension is sometimes described as the 10^{-1} dilution, but this is incorrect technically, because the weight of material collected by the swab is negligible and certainly very, very much less than 10% of the weight of diluent. An alternative approach is to describe the initial suspension as the 'primary suspension'; any dilutions prepared from this would then be designated 10^{-1}, 10^{-2}, etc. (see Plate 27).

10.4.2   *Contact plates, dip slides and exposure (settle) plates*

Contact plates (see Plate 13b) are used to estimate the numbers of microorganisms on surfaces by placing agar in direct contact with the surface. Similarly dip slides are 'paddles' coated with agar that can be dipped into a liquid so that a small amount of liquid is absorbed into / onto the agar; some can be pressed onto surfaces in a similar way to contact plates. The media used in such devices can be general-purpose or selective agars. Inoculated contact plates and dip slides are incubated under conditions appropriate for the medium used / organisms sought and the resulting colonies counted; however, no 'sample' dilution is possible, so high levels of micro-organisms may simply produce confluent growth that cannot be estimated quantitatively. Equally, the number of organisms collected by a contact plate or dip slide is likely to be variable because of the conditions of the test, e.g. the pressure that is applied, or where access to the surface is impaired so only a part of the plate / slide is used, making it difficult to compare results between different sample sites and sampling occasions.

Exposure plates, or settle plates, are Petri dishes containing agar that may be a general-purpose medium or a selective diagnostic agar. They are taken into pre-selected production areas and the lids removed for a specified time, often 30 minutes or 1 hour, during which time organisms in the atmosphere settle onto the agar. After exposure, the lids are replaced, the plates incubated under appropriate conditions for an appropriate time and the number of colonies counted. This is a particularly useful method for assessing the level of mould contamination in the air of food production areas but the exposure time must be adapted to suit the concentration of organisms usually encountered in a particular environment. Results are usually used for 'trend analysis' purposes. All plates used for such procedures must be clearly labelled with the medium's unique identifier, exposure location and the date and time of test.

Note: Swabs, contact plates and exposure plates are also used to monitor the hygienic status of laboratory environments and form an important part of laboratory quality assurance programmes.

10.5 Recognition of microbial growth after incubation

For all microbiological tests, it is important that laboratory staff can accurately recognise the microorganisms and / or their reactions that have developed during incubation. A false negative result, i.e. a failure to detect a target microorganism when it is present, is always unacceptable as this could result in a food product being a danger to public health. False positive reactions can occur when a test system is not selective enough and allows the growth of non-target microorganisms that are indistinguishable from the target organism without carrying out much further work; false positives are preferable to false negatives but still undesirable because they incur additional costs and sometimes considerable delays in confirming the presumptive result. To minimise the occurrence of false results, it is important that laboratory staff are adequately trained and their competence demonstrated before being allowed to interpret test results, especially as in microbiology, more so perhaps than in other sciences, day-to-day experience with organisms is necessary for the microbiologist to develop a practical 'sense' for the organism sought.

10.5.1 *Assessment of microbial growth and reactions*

When test plates and broths are removed from the incubator, a range of growth indicators and reactions may need to be assessed. For colony counts on general-purpose media, inspection of colonies must take into account the possible presence of particles in or on an agar that are not of microbiological origin so that only microbial colonies are counted. Diagnostic broths generally became turbid and exhibit colour changes during incubation if a target organism has grown, or gas may be produced. Microbial growth on diagnostic agars may exhibit up to 4 or 5 different reactions, all associated with a single colony; thus it is essential that food microbiologists who assess plates and broths after incubation are able to recognise the different reactions that the media used in their laboratory can display both for target and non-target organisms. Test reactions are also discussed in Chapter 11 ('Confirmation tests').

10.5.2 Growth on non-selective media

Usually, colonies that have grown on non-selective agars require no more than to be counted accurately, since tests such as total colony counts are generally intended to establish only the number, and not the identity, of organisms in a sample. Occasionally, however, some simple identification procedures can provide useful information, for example, on the possible source of a contaminant or the reason for its presence at higher numbers than expected. So, a microbiology technician needs to understand what *can* grow on a medium and what *has* grown on the medium.

In the case of non-selective agars, the colony morphology (size, shape, structure, colour) provides little indication of an organism's identity, perhaps the only relatively easily recognisable colony morphologies being those of some (but not all) *Bacillus* species that produce colonies variously described as dry, wrinkled, irregular, rough, spreading or rhizoid and often with a dull surface. Similarly, in broth culture, some *Bacillus* species form a pellicle (skin) due to aerobic growth on the liquid surface. These descriptions, however, must always be regarded as no more than guidelines that provide only a suggestion as to an organism's identity. Most microorganisms, however, produce nondescript colonies and, indeed, it is often impossible to distinguish between bacteria and yeasts by visual inspection of a colony. Although most moulds produce spreading, filamentous or 'furry' colonies that are easily distinguished from bacteria or yeasts, some do produce yeast-like forms when growing on agar; indeed, some yeasts produce rhizoid colonies that have an appearance similar to moulds.

Some particulate food components can produce physical or chemical reactions in microbiological media that are sometimes very difficult to distinguish from microbial colonies. In the case of colony counts, where serial dilutions have been plated and the higher dilutions are counted, this is unlikely to pose a problem because the food particles are diluted out. However, the samples encountered in a laboratory often require only low dilutions to be plated and, under these circumstances, food particles can cause considerable interference. Perhaps the best-known example is hard boiled egg, in which the solidified yolk is dispersed into tiny particles during blending and these can be very difficult to distinguish from bacterial colonies; thus, it is possible for a sample to be reported as having a TCC in excess of 10^5 cfu/g when the correct result should be < 10 cfu/g. In this example, the particles of egg yolk can very easily be distinguished from microbial colonies by means of a hand lens or a low power binocular microscope (see Chapter 11 and Plate 28).

Colonial morphology alone, therefore, provides only limited information, which should still be recorded, and much more useful information can be obtained by the use of a few simple, basic tests, such as microscopy, to observe the microbial cells directly. Microscopy can reveal the cell morphology of an organism, the Gram reaction (after staining), presence of spores and motility all of which give the food microbiologist much useful information about the type of organism grown. (For a more detailed description of the use of a light microscope, see Chapter 11.) However, because most commercial food microbiology laboratories rely heavily on modern selective media to grow and provide presumptive information about an organism's identity, microscopic examination of colonies or indeed broth cultures has become less commonplace in recent decades. Nevertheless, the use of microscopy is still strongly recommended as it remains one of the most simple and effective means of providing direct evidence of the identity of microbial cultures and is, perhaps, the most fundamental of confirmation tests.

10.5.3   *Reactions on selective-diagnostic media*

Tests using non-selective media can be made more selective by adjusting the incubation conditions, e.g. by incubating in an anaerobic atmosphere or by incubating at a higher or lower temperature to encourage the development of less diverse groups of microorganisms such as psychrotrophic or thermophilic aerobes. The use of selective, and usually diagnostic, media, however, generally enables more effective isolation of a particular genus, species or group of closely-related organisms and, for these test media, a sound knowledge of the diagnostic features and reactions associated with specific media used is important.

Selectivity of isolation media

Many modern media are reasonably selective and therefore limit the range of organisms that can grow on them to only a few genera, and selectivity may be enhanced when a selective agar is used to isolate target organisms from a selective broth because competing organisms are suppressed both in the broth and on the agar. In tests to detect *Salmonella*, for example, xylose lysine desoxycholate (XLD) medium is widely used as one of several selective and diagnostic media for isolating suspect organisms (presumptive positives) from the selective enrichment broths. This is a highly selective medium and colonies capable of growth on XLD are likely to be Gram-negative bacteria as Gram-positive bacteria and yeasts and moulds are effectively suppressed.

Nonetheless, because of the diverse capabilities of microbial populations, different microorganisms capable of growth on the selective medium used may be encountered from time to time and it should not be regarded as inevitable for example that XLD medium yields only genera of the family Enterobacteriaceae. In this context, some knowledge of the type of sample under examination can be helpful for understanding the types of microflora that may be present and that could confuse the interpretation of growth on a selective medium. A heat-processed, dried product such as milk powder, for example, is unlikely to be contaminated with pseudomonads, whilst a high-moisture, raw food such as fresh meat is much more likely to contain these organisms, and because some *Pseudomonas* spp. can produce red colonies on XLD they may contribute to false suspect positive isolations from these latter product types.

Some isolation media, however, are not very selective and can permit the growth of a significant range of non-target organisms, two examples being rose bengal chloramphenicol agar (RBCA) and Baird-Parker agar (BPA). The growth of bacteria on

RBCA is inhibited by chloramphenicol, which is a wide-spectrum antibacterial agent, and 'non-furry' colonies might therefore be considered to be yeasts; however, chloramphenicol-resistant bacteria are encountered occasionally, especially in some animal-derived products, and their colonies can be misinterpreted unless confirmatory tests are undertaken e.g. microscopy. Similarly, in addition to *Staphylococcus aureus*, BPA can support the growth of a range of Gram-positive and Gram-negative bacteria, e.g. other staphylococci, micrococci, *Bacillus*, *Escherichia* and *Proteus* as well as yeasts, and confirmation of suspect colonies that develop on this medium is essential.

Diagnostic reactions

Most selective media, both liquid and solid, also incorporate substrates, the utilisation of which facilitates the recognition of target organisms, and helps to distinguish target from non-target organisms; these media are also known as diagnostic media. In their simplest form, diagnostic reactions consist of a colour change brought about by the growth of organisms. MacConkey broth, for example, contains bile salts; when incubated at 37°C, growth of microorganisms is restricted largely to members of the Enterobacteriaceae. It also contains lactose which the coliform sub-group is able to ferment into lactic acid and the resulting change in pH value is detected by means of the pH indicator bromocresol purple, which changes colour from purple to yellow in acid conditions. Thus, a colour change in the medium from purple to yellow provides indirect evidence of the growth of coliforms (see Plate 26).

Selective media can incorporate the substrates for two, three or even four or more different diagnostic reactions, e.g. Baird-Parker agar (BPA) for *Staphylococcus aureus* (see Plate 29). In theory, this approach enhances the specificity of a selective medium but it can also create difficulties when choosing colonies for further confirmation. Colonies of *Bacillus*, *Escherichia*, *Proteus* and yeasts that grow on BPA can be distinguished easily from staphylococci on the basis of colony size, shape, colour and the presence or absence of reactions on the agar itself, but it is not possible to eliminate all false positives on this basis alone because micrococci and *Staphylococcus* spp., other than *Staph. aureus*, may be indistinguishable from the target organism. The problem is compounded because different strains of *Staphylococcus aureus* have diverse colonial characteristics; not all strains produce the full range of characteristic reactions on BPA and therefore colonies do not conform to the 'textbook' description. Thus, colonies showing either 'typical' or 'atypical' reactions may need to be confirmed in order to avoid false negative results. The result of all these complications is that the confirmation of *Staph. aureus* can be a particularly onerous task.

A further complication is that many microbiological standards refer to '*Staphylococcus aureus*', whereas most food microbiological methods describe confirmed positive colonies as 'coagulase-positive staphylococci' because the result of the coagulase reaction provides the most reliable indication of the organism's potential pathogenicity. However, most, but not all, strains of *Staph. aureus* produce coagulase whilst a few *Staphylococcus* species, other than *Staph. aureus*, also produce coagulase; thus, even an internationally used test may not always be able to provide the complete information required that allows the microbiological status of food products to be assessed against a legal standard. (This is discussed further in Chapter 11.)

10.6   Automation and proprietary tests

Practical food microbiology is a labour-intensive activity and a number of attempts have been made to automate some of the repetitive procedures in conventional test methods or, to make the tests more convenient to use through the development of proprietary test kits.

10.6.1   *Automation of repetitive procedures*

The gravimetric diluter is a machine that automatically dispenses the correct volume of diluent according to the amount of test sample weighed out. It consists of a balance and a peristaltic pump connected to a reservoir of sterile diluent. A microprocessor records the test sample weight and automatically operates the pump to dispense the correct weight of diluent (Plate 30).

The spiral plater (Plate 31) is an automated device for preparing spread plates without the need to prepare and plate a large number of test sample dilutions. It consists of a rotating turntable that holds a Petri dish containing agar. A stylus is connected to a syringe filled with the test sample suspension (normally a 10^{-1} dilution). The stylus is placed in contact with the surface of the medium and the turntable rotated. As the stylus moves from the centre of the plate to the edge, the syringe delivers a decreasing amount of the suspension, so that a spiral of inoculum of varying concentration is formed. The plate is then incubated under appropriate conditions and colonies are counted using a colony counter fitted with a template that is divided into sectors, each of which defines a different volume of inoculum. Sectors containing 'countable colonies' are selected, the colonies are counted and the count per gram of test sample calculated according to the equipment manufacturer's instructions.

A number of instruments has been developed for counting colonies on / in agar automatically. The first automatic colony counters to be produced counted the number of colonies on an agar plate by scanning the Petri dish using laser light. More recent instruments employ visible light that is detected by a camera linked to image analysis software. Automatic colony counters work well when many tests are carried out on similar samples, especially if the sample is liquid and the colonies are grown on a general-purpose agar, e.g. to estimate the TCC of raw milk samples as used in farmer quality payment scheme laboratories worldwide. However, automatic colony counters may have difficulty in discriminating between particulate matter from the test sample and microbial colonies or in providing an accurate count from pour plates in which large numbers of small colonies overlay others or in which large surface colonies are also present. They also may not work well with coloured, diagnostic media. It is essential to ensure that the preliminary setting up procedure for these instruments is thorough for each type of medium and sample type and that regular check monitoring of performance is carried out and results documented.

10.6.2   *Proprietary tests*

A growing number of proprietary tests is available that employ conventional microbiological test principles but are often more convenient to prepare and use than traditional broths and agars. There are now hundreds of commercially available proprietary tests based on a variety of operating principles including those of conventional microbiology and it is not the purpose of this introductory text to practical food microbiology to detail these. Baylis (2000) provides a useful overview and summary of many of these types of test. In common, however, with any item of equipment, medium, test method, etc. in the laboratory, it is essential that the aim and principles of the proprietary test as well as the manufacturer's instructions are understood and that the test is operated in full to ensure the achievement of correct results from the specific test. In addition, the laboratory's Standard Operating Procedures should detail, as necessary, the application of, and key steps in the operation of and interpretation of, results from these tests.

Examples of conventional method based proprietary tests include a variety of Petrifilm™ (3M Healthcare, UK) products which have been developed to replace some conventional agar plate methods. They comprise a dehydrated gel deposited onto a cardboard base with a thin plastic film overlay. The test sample dilution is inoculated onto the gel beneath the film overlay and the liquid in the inoculum rehydrates the gel, the film is replaced carefully over the gel to exclude air bubbles and this protects the gel and minimises moisture evaporation during incubation. The Petrifilm™ 'plate' is incubated in the normal way at a temperature appropriate to the target organism, e.g. coliforms at 37°C, total colony count at 30°C, and, following incubation, colonies are counted and results recorded.

Another test that has been developed for the enumeration of microorganisms is the Simplate (Biocontrol, UK), which is based on the MPN technique. It is a 'plate' divided into small compartments that contain growth media into which the test sample dilutions are inoculated; after the 'plate' has been incubated, the count of organisms per gram is estimated from the proportion of positive compartments.

Proprietary devices have also been developed for pathogen tests and new types of such test are regularly becoming available. Baylis (2000) provides a review of such commercially available proprietary methods and updates this regularly. A different time and labour saving approach to the conduct of *Salmonella* tests in particular is Sprint (Oxoid Ltd, UK). This comprises a timed-release capsule that contains selective agents. It is added to a pre-enrichment broth before incubation and automatically dissolves after several hours, slowly releasing the inhibitors and 'transforming' the pre-enrichment broth into a selective medium; thus, no sub-culture is required from the pre-enrichment broth to selective broths. At the end of incubation, the resultant selective enrichment broth is sub-cultured onto agar in the conventional way. Proprietary tests are also discussed in Chapters 11 and 12.

10.7   Further reading

Anon (2000) *Compendium of Microbiological Methods for the Analysis of Food and Agricultural Products.* (CD-ROM) AOAC International, Gaithersburg, USA.

Baylis, C.L. (ed.) (2003) *Manual of Microbiological methods for the Food and Drinks Industry, Guideline No. 43*, 4th edn. Campden & Chorleywood Food Research Association, Chipping Campden, UK.

Corry, J.E.L., Curtis, G.D.W. & Baird, R.M. (2003) *Handbook of Culture Media for Food Microbiology. Progress in Industrial Microbiology, Volume 37*. Elsevier Science, B.V., Amsterdam, The Netherlands.

Downes, F.P. & Ito, K. (eds) (2001) *Compendium of Methods for the Microbiological Examination of Foods.* Compiled by the APHA Technical Committee on Microbiological Methods for Foods, American Public Health Association, Washington D.C., USA.

Environment Agency (2002) *The Microbiology of Drinking Water – Parts 1–10, Water Quality and Public Health. Methods for the Examination of Waters and Associated Materials.* Environment Agency, Nottingham, UK.

Food and Drug Administration (1995) and Revision A (1998) *Bacteriological Analytical Manual*, 8th edn. Association of Official Analytical Chemists International, Arlington, Virginia, USA.

Hocking, A.D., Arnold, G., Jenson, I., Newton, K. & Sutherland, P. (eds) (1997) *Foodborne Microorganisms of Public Health Significance*, 5th edn. Australian Institute of Food Science and Technology, NSW Branch, Food Microbiology Group, Sydney, Australia.

Holbrook, R. (2000) Detection of Microorganisms in Foods – Principles and Culture Methods. In: *The Microbiological Safety and Quality of Food Volume II* (eds B.M. Lund, T.C. Baird-Parker, G.W. Gould) Aspen Publishers Inc., Maryland, USA. pp. 1761–1790.

Jarvis, B. (2000) Sampling for Microbiological Analysis. In: *The Microbiological Safety and Quality of Food Volume II* (eds B.M. Lund, T.C. Baird-Parker, G.W. Gould) Aspen Publishers Inc., Maryland, USA. pp. 1691–1733.

Kyriakides, A., Bell, C. & Jones, K. (eds) (1996) *A Code of Practice for Microbiology Laboratories Handling Food Samples – (Incorporating Guidelines for the Preparation, Storage and Handling of Microbiological Media), Guideline No. 9.* Campden & Chorleywood Food Research Association, Chipping Campden, UK.

Roberts, D. & Greenwood, M. (eds) (2003) *Practical Food Microbiology* 3rd edn. Blackwell Publishing Ltd., Oxford, UK.

11 Confirmation Tests

11.1 Introduction

Although there are accepted common biochemical characteristics ascribed to each microbial genus and species, there can also be differences from those normal profiles exhibited by different strains within a species of microorganism e.g. *Escherichia coli* comprises numerous strains distinguishable both biochemically and antigenically. Conversely, species within different genera may show remarkable similarities e.g. *Salmonella* and *Citrobacter*. Together, these variations from the 'norm' make the task of confirming a presumptive positive result a particular challenge to the food microbiologist. The need for accurate identification of a microorganism in food microbiology is especially important because the confirmed identification of a pathogen in a food sample raises concerns for the safety of the consumer that, depending on the food type, shelf life and consumer handling instructions, could lead to commercial losses associated with stopping production, launching a production investigation and, possibly even, a product recall.

Microbiological media are rarely (if ever) perfectly selective or diagnostic. They often allow some species to grow that are of no interest to the food microbiologist; these organisms may either produce dissimilar reactions or they may produce reactions similar to those of the organism sought. Non-target organisms displaying non-characteristic reactions of the target organism can quickly be eliminated but any showing reactions similar to the organism sought, i.e. showing 'false positive' reactions, require further work to eliminate them as target organisms. Conversely, organisms that are of interest sometimes produce atypical (but not negative) reactions and may therefore be overlooked if the 'textbook definitions' for colonial morphology and characteristic reaction are followed too strictly; this can lead to a 'false negative' result.

As a result of these complications, when colonies typical of the target organism have developed on selective / diagnostic agar or when typical reactions are observed in diagnostic broths, the results are usually considered to be **presumptive** and further confirmation tests are then applied to eliminate any false positive results before a **confirmed** test result is recorded and reported. It is usual to base an identification upon the results from a series of tests that provide a 'profile' for a particular organism supplemented, where available, by a test demonstrating a characteristic that is unique to the organism. Because of strain to strain differences, several similar but indivi-

Plates 1, 2, 3, 4 and 11 are located in the colour plate section at p. 100.
Plates 32–34 are located in the colour plate section at p. 228.

dually unique profiles often exist for any particular species, so the greater the number of tests that can be incorporated into a profile the greater is the confidence in the final identity that can be ascribed to the organism.

Confirmation tests can be much more labour-intensive, more expensive and take considerably longer to complete than the preceding work involved in producing the presumptive information. Therefore, the number of tests undertaken to produce a satisfactory profile for any given isolate depends, to some extent, on the importance of the result, as it is unnecessary for laboratories to spend considerable time and effort obtaining an unequivocal result when all the customer requires is the presumptive result or a simple indication that the presumptive result is genuine. For example, only simple confirmation tests may be required for a presumptive (suspect) coliform isolate or broth culture, whilst more numerous and complex tests to identify and characterise an isolate are more likely to be applied to presumptive isolates of organisms such as *Salmonella* spp. or *Listeria* spp. However, for many microorganisms, the choice of confirmation and identification tests is well-established internationally so that for these, similar tests are used in different laboratories throughout the world.

It is important therefore that the staff in food microbiology laboratories are properly trained to recognise and objectively assess the microbial growth and diagnostic reactions that develop during the incubation of agar plate and broth cultures and that only relevant confirmation tests that are recognised and accepted by all parties involved in the application of the results are used. This helps to ensure that the interpretation of test results has a sound scientific basis. To ensure conformity between laboratories, the interpretation of diagnostic reactions both on selective media and in the subsequent confirmation tests applied should be based upon information published in reference methods.

11.2   Preliminary confirmation tests

11.2.1   *General*

The appearance of colonies (colony morphology) grown under specified conditions including medium, atmosphere, temperature and time, is often important as it can provide useful information about the nature of the organisms that have been isolated. This is especially the case for microorganisms grown on diagnostic agars that have been especially designed to select for the growth of a specific organism and facilitate its production of colonies exhibiting one or more characteristic reactions (see Chapter 10). It can also be useful in tests such as the TCC, that do not yield diagnostic reactions, especially if a high count has been obtained, indicating that the sample is 'out of specification'. In such cases, separate counts of colonies showing similar morphologies can provide an indication of the most dominant organism in the population and, if required, a basis upon which to develop a production investigation.

Probably the most useful basic confirmation test is the direct observation of individual cells from a culture by light microscopy, of which there are two commonly used forms.

(1)   Use of a light microscope to observe stained microorganisms allows cell shape and size to be recorded, as well as (in the case of bacteria) to provide information

about the cell wall composition, i.e. from the Gram stain reaction, but the organisms are killed during the preparation of the specimen (see Plates 1(a), 2 and 3. See colour plate sections for all plates).

(2) Phase contrast microscopy is a form of light microscopy that permits live organisms to be observed so that their motility can be observed (see Figure 11.1, the 'hanging drop' technique); this method can also emphasise cell inclusions such as bacterial spores that may sometimes be difficult to observe by routine cell staining methods.

Figure 11.1   A diagram of a 'hanging drop' preparation used to observe motility in bacteria.

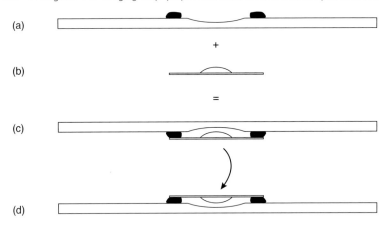

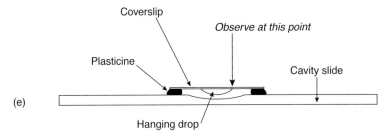

Method of Preparation:
(a) Place very small pieces of plasticine or similar material on either side of the cavity on a clean cavity slide.
(b) Place a drop, e.g. 10 μl, of a 'young', e.g. 18 hour, broth culture in the centre of a well-cleaned coverslip.
(c) Invert the cavity slide over the coverslip so that the drop of culture is central in the cavity and press the slide down gently to stick the coverslip and slide together by means of the plasticine.
(d) Turn the slide / coverslip unit over, quickly and smoothly, in an action that does not break the surface-tension of the drop which will now hang from the coverslip into the well of the slide.
(e) Focus on the edge of the drop, using the lower power objectives of the microscope, then place a drop of immersion oil onto the coverslip at the edge of the hanging drop and use an oil immersion objective, e.g. × 100 magnification, to observe the form of motion of the cells in the culture. If available, phase contrast microscopy (see 11.2.2) should be used; this allows motility to be observed more clearly.
Note: Coverslips used in the preparation of 'hanging drops' must be thin enough (approx. 0.2 mm) to allow the objective to be clearly focused.

Use of light microscopy can eliminate an isolate quickly from further study: for example, if a presumptive coliform has been isolated then the isolate should be a Gram-negative rod and not a Gram-positive coccus; similarly, if a yeast has been isolated then large oval cells might be expected and small bacterial rods can be disregarded.

In most routine food microbiology laboratories, light microscopy is used mainly for the observation of bacteria, and sometimes of yeasts and moulds. However, the confirmation and identification of yeasts and moulds generally requires specialist knowledge and would normally be referred to a recognised authority and specialist who has specific knowledge of, and relevant experience with, these organisms. Therefore, the tests described in the remainder of this section apply to bacteria and confirmation methods for yeasts and moulds will be described only briefly.

In addition to microscopy, the responses of bacteria in biochemical tests, such as the catalase reaction, oxidase reaction and breakdown of specific carbohydrates, are quite straightforward to obtain. The results of these tests can be used in conjunction with the results of the examination of isolates by light microscopy to categorise an isolate quickly and effectively, usually to family level, e.g. Enterobacteriaceae, which are generally catalase-positive, oxidase-negative, Gram-negative rods that produce acid and gas from glucose and are resistant to bile salts.

Additional biochemical tests are employed to study the more complex metabolic capabilities of an isolate. Not all such tests are applicable to all organisms and the range of 'complex' biochemical tests to be applied is best selected when the results of the more basic tests have become available. The following biochemical characteristics of microorganisms are probably those most commonly investigated when identifying a presumptive positive culture:

- production of indole from tryptophan
- acid production from glucose metabolism to pH <4.4
  (Methyl Red test),   ⎫
- production of acetylmethylcarbinol (Voges-Proskauer (VP) test)   ⎬ IMViC reactions
- use of citrate as sole carbon source   ⎭
- hydrolysis of *o*-nitrophenyl-β-D-galactopyranoside (ONPG) by β-galactosidase
- decarboxylation of amino acids, e.g. lysine, ornithine
- production of hydrogen sulphide
- production of the enzyme urease
- liquefaction of gelatin
- reduction of nitrate to nitrite or nitrogen.

Most of the reactions listed above may be applied to either Gram-positive or Gram-negative bacteria but the first four tests form the basis for the characterisation of Enterobacteriaceae, particularly the coliforms to distinguish the 'coli' sub-group from the 'aerogenes' sub-group. Collectively, they are known as the IMViC tests (indole reaction, methyl red reaction, Voges-Proskauer reaction, citrate utilisation); the addition of a fifth test to the series, the Eijkman test (production of acid and gas from lactose at 44°C), enables the identification of *Escherichia coli*, and the tests are then known as IMVEC tests.

For some organisms, the common range of tests does not provide adequate discrimination and more specific characteristics may need to be examined, for example:

- growth in an aerobic or microaerobic atmosphere, e.g. for *Campylobacter*
- resistance to heat, e.g. 60°C for 30 minutes, for some *Streptococcus* spp.
- production of the enzyme coagulase, e.g. for *Staphylococcus aureus* (coagulase-producing staphylococci)
- fermentation of lactose at elevated incubation temperatures (the Eijkman test for *E. coli*)
- identification of specific surface antigens (serology), e.g. for *Salmonella* spp. or *E. coli* O157
- enhanced haemolysis in the presence of other microorganisms, Christie–Atkins–Munch–Petersen (CAMP) test, e.g. for *Listeria monocytogenes*.

Most microorganisms cannot be identified with certainty using only one or two tests and a range of tests is generally applied. The test results based on a broad range of metabolic and / or physiological properties of the organism help to define the biotype of the organism allowing it to be distinguished from other types of bacteria and providing a higher level of confidence in the identity ascribed to the organism than the results of just very few tests.

Traditionally, confirmation test media and reagents were prepared in the laboratory from the individual raw materials, but in recent decades, proprietary test kits have become available that provide test materials in a ready-prepared format, probably the best-known of which is the API series of biochemical test strips manufactured by bioMérieux, SA, France (see Section 11.7.1).

11.2.2   *The microscope and microscopy*

The compound light microscope is so-called because the image of the specimen is first magnified by an 'objective' lens and then by an 'ocular' or eyepiece lens, so the magnification viewed by the operator is 'compound'; for example, a × 40 magnification objective and a × 10 magnification eyepiece would together provide the viewer with × 400 magnification. A typical, simple bright-field (light) laboratory microscope is shown in Figure 11.2 and its components are shown diagrammatically in Figure 11.3. Its essential components are listed below and their functions are explained in Table 11.1.

- a light source, which may or may not include an adjustable (variable aperture) iris diaphragm ('field' diaphragm),
- a mechanical stage on which to mount slides or specimens,
- a sub-stage condenser and iris diaphragm,
- one or more objectives, mounted on a revolving nosepiece,
- one or more eyepieces,
- coarse and fine focus adjustment 'wheels'.

Most bright-field microscopes suitable for use in food microbiology have several objectives of different magnifications; a × 10, or 'low-power' objective is useful for

Figure 11.2 An example of a compound light microscope.

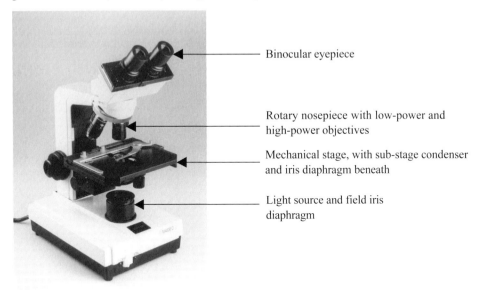

         Binocular eyepiece

         Rotary nosepiece with low-power and
high-power objectives

         Mechanical stage, with sub-stage condenser
and iris diaphragm beneath

         Light source and field iris
diaphragm

(Courtesy of Psyer-SGI Ltd.)
Key points:
- The condenser, eyepieces and objective lenses must be maintained in a clean, dry condition at all times always using proper lens cleaners and lens tissues.
- The microscope must be properly set up before use.
- Care must always be taken to avoid damaging the high-power objective lens by contact with the specimen slide.
(See also Table 6.2.)

the initial focussing of a specimen whilst 'high-power' objectives ($\times 40$ and $\times 100$) are required to be able to observe moulds, yeasts and bacteria in detail.

    Phase contrast microscopy is a specialised form of light microscopy that enhances the contrast between the specimen and the background so that the specimen becomes visible without the need for staining. The technique makes use of the fact that light takes the form of a sine wave that has 'phases'. When light is diffracted by microbial cells, its phase is retarded by about $\frac{1}{4}$ wavelength compared with light that is not diffracted (direct light), i.e. the diffracted and direct light become 'out of phase'. In addition, it is possible to advance the phase of the direct light by $\frac{1}{4}$ wavelength so that the two waves become $\frac{1}{2}$ wavelength apart, which produces destructive interference; in other words, the 'peaks' of the waves add together, and the 'troughs' also add together but in the 'opposite direction'; this darkens the image against a light background and creates an increase in the contrast about 4-fold.

    In phase contrast microscopy, a sub-stage phase contrast condenser and a set of phase contrast objectives are fitted to the microscope. These are set up and used in such as way as to ensure that the direct and specimen-diffracted light waves become $\frac{1}{2}$ wavelength out of phase making the specimen (microbial cells) visible. Further

Figure 11.3 Diagram of a compound light microscope indicating the key components, Kyriakides *et al.* (1996).

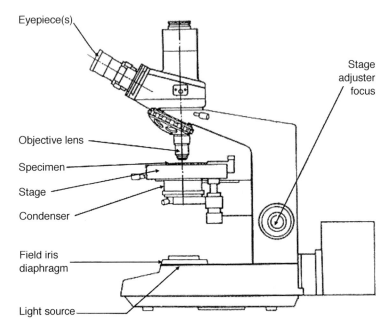

Table 11.1 The functional components of a compound light microscope.

Component	Function
Light source	Produces a parallel beam of light
Field iris diaphragm	Directs the beam of light towards the sub-stage condenser
Sub-stage condenser and iris diaphragm	For focusing the light onto the specimen
Mechanical stage	Provides a means of holding a microscope slide on which the specimen has been prepared
Objective(s): • low-power objectives, usually ×10 magnification • high-power objectives, usually ×40 and ×100 magnification (×100 objectives normally require 'oil immersion')	Mounted on a rotating nosepiece and are selected according to the magnification required: • provide low magnification of the specimen • provide high magnification of the specimen and may require 'oil immersion' in order to achieve a high resolution
Eyepiece(s), usually either ×5 or ×10 magnification	For viewing the specimen. Also, provide further magnification that 'compounds' the magnification by the objective Almost all modern microscopes are fitted with binocular eyepieces
Coarse and fine focus adjustments	For accurately focusing the specimen

technical and practical information can be obtained from microscope suppliers and their specific instructions should be followed for the operation of the equipment supplied.

Phase contrast microscopy is very useful for observing cell shape and motility in living microorganisms, but especially for observing bacterial spores which, when dormant, appear bright with a dark ring around their perimeters (known as 'phase bright spores', see Plate 4) but become uniformly grey ('phase dark') when they start to germinate into a new vegetative cell.

Light microscopy – setting up and using a microscope

For reliable information to be gained from the use of a light microscope it is essential to ensure that the microscope is properly set up and clean before use. Laboratory staff often consider the set-up procedure to be difficult and tedious, and it therefore tends to be overlooked, but with most modern microscopes, it is generally quite straight-forward and rapid, and it must be done. For instance, the first thing to do when using a light microscope is to adjust the lighting to provide the correct uniform intensity of illumination (known as Köhler illumination), since poor lighting causes aberrations that distort the image and poor illumination can very easily make Gram-negative bacteria appear Gram-positive.

All microscopes differ to some extent in their design, some being simpler than others. It is important that among the staff, there is the expertise to use the microscope correctly and the manufacturer's manual should always be the first point of reference for training. The following therefore, describes the set up operation of a microscope in general terms only.

(1) With the $\times 10$ objective in place and the light switched on, adjust the light source to a comfortable level of illumination and adjust the distance between the eyepieces to suit your own eyes.

(2) Place a prepared microscope slide, with the specimen uppermost, on the stage, using the clips to secure it in place and focus on the specimen using the coarse and fine focus controls. Focus adjustable eyepieces so that a sharp image is seen by both eyes and then move the slide to a clear area away from the specimen.

(3) Partially close the field-iris diaphragm and fully open the iris diaphragm on the sub-stage condenser.

(4) Move the condenser up or down until the image of the field-iris diaphragm is in sharp focus.

(5) Centre the field-iris diaphragm. If the sub-stage condenser has centring screws, adjust these until the image of the field-iris diaphragm is central in the field of view or use the field-iris diaphragm's centring mechanism.

(6) Open the field-iris diaphragm until the image of the iris just disappears.

(7) Remove an eyepiece, look into the light tube and close the sub-stage condenser iris until it obscures about $\frac{1}{3}$ of the view. If there is a means for centring the condenser iris then this should be done if required; replace the eyepiece. Move the specimen back into the field of view.

(8) Examine the specimen on the slide, adjusting the focus as necessary; it should only be necessary to use the fine focus control.

(9) To increase the magnification, rotate the nosepiece so that the required objective is in the light path; if necessary, re-adjust the condenser iris as in (7), and the level of illumination.

High-power objectives, e.g. $\times 100$ magnification, usually require the use of 'oil immersion' i.e. a drop of immersion oil is placed on the specimen and the objective is gently adjusted using the focusing mechanism until it makes contact with the oil, then the specimen is focused in the normal way. Thus, between the specimen and the objective lens, the light passes through oil instead of air and, because immersion oil has a higher refractive index than air, the technique allows a higher resolution of the image of the specimen to be obtained from the microscope optical system with enhanced brightness and clarity of the image. Objectives are designed for use either with immersion oil or without and this is usually marked on the side of the lens. After using an oil immersion objective, it is normally not possible to revert to using a lower magnification because of the oil now covering the specimen so any observations requiring use of low magnification objectives should be done before oil is used. After use, oil remaining on an oil immersion objective should always be immediately and gently removed using lens tissue.

11.2.3 *Staining techniques*

Many living microorganisms are translucent and therefore almost invisible using conventional light microscopy. Staining techniques have therefore been developed so that individual cells can be seen using a light microscope. Most staining techniques first require the preparation of a heat-fixed smear of the specimen culture. This is a method of spreading a culture of a microorganism onto a microscope slide and attaching or 'fixing' it so that it is not washed off during the staining procedure.

Note that the staining procedures applied to the heat-fixed specimen do not require the operation of aseptic practices, but it is important not to contaminate the culture from which the smear is prepared in case it is required for further sub-culture for tests that are to be incubated, and also, not to contaminate the working environment with the culture. A smear for staining should therefore be prepared using sterile equipment and it is essential to handle the live cultures aseptically to avoid contaminating the staff and the environment; indeed this is **vital** when handling pathogens.

Preparation of a heat-fixed smear

(1) To the centre of a clean and labelled microscope slide, add a drop of sterile saline solution (usually 0.85% w/v sodium chloride is used).
(2) Using a sterile loop, pick up a small quantity of culture from a single, isolated colony and emulsify it in the saline to produce an evenly distributed and smooth suspension. The quantity of culture to use needs to be determined by experience: too small a quantity can result in only a few cells, or no cells, being visible in any field of view, whilst too much can result in a densely packed smear in which individual cells cannot be distinguished.
(3) Allow the smear to dry, preferably slowly by natural evaporation in air as bacterial cells can become distorted if the culture dries too quickly.

(4)   Heat-fix the dried smear by rapidly passing the slide, inverted, 6 times through a 'blue' Bunsen flame (see Chapter 9); for safety, it is advisable to hold the slide with forceps. This procedure makes the cells adhere to the glass so that they do not wash off during subsequent steps. Allow to cool for 1 minute.

The Gram stain

The Gram reaction allows Gram-positive and Gram-negative bacteria to be distinguished and its principles are described in Chapter 1. The Gram staining technique not only differentiates the two groups of bacteria but also enables the food microbiologist to observe the cell morphology of the organisms in a culture. The Gram staining technique is undertaken as follows and is also shown diagrammatically in Figure 11.4.

(1)   Place the slide on a staining rack over a cleanable drip tray or laboratory sink.
(2)   Completely flood the slide with crystal violet stain (crystal violet-2 g, 95% ethanol-20 ml, ammonium oxalate-0.8 g, distilled water-80 ml*) for 1 minute.
(3)   Wash off with tap water.
(4)   Completely flood the slide with Lugol's iodine (solution containing 1% w/v I_2 + 2% w/v KI*) and leave for 1 minute. Iodine is a mordant that binds the crystal violet stain irreversibly to the cells of Gram-positive organisms.
(5)   Wash off with tap water.
(6)   Decolourise the smear by holding the slide vertically and rinsing with 100% ethyl alcohol until the blue stain ceases to be washed off. **Immediately** remove the residual ethyl alcohol by washing in tap water. This stage is critical and it is very easy to over-decolourise a smear; the optimal technique for this step can only be achieved by practice and experience.
(7)   Counter-stain the smear by flooding the slide for 30 seconds with Safranin-O stain (Safranin-O-0.25 g, 95% ethanol-10 ml, distilled water-100 ml)
(8)   Wash off with tap water.
(9)   Dry by gently blotting on a paper towel and leaving to dry in air.
(10) Transfer the slide to the microscope stage and bring the specimen into focus using a low power ($\times 10$) objective, then increase the magnification and re-focus if necessary. Before using the oil immersion ($\times 100$) objective, place a drop of immersion oil on the smear, then change to the oil immersion objective and re-focus (see Section 11.2.2).
(11) Record the Gram reaction together with the shape of the cells and, if appropriate, their size and arrangement seen on the smear, e.g. chains, clumps, V-shapes, single cells (see Plates 2 and 3).

There are various modified versions of Gram's staining method including the concentrations of the solutions used and times used at each stage of the process.
   *The solutions required for Gram staining are commercially available. Where commercial preparations are used, the manufacturer's instructions for use must be followed.

Figure 11.4 An illustration of the preparation of a Gram stain.

(a) Preparation of a heat-fixed smear.

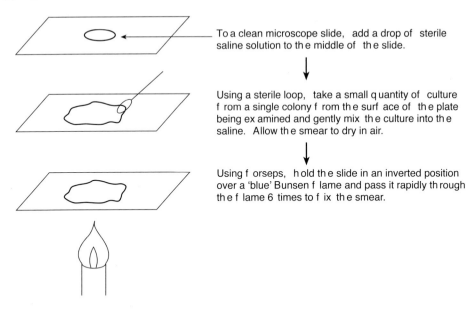

To a clean microscope slide, add a drop of sterile saline solution to the middle of the slide.

Using a sterile loop, take a small quantity of culture from a single colony from the surface of the plate being examined and gently mix the culture into the saline. Allow the smear to dry in air.

Using forseps, hold the slide in an inverted position over a 'blue' Bunsen flame and pass it rapidly through the flame 6 times to fix the smear.

(b) Gram staining method.

	Step
(1)	Place the slide onto a staining rack over a suitable reservoir.
(2)	Flood with crystal violet solution.*
(3)	Leave for 1 minute.
(4)	Wash gently with tap water and shake off any remaining water drops.
(5)	Flood with Lugol's iodine.*
(6)	Leave for 1 minute.
(7)	Wash gently with tap water and shake off any remaining water drops.
(8)	Wash with successive applications of ethyl alcohol (30 seconds to 1 minute) until no further dye is released from the smear then immediately wash with water.
(9)	Flood with safranin solution.*
(10)	Leave for 30 seconds.
(11)	Wash quickly with water, blot and dry.
(12)	The Gram stain is now complete and ready for observation by light microscopy.

*The solutions required for Gram staining are available commercially.

The bacterial spore stain

Bacterial endospores are resistant to staining by the Gram staining technique as they need to be heated to enable the stain to penetrate the spore coat. The spore stain (see Chapter 1) allows spores that have not been released from vegetative cells to be stained green whilst the vegetative cells are stained red; this enables spore shape, size and position within the vegetative cell to be determined, which is an essential part of the identification of *Bacillus* spp. The general procedure is as follows:

(1)   Prepare a heat-fixed smear of the culture using the method described in the Gram stain procedure. Ideally, pick the culture from the centre of a 1 day old colony or from the edge of a 2 day old colony.

(2)   Place the slide so that it is supported over a glass beaker containing boiling water and flood with malachite green dye (5% w/v aqueous solution). As the slide heats up, water vapour should be visibly rising from the stain, hold for 2 minutes.

(3)   Wash the slide with water and blot dry.

(4)   Place the slide on a staining rack over a cleanable drip tray or laboratory sink and counter-stain with safranin dye (0.5% w/v aqueous solution) for 20 seconds.

(5)   Wash off with tap water.

(6)   Dry by gently blotting on a paper towel and leaving to dry in air.

(7)   Transfer the slide to the microscope stage and bring the specimen into focus using a low power ($\times$ 10) objective, then increase the magnification and re-focus if necessary. Before using the oil immersion ($\times$ 100) objective, place a drop of immersion oil on the smear, then change to the oil immersion objective and re-focus (see Section 11.2.2).

(8)   Record the spore shape, position and size, i.e. whether or not the spore distends the vegetative cell that surrounds it.

A modification of the spore stain (Holbrook and Anderson, 1980) is used as a confirmation test for *Bacillus cereus*.

11.2.4   *Microscopy using live cultures*

Two techniques are used for preparing bacterial cultures for direct (unstained) observation by microscopy. The traditional method is the 'hanging drop' technique (Figure 11.1) that is used to determine motility, especially for obligate aerobes, but cannot be used for anaerobes because these organisms may become non-motile in the presence of oxygen. For these organisms, motility may be observed by growing the culture in a 'sloppy agar' medium, e.g. motility nitrate medium (see '11.4.11 Nitrate reduction'). A simpler method, the 'wet mount', can be used to observe bacterial spores or, with experience, to determine cell motility especially in facultative anaerobes such as *Listeria* spp., or as a confirmation method for the presence of yeast and moulds in colony counts, when only cell size and morphology need to be established.

Broth cultures should always be used for motility tests because many motile bacteria lose their motility (ability to move) on agar media. For motility tests, it is usual to use an 'overnight' culture as many motile bacteria lose their motility when the population reaches the stationary phase (see Chapter 1); also, some organisms lose their motility at certain incubation temperatures; *Listeria* spp., for example, are motile at temperatures between 20–25°C but not at 37°C even though they are able to multiply rapidly at this temperature. Note that all bacteria may appear to 'jiggle about' or oscillate in a liquid suspension; this is 'Brownian motion' caused by the molecular forces in the suspending liquid and is not true motility.

It is important to note that, by the production of aerosols during the removal of small volumes using pipettes or loops, broth cultures can easily contaminate hands, work surfaces and equipment and it is essential that good aseptic practice is applied consistently; indeed this is **vital** if pathogens are being handled.

The 'hanging drop' technique

(1) Preferably using a microscope cavity slide, place a clean slide on a clean, dust-free work surface and attach to it, centrally, two thin (about 3 mm in diameter) 'sausages' of plasticine or similar plastic material (Figure 11.1(a)).

(2) Place a clean coverslip on the work surface then, using a sterile loop, take a drop of a well-mixed broth culture and place it in a spot at the centre of the coverslip (Figure 11.1(b)).

(3) Invert the slide and press it gently on to the coverslip so that the plasticine attaches the two together (Figure 11.1(c)).

(4) Pick up the assembly and invert it in a swift but smooth motion, so that the drop hangs from the underside of the coverslip (Figure 11.1(d)).

(5) Following the manufacturer's instructions, set up the microscope for phase contrast illumination using a low power **phase contrast objective** and the *corresponding* **phase rings**.

(6) Transfer the assembly from (4) to the microscope stage and bring the specimen into focus.

(7) Move the slide so that the field of view corresponds to the edge of the hanging drop and focus on the organisms in this region (Figure 11.1(e)). This is important as motile organisms such as *Pseudomonas* are obligate aerobes and may cease to be highly active if the oxygen tension decreases.

(8) Increase the magnification and change the phase rings to those that correspond to the objective used; re-focus if necessary. Before using the oil immersion ($\times 100$) objective, place a drop of immersion oil on the coverslip, then change to the oil immersion objective and re-focus.
It is especially important not to break the coverslip as this may damage the microscope and will contaminate the environment.

(9) Observe the cells; motile bacteria may show rapid darting or tumbling movements.

(10) Record the presence or absence of motility and, if appropriate, the style of motion, as well as the cell morphology (see Plate 4 and Figure 1.5(a)).

The 'wet mount' technique

(1) Place a clean microscope slide on a clean, dust-free work surface.

(2) Using a sterile loop, take a drop of a well-mixed broth culture and place it in a spot at the centre of the slide.

(3) Place a clean coverslip over the drop of culture on the slide and taking care to avoid air bubbles, press down gently using a blunt instrument such as a loop handle, to 'squeeze the culture down'. **Take care not to break the coverslip or contaminate the environment with culture**.

(4) Following the manufacturer's instructions, set up the microscope for phase contrast illumination using a low power **phase contrast objective** and the *corresponding* **phase rings**.

(5) Transfer the slide from (3) to the microscope stage and bring the specimen into focus.

(6) Increase the magnification and change the phase rings to those that correspond to the objective used; re-focus if necessary. If an oil immersion ($\times 100$) objective

is used, place a drop of immersion oil on the coverslip, then change to the oil immersion objective and re-focus.

(7) Observe the organisms; endospores appear as bright 'holes' within rod-shaped cells (unless they have been released into the broth after cell lysis). Motile bacteria often show rapid darting or tumbling movements; yeasts are much larger than bacteria, are always non-motile and may show 'budding'. (Bacterial endospores can be seen in Plate 4 and yeast cells in Plate 1b.)

(8) Record, as appropriate, the presence of bacteria or yeasts, the presence or absence of motility, the style of motion and the cell morphology.

11.3   Basic biochemical tests

Just as when preparing slides for microscopy, it is important to note that broth cultures can very easily contaminate hands, work surfaces and equipment and, for all biochemical tests involving liquid cultures, it is essential that good aseptic practice is applied consistently; indeed this is **vital** if pathogens are being handled.

11.3.1   *The catalase test*

Most aerobic bacteria (and indeed some yeasts) produce the enzyme, catalase. This enzyme breaks down toxic peroxides that would otherwise accumulate during aerobic microbial metabolism and this is one means by which a microorganism protects itself from poisoning. The production of this enzyme (or lack of) by a bacterial culture, therefore, can be used as a simple test to help identify the isolate.

The principle of the test is the conversion of hydrogen peroxide to water and oxygen; when a culture (usually a colony) is mixed with hydrogen peroxide solution, the enzyme present in catalase-positive organisms, e.g. *Bacillus* spp., rapidly catalyses the breakdown reaction, with the release of oxygen that can be seen very easily as gas bubbles or 'fizzing'. The absence of gas bubbles produced within a few seconds indicates that the culture is catalase-negative, e.g. *Lactobacillus* spp.

A number of different techniques have been developed to test a culture for catalase production. **With all techniques, it is important to avoid the production of aerosols by catalase-positive organisms, especially if the isolate is pathogenic, as the vigorous evolution of oxygen during the reaction can easily spread contamination to the worker and to the environment**. Two commonly used techniques are described below.

Slide technique

(1) Using a sterile loop, pick up a small quantity of culture from a single, isolated colony and deposit it on the centre of a clean microscope slide. Alternatively, place a small drop of an overnight broth culture directly onto the slide.

(2) Place a clean coverslip over the culture.

(3) Using a sterile Pasteur pipette, add a small quantity of 3% aqueous solution ('10-volume') of hydrogen peroxide to the edge of the coverslip, so that it runs under the coverslip and on to the culture.

(4) Observe the mixture for the vigorous production of gas bubbles within 1 minute.

(5) Record vigorous gas production as 'positive' or no gas production as 'negative'.

Tube technique

This method can only be used for colonies that have grown on the surface of an agar medium.

(1) Dip one end of a capillary tube, e.g. 100 mm long × 0.9–1.0 mm internal diameter × 0.15 mm wall thickness, into a stock solution of 3% hydrogen peroxide so that the solution is taken up by capillary action and forms a column of liquid 2–3 cm high.
(2) Holding the capillary tube vertically, touch the end with the exposed liquid meniscus onto the surface of the colony to be examined, so that the hydrogen peroxide comes into contact with the colony.
(3) Observe the column of liquid in the capillary tube for the presence of vigorous gas bubbles rising through the liquid, within 30 seconds.
(4) Record vigorous gas production as 'positive' or no gas production as 'negative'.

It is important that all equipment used for catalase tests is very clean, as dirty equipment can cause a false positive result. Colonies should not be picked from blood-containing growth media as blood itself contains the enzyme catalase. Some bacteria produce 'pseudocatalase' in media containing low concentrations of glucose; colonies therefore should be picked from growth media that contain at least 1% w/v glucose. All used materials must be discarded safely.

11.3.2   *The oxidase test*

Those bacteria that have an aerobic respiratory metabolism possess cytochrome c that enables them to use free oxygen in their energy metabolism. Indophenol oxidase, also present in respiratory organisms, oxidises cytochrome c and it is the oxidised cytochrome c that oxidises the reagent in the oxidase test. The production of cytochrome c (or lack of) by a bacterial culture, therefore, can be used as a simple test to help identify the isolate.

The principle of the test is the oxidation of the colourless reagent, tetramethyl *p*-phenylene diamine dihydrochloride, to yield a purple compound. When a culture (usually a colony) is mixed with a solution of tetramethyl *p*-phenylene diamine dihydrochloride, the enzyme system present in oxidase-positive organisms, e.g. *Pseudomonas* spp., rapidly catalyses the reaction, usually within 10 seconds, with the production of a deep purple colour that can be seen very easily; the absence of a deep purple colour after 30 seconds indicates that the culture is oxidase-negative, e.g. Enterobacteriaceae.

(1) Prepare an aqueous solution of approximately 1% w/v tetramethyl *p*-phenylene diamine dihydrochloride. It is not normally necessary to weigh out the reagent very accurately and the solution must be used within 10 minutes, as the reagent breaks down naturally in solution.

(2) Place a drop of the solution on a piece of filter paper.
(3) Using a sterile plastic or platinum loop, or a glass rod (but not a nichrome wire loop), pick up a small quantity of culture from a single, isolated colony and rub it carefully onto the reagent.
(4) Observe the mixture for the development of deep purple colour within 30 seconds.
(5) Record a deep purple colour as 'positive'.

A number of proprietary reagents and kits are available for oxidase tests, e.g. a proprietary 'dipstick' test that comprises a plastic stick, one end of which is coated with a stabilised oxidase reagent; the end of the stick is touched onto a colony and observed for the development of a purple colour.

11.3.3   *Tests for carbohydrate utilisation*

Many microorganisms can utilise one or more carbohydrates in energy-yielding metabolic systems referred to generically as 'fermentation' reactions. As the range of carbohydrates utilised varies from organism to organism, the ability of a bacterial culture to break down carbohydrates can be used to help identify the isolate through the use of carbohydrate utilisation tests.

In carbohydrate utilisation tests microorganisms produce acid, often accompanied by gas, from a range of different carbohydrates, although some bacteria, e.g. *Listeria*, produce acid without gas. Bacteria that cannot utilise a carbohydrate usually grow utilising the peptones in the medium but produce neither acid nor gas. A basal medium, usually liquid, containing either peptone or yeast extract as a source of soluble nitrogen, salts and a pH indicator (usually Andrade's indicator, International Commission on Microbiological Specifications for Foods, 1978) is prepared and is dispensed into a series of culture tubes or bottles, usually in 5 ml or 10 ml quantities. If gas production needs to be observed, then a Durham tube (see Chapter 9) is added to each culture tube and the prepared basal broths are sterilised by autoclaving. Carbohydrates are then added by aseptically dispensing filter-sterilised carbohydrate solutions, each broth prepared containing a single, different carbohydrate, usually at a final concentration of 0.5–1.0% w/v.

The test culture is inoculated into each of the tubes in the series of carbohydrate broths and incubated at the organism's optimum growth temperature, often for 7 days or more. At intervals during incubation, the broths are observed for acid production, denoted by a colour change of the medium, and any gas produced which collects in the Durham tube, forming a bubble. Acid production in the medium, together with gas production if appropriate, are recorded as 'positive', but it is important not to record negative reactions until the full incubation period is complete as some organisms reduce the pH value of the medium only after extended incubation. The profile obtained is then used, together with the results of other tests, to identify the organism being examined (see 11.6 'Some examples of confirmation test profiles'). A similar approach is also used to identify yeasts except that non-utilisers of any particular carbohydrate usually fail to grow in the medium, as opposed to growing without the production of acid and / or gas.

It is possible to examine an isolate for the ability to break down a very wide range of carbohydrates, but this is generally considered too labour intensive and unnecessary for most routine purposes. Instead, different ranges of carbohydrates are usually employed for the differentiation of different groups of organisms. The range(s) to be employed for any particular isolate may be determined from the preliminary information obtained from the Gram reaction, cell morphology and catalase and oxidase reactions. Thus, the range of carbohydrates to be used can be restricted to those most likely to be of value in discriminating between the genera or species indicated by the results of the preliminary tests.

For example, the following range of carbohydrates has been suggested by Holmes and Costas (1992) for distinguishing members of the family Enterobacteriaceae:

glucose	inositol	salicin
adonitol	lactose	sorbitol
arabinose	maltose	sucrose
cellobiose	mannitol	trehalose
dulcitol	raffinose	xylose
glycerol	rhamnose	starch

Glucose is often included in these tests because most organisms capable of breaking down carbohydrates can utilise this sugar as it is central to respiratory metabolism. Thus, glucose is often included as a 'positive control' for such tests, even though the result of the glucose test may be of no use in discriminating between different species of the same genus. Additionally, although starch is not a simple sugar, being a composite D-glucan consisting of amylose and amylopectin, it can be utilised by a range of organisms that produce enzymes such as amylases and therefore, it can play a useful role in distinguishing one genus from another.

When required to examine an isolate's carbohydrate breakdown pattern, it is important to choose an appropriate medium formulation and culture technique as this can affect the results obtained. For example, some organisms are better able to use proteins than carbohydrates and can produce highly alkaline end-products from peptone that neutralise the small amount of acid they may produce from a carbohydrate leading to a false negative result. Some bacteria may not grow on simple media and need an enriched sugar medium to enable them to grow well, while some organisms that utilise sugars may show poor acid production in liquid media but may produce more reliable results when grown on an agar surface where the oxygen tension is higher. Additionally, aerobic organisms such as *Pseudomonas* spp. may not give reliable reactions on peptone-containing media and are best grown on media containing an ammonium salt as the main nitrogen source. Yet other organisms can actually decolourise the pH indicator so that the pH change is not visible; this problem is overcome by adding the indicator at the end of the incubation period so that it is not metabolised during incubation.

Details of different media formulations and culture methods, as well as their significance for different organisms, are described by Cowan (1974). Two contrasting examples used for food-borne organisms are:

(1) Carbohydrate breakdown by Enterobacteriaceae (Holmes and Costas, 1992)
   - Peptone water
     + Andrade's pH indicator (contains aqueous acid fuchsin and sodium hydroxide)
     + glucose added to one tube (final concentration, 0.75% w/v) and other carbohydrates added to separate tubes (1 tube per carbohydrate, final concentration, 0.5% w/v)
     + Durham tubes required, at least in the glucose tube
   - incubate all tubes at 37°C for 5 days
   - inspect for a yellow to pink colour change
   - inspect daily, as colour reversions may occur. Record results.

(2) Carbohydrate breakdown by *Listeria* spp. (Prentice and Neaves, 1992)
   - basal medium containing peptone, salt and beef extract
     + bromocresol purple pH indicator
     + α methyl-D-mannoside added to one tube (final concentration, 0.5% w/v), other carbohydrates added to separate tubes (1 tube per carbohydrate, final concentration, 1.0% w/v)
   - Durham tubes not required
   - incubate all tubes at 37°C for 2 days
   - inspect for a purple to yellow colour change
   - check tubes showing negative reactions for turbidity, to ensure that growth has occurred. Record results.

11.4   Further biochemical tests

Nowadays, the more complex biochemical tests used for differentiating genera and species of organisms are generally accomplished using proprietary kits and it is only in exceptional conditions that a laboratory may need to prepare the media and reagents required 'in-house' from the individual ingredients. Details of the following tests are therefore limited to their principles only, as comprehensive descriptions of the media and methods for these confirmation tests can be found in reference texts, e.g. Blazevic and Ederer, 1975; Cowan, 1974, Roberts *et al.*, 1995 and *Society for Applied Bacteriology Technical Series No. 1*, Gibbs and Skinner (1966); *No. 2*, Gibbs and Shapton (1968); *No. 17*, Corry *et al.* (1982) and *No. 29*, Board *et al.* (1992).

11.4.1   *Oxidation-Fermentation (O-F) test*

The O-F test demonstrates the metabolic pathway by which an organism breaks down simple carbohydrates, such as sugars. Bacteria that can utilise sugars do so either oxidatively or fermentatively whilst organisms that cannot break down carbohydrates are known as non-utilisers. Hugh and Leifson's medium is most often used for the O-F test for distinguishing these metabolic pathways.

   Hugh and Leifson's medium is an agar medium containing peptone, salts and a pH indicator (bromothymol blue), to which a suitable sugar, usually glucose, is added to a final concentration of 1% w/v. The prepared medium is dispensed into test tubes and sterilised by autoclaving; this also dispels oxygen from the medium which must be used quickly once the oxygen is dispelled. For each isolate, two tubes of oxygen-free media are inoculated by stabbing (see Chapter 9) and the medium in one tube is immediately overlaid with sterile liquid paraffin to provide anaerobic conditions; the tubes are then incubated at an appropriate temperature for up to 14 days.

Bacteria that utilise sugars oxidatively produce acid and therefore, a colour change of the pH indicator to yellow (due to the decrease in pH value) at the top of the non-overlaid tube only whilst acid (yellow colour) production throughout both tubes indicates sugar fermentation. Non-utilisers of the carbohydrate under test show no reaction in either tube; in this case, the tubes should be inspected for turbidity to ensure that growth has occurred, as an alternative basal medium may be required if the basal medium used proves unsuitable for the isolate under investigation (see 11.3.3).

11.4.2   *Indole test*

The production of indole from tryptophan by means of a tryptophanase is a characteristic possessed by many different microorganisms, but the indole test is used mainly for differentiating species within the family Enterobacteriaceae, e.g. to identify *E. coli* in general after incubation at 37°C, or to distinguish *E. coli* biotype 1 from other faecal coliforms using a test incubation temperature of 44°C.

The indole test can be done in various ways but, commonly, an isolate is inoculated into a tryptophan-containing medium, e.g. tryptone water (normally a 5 ml or 10 ml volume) and incubated for 24–48 hours. Various incubation temperatures have been reported in the literature, but for *E. coli* biotype 1, 44°C with an incubation time of 24 hours is usually used. After incubation, 0.5 ml of Kovács' reagent (*p*-dimethylaminobenzaldehyde dissolved in amyl alcohol and concentrated hydrochloric acid) is added and the culture is shaken then left to stand. The yellow reagent partitions from the medium and settles onto the surface of the tryptone water. Development of a deep red colour within 1 minute in this surface layer indicates a positive reaction (Plate 32(a)) i.e. the indole produced is 'seen' by its reaction with the reagent.

11.4.3   *Methyl red test*

The methyl red test is used to differentiate coliforms into 'coli' and 'aerogenes' subgroups. All coliforms ferment glucose, some more vigorously than others. After incubation for 48 hours, the 'coli' produce sufficient acid to acidify a peptone-phosphate growth medium containing 0.5% w/v glucose so that the pH value decreases to below pH 4.4 and remains stable. However, if 'aerogenes' organisms are present, e.g. *Enterobacter*, they do not produce so much acid but also produce large amounts of neutral products and use-up their fermentation end-products so raising the pH value of the medium to pH 6–7.

The addition of methyl red pH indicator to a culture after 48 hours' incubation therefore distinguishes between those organisms that have produced copious quantities of acid (a red colour is interpreted as MR positive) and those that have not, but cause the pH value to revert towards neutral values (a yellow colour is interpreted as MR negative). It is important that cultures should be given a full incubation period, normally 2 days at 37°C or 3 to 5 days at 30°C, as 'false positives' may occur if the incubation time is shortened.

11.4.4　*Voges-Proskauer test*

Although acetylmethylcarbinol (acetoin) is produced by a wide range of micro-organisms, the Voges-Proskauer (VP) test is used mainly to differentiate coliforms and is usually combined with the methyl red test, as the two tests show opposite reactions (negative correlation) for lactose-fermenting coliforms, i.e. MR-negative coliforms are always VP-positive and *vice versa*. A negative reaction to the methyl red test simply shows that the pH value of the medium is above pH 4.4 whilst a positive VP reaction gives a specific indication that acetoin has been produced.

Several VP test procedures have been developed, varying mainly in their incubation time and temperature, as well as the indicator system used to produce the final colour reaction. However, the medium described for the methyl red test (often known as MR–VP medium) is often used, and a 'standard' temperature / time for incubation applied, i.e. 37°C for a minimum of 48 hours. However, it should be noted that the reactions of organisms such as *Hafnia* may be negative at 37°C but positive at 30°C or below and 3 to 5 days at 30°C remains the better option for the incubation of this test to ensure reliable results. Several rapid methods have been proposed in which the production of acetoin can be accelerated by using a heavy inoculum from an agar culture and by reducing the volume of the broth medium to 0.5 ml or less.

After incubation, acetoin production is visualised by adding, either (a) α-naphthol followed by a potassium hydroxide / creatine solution or (b) creatine solution followed by potassium hydroxide solution. For both methods, the culture is shaken vigorously to increase the exposure to oxygen and then allowed to stand; a positive reaction is denoted by a strong red (method (a) above) (Plate 32(b)) or eosin-pink (method (b) above) colour. Various times have been reported for which the test mixture should be allowed to stand before declaring the test negative, i.e. no colour development; these range from 30 seconds to 1 hour. It is important to ensure that the time allowed is that specified in the actual method employed.

11.4.5　*Citrate utilisation*

The test for microbial utilisation of citrate as a sole carbon source is the fourth of the IMViC tests used to differentiate Gram-negative rods, particularly members of the Enterobacteriaceae and the test may also be useful for differentiating some of the more unusual *Salmonella* serotypes. Three media have been developed, and these may give conflicting results for the same organism. Any results from these tests of an organism's ability to utilise citrate should always therefore, be accompanied by the name of the test medium. The test principle is based on the ability of some organisms to convert the salts of organic acids into alkaline carbonates, producing an alkaline reaction in the test medium.

The first medium to be developed, Koser's citrate medium, is a liquid medium containing only simple salts to which citric acid is added. This medium was modified by Simmons who added agar and a pH indicator, bromothymol blue. In both of these, citrate is the only source of carbon. Christensen's medium is more complex and contains yeast extract and cysteine hydrochloride as well as sodium citrate. Thus, this is a medium rich in nutrients including additional sources of carbon, so organisms that grow on Christensen's medium may not grow on Koser's or Simmons' media.

Of the three media, Simmons' citrate agar is probably the most widely used. The medium is often prepared as an agar slope in a test tube and is inoculated by stabbing the butt and / or streaking the surface. It is important that a light inoculum is used, taken from a cell suspension in saline or buffer, and applied using a sterile inoculating needle not a loop, to avoid any carry-over of nutrients with the inoculum as these may lead to false results; for the same reason, all glassware used must be particularly clean. Cultures are incubated for 4 to 7 days at the organism's optimum temperature and examined for colour reactions; a blue colour (alkaline) is recorded as positive whilst the original green colour of the medium indicates a negative reaction; tests showing positive reactions should always be sub-cultured to a second tube of the same medium, re-incubated and examined to ensure that the reaction was not due to carry-over of nutrients.

11.4.6   *ONPG test*

In conventional fermentation tests, the fermentation of lactose only gives a positive result if the isolate possesses two enzymes, a lactose permease and β-galactosidase. The permease facilitates the entry of lactose into the bacterial cell whilst the intra-cellular β-galactosidase hydrolyses lactose into galactose and glucose, and organisms that lack a lactose permease show negative reactions even though they possess the ability to break down lactose. The presence of β-galactosidase in lactose negative organisms (lack lactose permease) such as *Yersinia enterocolitica* is therefore detected by the ONPG test in which the colourless substance, *o*-nitrophenyl-β-D-galactoside (ONPG) which does not need a specific permease to enter cells, is hydrolysed by β-galactosidase to produce galactose and also *o*-nitrophenol, which is yellow.

Traditionally, the ONPG test is undertaken by inoculating an isolate into ONPG broth (containing ONPG, peptone and phosphate buffer), incubating at an appro-priate temperature, usually 37°C for 24 hours, and examining the culture for the development of a yellow colour. However, the ONPG reaction can take place very quickly and the yellow colour may develop in less than 6 hours. An alternative, proprietary technique is to use paper discs impregnated with ONPG; these are placed into a small sterile tube, sterile saline solution (0.1 ml) is added and a colony of the test organism is emulsified in the liquid. The suspension is incubated at 37°C and observed hourly for up to 6 hours for the development of a yellow colour in the disc; cultures displaying a negative reaction are then further incubated and inspected after 24 hours.

11.4.7   *Decarboxylation of lysine and ornithine*

The decarboxylation of amino acids provides an important means of differentiating related microorganisms. Several amino acids may be attacked by decarboxylase enzymes but the decarboxylation of lysine to cadaverine and ornithine to putrescine in particular are two of the most useful characteristics for discriminating between different genera, particularly the genera of the family Enterobacteriaceae.

Møller's medium and method are recognised as the reference method for decarboxylase tests. The medium is a peptone-glucose broth with beef extract and pyridoxal, containing bromocresol purple and cresol red to which the required amino acid, e.g. lysine or ornithine, is added at a final concentration of 1% w/v; a negative control is also prepared which consists of the basal medium without the amino acid. The medium is prepared and dispensed in small volumes into tubes containing sterile liquid paraffin, then sterilised by autoclaving (Cowan, 1974). To prevent alkaline oxidation reactions from producing spurious results, the tubes are of narrow diameter to keep the surface area of the medium small in relation to the total volume in the tube. Using a needle, the medium is inoculated, through the paraffin, with a heavy inoculum from a culture grown on agar and incubated at 37°C for up to 4 days, although reactions usually occur within 24–28 hours. The fermentation of glucose in both tubes initially produces an acid, yellow reaction. Then, decarboxylation of the amino acid raises the pH value and the medium becomes purple while the control tube remains yellow; a yellow reaction in the amino acid-containing tube indicates a negative reaction.

An agar version of Møller's medium has been developed. The medium is dispensed into tubes and solidified with a deep butt and short slope. It is inoculated by stabbing the butt and streaking the slope, then incubated at 37°C for 24 hours. A purple slope with a yellow butt indicates a negative reaction whilst a purple slope with a purple butt is interpreted as positive. This medium has two other advantages: it contains indicators of hydrogen sulphide production (a characteristic of *Salmonella*), resulting in an intense blackening of the medium, whilst lysine deamination, a characteristic of *Proteus*, can be seen as a red slope with a yellow butt.

Later modifications of Møller's medium and method have been used and a simpler version of the medium that does not require such stringent control of the growth conditions is used in routine testing laboratories. This medium contains yeast extract and glucose but does not include peptone, so organisms that can utilise peptone cannot use this as a nitrogen source and produce false positive alkaline reactions that can obscure an otherwise negative result for the decarboxylation of an amino acid, as occurred with the earlier media formulations.

11.4.8 *Hydrogen sulphide production*

The production of hydrogen sulphide detects the ability of an organism to reduce sulphur compounds to sulphide. This ability is found in some members of the Enterobacteriaceae and some clostridia. Hydrogen sulphide may be produced from sulphur-containing organic substances found in peptones, e.g. cysteine, or from inorganic sulphur compounds, e.g. thiosulphate, although different enzyme pathways are used for the two types of substrate.

Thus, test results may differ depending on the medium used; *E. coli*, for example, does not produce hydrogen sulphide when grown on triple sugar iron (TSI) agar but may yield a positive reaction in media containing large amounts of cysteine. The reactions observed are also affected by the indicator used to detect the hydrogen sulphide produced; some heavy metal salts e.g. of lead, iron and bismuth are used to detect hydrogen sulphide production because their sulphide salts are black compounds and easily detectable, but some, e.g. of bismuth, may inhibit growth of some organisms. In media that contain ferrous (iron) salts, e.g. TSI or Kligler's iron agar, hydrogen sulphide production can be detected with great sensitivity. It is usual therefore, to state the medium used alongside a test result. In practice, TSI is the most commonly used medium for specifically detecting hydrogen sulphide production and, when necessary, this allows the ready comparison of results from different laboratories.

Different substrates for and indicators of hydrogen sulphide production are incorporated into a number of selective media used to detect a range of different organisms; thus different incubation conditions are employed and varied test reactions are obtained (though a blackening (for H_2S positive) reaction is always observed). Some well-known selective-diagnostic isolation media for Enterobacteriaceae that incorporate substrates for and indicators of hydrogen sulphide production are xylose lysine desoxycholate medium (XLD), hektoen enteric agar (HEA) and bismuth sulphite agar (BSA); for clostridia, differential reinforced clostridial medium (DRCM), tryptose sulphite cycloserine (TSC) agar and oleandomycin polymyxin sulphadiazine perfringens agar (OPSP) are commonly used.

11.4.9 *Urease test*

Production of the enzyme urease, is a characteristic of the genus *Proteus* (although other bacteria also possess this enzyme) and the urease test is primarily of use to distinguish this genus from *Salmonella* and *Shigella* which do not produce urease. Urease hydrolyses urea to form ammonia and carbon dioxide, and the resulting alkaline reaction is detected by means of a pH indicator, phenol red, included in the medium.

Urea is added to the peptone-glucose-salt basal medium of Christensen which is then inoculated from a single isolated colony, incubated at the organism's optimum temperature, examined after 4 hours and then, if negative, daily for 5 days. A colour change from yellow to pink in the test medium but not in the control indicates a positive reaction.

If the medium is highly buffered, then positive reactions are restricted to *Proteus* but a result can be obtained more quickly, in 2–4 hours, by decreasing the buffering capacity of the medium. Some organisms, e.g. *Pseudomonas aeruginosa*, can create an alkaline reaction by producing ammonia from peptones in the medium, giving rise to false results; a negative control should therefore always be set up, containing an inoculated broth from which the urea has been omitted.

11.4.10 *Gelatin liquefaction*

Hydrolysis of gelatin, a protein, results in a loss of its gel strength, i.e. the gel liquefies. Gelatin is derived from collagen, an important component of animal connective tissue, and the enzymes capable of its hydrolysis are known as gelatinases or, alternatively, collagenases. Gelatinase production is found widely among bacteria but is mostly useful for differentiating the Enterobacteriaceae, pseudomonads and clostridia, although some other organisms, e.g. micrococci and species of *Bacillus*, also possess the ability to liquefy gelatin.

Usually, a peptone gelatin medium is used, e.g. nutrient gelatin that is a peptone-beef extract broth to which 12% w/v gelatin is added; this is dispensed into test tubes and sterilised by autoclaving at 115°C for 20 minutes. The culture for examination is then inoculated by stabbing and the tubes are incubated. Gelatinase activity is often greatest at incubation temperatures lower than 37°C, so incubation at 22°C is often used, the cultures being examined for liquefaction, at intervals, for up to 30 days. If cultures are incubated at 37°C, the tubes must be refrigerated for 2 hours before examination as gelatin is liquid at the incubation temperature and it is also important that an uninoculated, incubated control tube is examined. The rate of gelatin liquefaction can also be of interest and organisms may be classified as weak-, strong- or non-liquefiers.

Lactose-gelatin medium is a confirmatory medium for *Clostridium perfringens* that combines a lactose fermentation test with a test for gelatin liquefaction. It is a nutrient medium containing lactose, 12% w/v gelatin and phenol red as a pH indicator dispensed into tubes that are stab-inoculated immediately following steaming and cooling (to dispel oxygen), then incubated at 37°C. After 20 hours, the medium is examined for lactose fermentation (acid production changing the colour from red to yellow, accompanied by gas bubbles); after 44 hours, the tubes are refrigerated and inspected for gelatin liquefaction. *Clostridium perfringens* gives positive reactions in both tests.

11.4.11 *Nitrate reduction*

In those organisms that reduce nitrate, there is a variety of chemical end points in the breakdown pathway; nitrate is always reduced to nitrite, but nitrite may be further reduced to ammonia, nitrogen or other nitrogen-containing compounds. The test for nitrate reduction therefore, has several possible outcomes and each must be examined.

The organism under investigation is grown, at its optimum incubation temperature, in nitrate broth, which is a nutrient medium containing 0.1%w/v potassium nitrate. Normally, incubation is for 2 to 5 days, but several, more rapid, tests are available. After incubation, the presence of nitrite in the medium is detected by the addition of sulphanilic acid and 5-amino-2-naphthylene-sulfonic acid, that all together produce a water-soluble, red dye. A positive (red) reaction indicates that the organism has reduced the nitrate to nitrite, but no further. If the reaction is negative (no red colour), then either the microorganism has not reduced nitrate, or the nitrate reduction has continued beyond nitrite to ammonia or nitrogen. These two possibilities can be distinguished by the addition of a small amount of zinc dust that chemically reduces nitrate to nitrite; if a red colour develops on the addition of zinc dust, then nitrate was still present in the medium and was not reduced by the microorganism, but if no red colour develops, then the nitrate was reduced beyond nitrite. The production of nitrogen from nitrite can be confirmed by the inclusion in the medium of a Durham tube, in which a bubble of nitrogen gas collects. It is important to test an uninoculated broth for the presence of nitrite to demonstrate that it is absent from the sterile medium.

Motility-nitrate medium is a useful confirmation test for *Clostridium perfringens* that combines the nitrate reduction test with a test for motility. The de-aerated nutrient medium, containing potassium nitrate, 0.3% agar and 0.5% w/v glycerol, is

stab-inoculated and incubated at 37°C for 20 hours. Motile organisms produce turbidity throughout the medium whilst non-motile species, including *C. perfringens*,
grow in a discrete line along the stab; *C. perfringens* reduces nitrate to nitrite and so
the addition of the nitrite detection reagents yields a red colour in the medium.

11.5   Some confirmation tests for specific organisms or groups of organisms

For some organisms, specific additional tests can provide useful information to aid
their identification. The following describes some of these.

11.5.1   *Growth in different atmospheres*

The ability, or inability or an organism to grow in an aerobic, anaerobic or microaerobic atmosphere is often used in confirmation tests, e.g. for obligate anaerobes
such as *Clostridium* spp. or for micro-aerobes such as *Campylobacter* spp. If the
objective is to establish an organism's ability to grow in air, then the culture simply
has to be inoculated onto an appropriate agar and incubated in air, then the agar is
observed for growth / no growth. However, if an organism's ability to grow in an
atmosphere containing an oxygen concentration lower than the approximately 21%
normally found in air needs to be established, then a suitable atmosphere composition
must be provided.

   To obtain anaerobic or microaerobic conditions, it is usual to inoculate an isolate
onto an appropriate agar and incubate the culture in a sealed jar (an 'anaerobe jar',
see Plate 11) containing the appropriate atmosphere. For obligate anaerobes, almost
all traces of oxygen must be removed whilst for micro-aerobe organisms, the oxygen
concentration needs to be reduced to around 6%; in both cases, the atmosphere also
contains around 10% carbon dioxide, the remainder comprising nitrogen and
hydrogen.

   There are a number of commercially available systems / kits for achieving the
required gas mixture in an anaerobe jar; one proprietary kit, when activated,
generates a mixture of hydrogen and carbon dioxide. For safety reasons, such kits
must always be used in accordance with the manufacturer's instructions. Generally,
after the cultures have been placed in the jar, the gas generating kit is activated by
adding water, and is placed in the jar and then the jar is sealed quickly. The jar must
also contain an active palladium catalyst to catalyse a reaction between the evolved
hydrogen from the kit and the atmospheric oxygen remaining in the jar, producing
water and so removing oxygen from the atmosphere and creating an anaerobic
environment surrounding the cultures. When setting up the jar for anaerobes, it is
important to include an indicator of anaerobic conditions; usually this is a paper strip
soaked in a redox dye, the colour of which is sensitive to the oxygen concentration so
that when the jar is opened it is easy to determine whether or not the correct conditions have been maintained. Other commercial kits are available for generating the
different atmospheres that may be required and there is a variety of procedures from
which to select. The laboratory should have Standard Operating Procedures that
describe the types of system to be used and the required method of operation.

11.5.2 *Heat-resistance*

A microorganism's resistance to heat can help to distinguish it from related species. The '*Streptococcus* group', for example, contains a number of species, most of which are easily inactivated by a mild heat treatment, except for the enterococci (also known as Lancefield's Group D or 'faecal' streptococci and now ascribed to the genus *Enterococcus*) that survive a heat process of 60°C for 30 minutes. Therefore, if a culture is heat-treated under accurately controlled conditions, before inoculation on a suitable growth medium and incubation, the presence of growth indicates that the organism survived the process.

On a more practical note, those organisms that may survive a traditional milk pasteurisation process, e.g. 63.5°C for 30 minutes, can be isolated and enumerated by the 'thermoduric plate count'. For this test, the growth medium and incubation conditions are similar to those used for the TCC (see Chapter 10) but, before plating, the primary raw milk test sample and / or dilutions are 'pasteurised' in the laboratory using the same factory pasteurisation process temperature and time conditions so that only the surviving organisms grow on the agar. This test can help to establish the spoilage potential of the heat-treated production material.

11.5.3 *Coagulase test*

The production of coagulase, an enzyme that causes clotting of plasma (usually rabbit plasma) in laboratory tests, is a characteristic strongly associated with pathogenic strains of staphylococci, usually *Staphylococcus aureus*. To demonstrate the production of 'free' coagulase, a presumptive positive isolate (commonly a well-separated characteristic colony on Baird-Parker agar) is first grown as a broth culture, e.g. in brain-heart infusion broth, incubated overnight at 37°C. A small quantity of the incubated culture (usually 0.1 ml) is added aseptically to rabbit plasma (usually 0.3 ml) in a small, sterile test tube (10 mm in diameter × 75 mm in length); the size of the tube is important as it enables the reaction to be observed correctly. The culture is then incubated at 37°C and examined for clotting of the plasma after 4–6 hours and again after 24 hours if it is negative after the initial incubation period. To observe clotting, the tube is tilted, very carefully; if the culture has remained liquid the test is negative, but if a clot or 'gel' has formed that occupies more than half of the original liquid volume (Figure 11.5), then a positive result is recorded.

It is important that negative and positive controls are also examined; the negative control should consist of plasma to which sterile brain-heart infusion broth has been added whilst the positive control should comprise a known, coagulase-producing strain of *Staph. aureus*.

11.5.4 *Eijkman test*

Escherichia coli can be distinguished from other Enterobacteriaceae by its ability to produce acid and gas by the fermentation of lactose, during incubation at an elevated temperature. The culture to be examined is inoculated into a medium containing bile salts, lactose and a pH indicator, e.g. MacConkey broth, and incubated at 44°C for 48

Figure 11.5 An illustration of the coagulase test reaction.

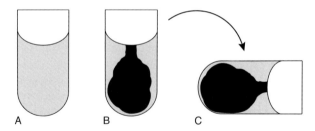

A   B   C

Notes:
A = a negative reaction.
B and C = a strong positive reaction. A strong straw-coloured gel (clot) forms and occupies a volume more than half the original liquid volume. When tilted gently, the clot remains firm at the bottom of the tube.

hours. A colour change in the medium, indicating the production of acid, together with the production of gas (detected using a Durham tube), is interpreted as a 'positive' result.

11.5.5 *Slide agglutination tests*

In food microbiology, serology is the study of reactions involving antibodies and antigens in which specific antibodies are used to help characterise microorganisms according to the antigens they possess. The simplest serological test, the slide agglutination test, is used for example, to sub-divide *Salmonella* into different 'serotypes'. It is based on the principle that salmonellae possess a range of different surface antigens including those in their cell walls ('O' or 'somatic' antigens) which are lipopolysaccharide in nature and / or those associated with their flagella i.e. flagellar proteins ('H' antigens). When an homologous antiserum, i.e. a solution of purified antibody that has been produced against a specific antigen, is mixed with a suspension of the organism from a pure culture, the antibody binds to its corresponding antigen (if present) and makes the bacterial cells stick together in clumps that can be seen with the naked eye; this effect is known as agglutination (see also 12.3.4 'Serology').

The slide agglutination test is often very time-consuming, and a complete serological characterisation of a *Salmonella* serotype may require a wide range of purified antisera, so it is only undertaken by specialist laboratories. However, some basic serological grouping can be done, and is often undertaken, by some of the larger food microbiology laboratories. To do the test, a drop of saline is placed on a clean microscope slide and a small amount of the purified isolate (usually picked from a freshly grown colony) is emulsified in it to make a smooth suspension. A drop of antiserum is then added and the slide is gently rocked back and forth for a few minutes. The development of agglutination (clumping) indicates a positive reaction but if the bacteria remain as a smooth suspension, the reaction is considered negative. The pattern of antigens (antigenic formula) of an isolate characterises its serotype and the majority of serotypes of *Salmonella enterica* subsp.

enterica are given 'names', e.g. *Salmonella enterica* subsp. *enterica* serotype Typhi-murium, shortened to *Salmonella* Typhimurium, but slide agglutination tests can also be used to identify strains of *E. coli*, e.g. *E. coli* O157 : H7 or *E. coli* O26 : H11.

11.5.6   *CAMP test*

The CAMP (Christie, Atkins, Munch-Petersen) reaction demonstrates the ability of an organism to produce a zone of 'enhanced haemolysis' when grown on a blood-containing medium in close proximity to another microorganism; it is used in the identification of *Listeria* spp. and especially in the differentiation of *Listeria mono-cytogenes*. The method is described in detail by Prentice and Neaves (1992).

Double-layer agar plates are prepared with a very thin upper layer containing sheep blood. Cultures of *Staphylococcus aureus* and *Rhodococcus equi* are picked from slopes (see Chapter 9), using needles, and streaked in parallel lines across the agar. The *Listeria* isolate is then carefully streaked at right angles to these organisms so that it is very close to, but not actually touching, the *Staphylococcus* and *Rhodococcus* cultures. The plate is incubated at 37°C for no longer than 12–18 hours and examined for enhanced zones of haemolysis in the regions where the streaks are close to each other (see Figure 11.6). The presence or absence of a CAMP reaction with either *Staphylococcus aureus* or *Rhodococcus equi* and the size of the zone, together with the results of reactions in other tests, identifies the species of *Listeria* present.

Figure 11.6   An illustration of the reactions given by *Listeria* spp. in the CAMP test, adapted from Prentice and Neaves (1992).

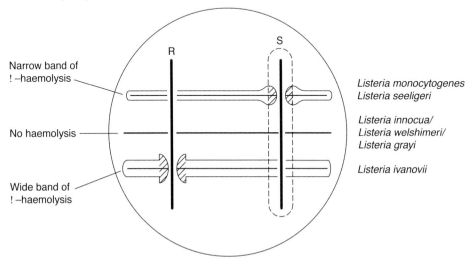

R = *Rhodococcus equi*, S = *Staphylococcus aureus*.
Note the bulb-like zones of enhanced β-haemolysis produced by *Listeria monocytogenes* and *L. seeligeri* in proximity with *Staph. aureus* and the mushroom-shaped zone of β-haemolysis produced by *L. ivanovii* when growing in proximity with *Rhodococcus equi*. *Listeria innocua*, *L. welshimeri* and *L. grayi* do not produce haemolytic reactions.

11.6 Some examples of confirmation test profiles

Many selective agar media allow some initial diagnostic characteristics of the organisms that grow to be noted, which means that some general assumptions may be made about a presumptive positive isolate before confirmation tests are initiated. For example, XLD medium, used widely to isolate *Salmonella*, is highly selective and it is very likely that organisms capable of forming colonies on this medium would be Gram-negative rods; the colours of a colony and the surrounding medium also allow the organism's ability to decarboxylate lysine, ferment certain carbohydrates and produce hydrogen sulphide to be determined. Thus, typical colonies on this medium are likely to be members of the Enterobacteriaceae although some *Pseudomonas* spp. can produce transparent colonies on a red medium, similar to those of some salmonellae; this is why colonies exhibiting the characteristics of *Salmonella* are termed 'suspect' and more information would be needed to provide increased confidence in an identification of colonies growing on a medium.

Thus, a profile for the identification of a particular organism is built up from information gathered from a variety of sources, the diagnostic reactions on isolation media being confirmed and extended by the organism's reactions in confirmation tests. Figures 11.7 to 11.9 show some examples of simple profiles that allow some groups of organisms that are important in food microbiology to be differentiated. However, it cannot be over-emphasised that different strains of a particular species can give different profiles whilst organisms of different species can give similar patterns of reactions.

It is important that during training, staff learn the primary characteristics of the organisms sought and the more usual of the atypical forms encountered in the sample types examined.

11.7 Proprietary kits and reagents

The preparation of large numbers of tubes containing small volumes of sometimes complex microbiological media for microbial identification purposes is laborious and can therefore be expensive. Consequently, since the 1960s, many proprietary test kits and reagents have been developed that reduce the test preparation time and the time to obtain a result. In addition, the interpretation of test reactions has been greatly simplified. However, whilst the use of proprietary test kits to simplify the routine identification of organisms may be cost-effective, the result has been that many technicians have not learned the fundamental principles of confirmation tests. Thus, if a test has provided incorrect results or if an unusual strain of an organism has been encountered, the staff of a laboratory may find it difficult to recognise and / or resolve the problem. It is important therefore, even when proprietary kits are used, that laboratory staff understand the basic principles of the tests incorporated into the proprietary kit, as well as the characteristics of the organisms they are designed to identify.

Figure 11.7 Basic profiles for some Gram-negative bacteria.

Pseudomonas

- Gram-negative
- rods
- catalase positive
- oxidase positive
- motile (but may be non-motile)
- obligate aerobe
- oxidative glucose metabolism, or does not utilise glucose (O-F test)
- never produce endospores.

Note that *Pseudomonas* is only one genus within the technologically significant group, Gram-negative psychrotrophs, which also contains oxidase-negative and non-motile organisms.

Salmonella

- conforms to the profile for Enterobacteriaceae
- lactose negative, i.e. not a coliform ⎫
- lysine decarboxylase positive ⎬ can be seen on XLD medium
- produces hydrogen sulphide ⎭
- ONPG negative
- indole negative
- methyl red positive ⎫
- Voges-Proskauer (VP) negative ⎬ IMViC reactions
- citrate positive (Simmon's) ⎭
- agglutinated by specific 'O' (somatic) and 'H' (flagella) antisera (see also Chapter 12).

Enterobacteriaceae

- Gram-negative
- rods
- catalase positive
- oxidase negative
- motile
- facultative anaerobes
- oxidative and fermentative glucose metabolism (O-F test)
- grows on bile salt-containing media, e.g. VRBGA
- never produce endospores.

Note that the Enterobacteriaceae is a large family of bacteria containing a number of genera of interest to food microbiologists including food-borne pathogens. Within this family, a number of artificial sub-divisions is generally accepted and used in food and water microbiology, for example,

- coliforms are Enterobacteriaceae that ferment lactose
- 'faecal coliforms' are coliforms capable of growth at 44°C.
- *E. coli* biotype 1 is a faecal coliform that produces indole from tryptophan at 44°C.

Escherichia coli

- conforms to the profile for Enterobacteriaceae
- lactose positive, i.e. a coliform
- grows at 44°C (except Vero cytotoxigenic *E. coli*, e.g. *E. coli* O157 : H7)
- does not produce hydrogen sulphide
- ONPG positive
- indole positive
- methyl red positive ⎫
- Voges-Proskauer (VP) negative ⎬ IMViC reactions
- citrate negative. ⎭

Figure 11.8 Basic profiles for some Gram-positive bacteria.

Listeria spp.

- Gram-positive
- short rods
- catalase positive
- oxidase negative
- motile, with a distinctive 'tumbling' motion when incubated at temperatures of between 20–25°C
- produce acid, but no gas, from glucose
- aesculin hydrolysis positive
- facultative anaerobes
- never produce endospores.

Listeria monocytogenes

- conforms to the profile for *Listeria* spp.
- produces β-haemolysis on media containing horse or sheep blood
- rhamnose positive
- xylose negative
- CAMP reaction positive against *Staphlococcus aureus* on sheep blood agar
- CAMP reaction negative against *Rhodococcus equi* on sheep blood agar.

Staphylococcus spp.

- Gram-positive
- cocci, often in grape-like clusters
- catalase positive
- oxidase negative
- non-motile
- break down carbohydrates by fermentation (O-F test)
- facultative anaerobes
- never produce endospores.

Staphylococcus aureus

- conforms to the profile for *Staphlococcus* spp.
- reduces tellurite to tellurium (black) } can be seen on Baird-Parker agar
- most strains produce the 'egg-yolk' reaction causing clear zones to form around the colony when grown in media containing egg-yolk; also some strains hydrolyse the fat in egg-yolk causing precipitation of calcium and magnesium salts of fatty acids in the clear zone.
- coagulase positive

Note 1: Some strains of *Staph. aureus* e.g. from dairy products and environments do not produce typical (characteristic) colonies on Baird-Parker agar (one of the most common media used for detecting the organism) and for these sample types, care is required in both the selection of media to be used and the interpretation of growth on the selected medium.

Note 2: Within the genus *Staphylococcus*, *Staph. aureus* is the principal species that produces the enzyme coagulase; however, other species, notably *Staphylococcus intermedius* and some members of *Staphylococcus hyicus*, also produce free coagulase. The coagulase test is widely used as the main confirmation test because it is strongly associated with pathogenicity. Strictly, therefore, the test is not definitive for *Staphylococcus aureus* but confirms the presence of any coagulase-positive staphylococci.

Note 3: A control culture exhibiting a weak coagulase positive reaction may also be used by laboratories that frequently encounter such organisms as food isolates.

Figure 11.9 Basic profiles for some spore-forming bacteria.

Bacillus spp.

- Gram-positive
- rods
- capable of producing endospores in the presence of oxygen
- catalase positive
- oxidase variable (species differ; can be positive or negative)
- motile (most species)
- species differ in the way they break down carbohydrates, some do attack them and some do not attack them at all
- aerobes or facultative anaerobes.

Bacillus cereus

- conforms to the profile for *Bacillus* spp.
- mannitol negative ⎫ can be seen on *Bacillus cereus*
- lecithinase positive (egg yolk) ⎬ isolation media
- spores central, oval and do not distend the cell – can be seen from a spore stain
- capable of producing fat globules within the vegetative cell – can be seen from a Gram or spore stain.

Clostridium spp.

- Gram-positive
- rods
- capable of producing endospores in the absence of oxygen; spore may or may not distend the cell
- catalase negative
- oxidase negative
- motile (most species)
- break down carbohydrates fermentatively or not at all
- obligate anaerobes.

Clostridium perfringens

- conforms to the profile for *Clostridium* spp., except non-motile
- reduces sulphite to sulphide ⎫ can be seen on many
- produces lecithinase C (α-toxin) ⎬ *C. perfringens* isolation media
- lecithinase production inhibited by *C. perfringens* antitoxin (Nagler reaction)
- reduces nitrate to nitrite (most strains)
- liquefies gelatin
- lactose positive
- spores central or sub-terminal and distend the cell
- rapid growth at 45°C.

11.7.1   *Miniaturised, conventional tests*

Probably the most widely known of proprietary products for undertaking confirmation tests is the 'API' range of kits (bioMérieux SA, France). API kits comprise a set of miniaturised, conventional confirmation tests, i.e. a set of cupules in a plastic strip contain the relevant (dehydrated) growth media and diagnostic reagents for a range of tests appropriate to a particular group of organisms (Plate 33). Kits are available that contain different ranges of tests for the identification of different groups of organisms, of which the best-known kit is API 20E that contains 21 tests for identifying members of the Enterobacteriaceae. Most of the tests are proprietary versions of the conventional (traditional) biochemical tests described above; hence the test principles are similar to, but not identical with, those of the 'traditional' versions.

All of the cupules in a test strip are inoculated with a suspension of the purified test culture and the strip is then incubated at the organism's optimum growth temperature, usually for 24 hours. After incubation, the reactions are interpreted as positive or negative according to the manufacturer's instructions and the results recorded; mostly these are straightforward colour reactions but some cupules require reagents to be added before the result can be seen. The positive reactions are assigned a number whilst the negative reactions are scored as zero. Then, the numbers are added together in groups to form a numerical profile, (Plates 33 and 34). This profile is then compared with a database of 'scores' obtained from known organisms, from which the most likely identity can be derived; for example, the API 20E code 6704753 is one of many profiles produced by strains of *Salmonella* (Plate 34).

Other manufacturers produce miniaturised identification systems based upon similar principles to API, e.g. 'Enterotube® II' (Becton Dickinson, USA). Enterotube® II comprises a plastic tube divided into 12 compartments that contain diagnostic agars for 15 test reactions. A metal needle runs through all of the compartments and is protected by a cap at each end to maintain sterility. After removing the end caps, the agars are inoculated by touching the end of the needle on an isolated colony, then withdrawing the needle from the tube so that the inoculum is drawn through all of the compartments. The needle is then replaced, the tube is incubated and the test results are read and scored in a manner similar to that used for API 20E.

Oxoid Ltd, UK, market a range of kits (Microbact™ products) that is similar to the API galleries except that in some of the kits, the reagents are dispensed into the wells of microtitre plates. As with the API, after inoculation and incubation, diagnostic reagents are added to the relevant test wells and the test reactions are then read, recorded and compared with a computerised database of reactions for known organisms.

Whichever test kits are used, it is essential that the manufacturer's specific instructions are followed for handling and setting up the test as well as reading / interpreting the results.

Kits from different manufacturers, for the identification of the same group of organisms, often contain slightly different ranges of tests, which when used in combination, can be useful for characterising strains of organisms that sometimes are difficult to differentiate / identify using a single proprietary kit. In these circumstances, it may be possible for the experienced microbiologist to use the results of individual test reactions and make a professional judgement about the identity of such isolates or determine which further tests may be required to complete its identification. For example, Enterotube® II contains a test for dulcitol fermentation that does not form part of the API 20E test range. Dulcitol fermentation, together with other tests such as growth in the presence of potassium cyanide, can be useful for differentiating some strains of *Salmonella* and *Citrobacter freundii*, as these organisms are very closely related and can have almost identical biochemical reactions but differ in their ability to cause disease. By combining the test reactions from more than one test kit, or performing additional individual tests it may therefore be possible to ascertain the probable identity of a 'difficult' isolate by reference to textbooks on microbial taxonomy. However, it must be emphasised that such a judgement should be undertaken only by those in the laboratory who have sufficient knowledge, experience and authority.

11.7.2　*Other proprietary tests*

A wide range of proprietary tests targeting single or a selected few biochemical or serological characteristics is available for assisting the identification of specific food-borne pathogens, for example serologically based latex particle agglutination tests for identifying *Salmonella* and *E. coli* O157.

Also, specific individual biochemical tests e.g. the O.B.I.S. *Salmonella* kit' (Oxoid Ltd, UK) which is a disposable test card-based kit that allows the differentiation of *Salmonella* from other organisms that have similar colony morphologies on common selective media particularly *Citrobacter* spp. Test cards are supplied with L-pyroglutamic acid 7-amino-4-methyl coumarin (7AMC), a substrate which detects PYRase activity and Nitrophenylalanine (NPA) that detects deaminase activity. If these enzymes are present in bacteria, the respective substrates are hydrolysed by the enzymes and, on addition of the appropriate colour-developing reagent, a purple colour indicates a positive PYRase test and an orange-brown colour indicates positive NPA activity in the respective tests. *Salmonella* is negative in both tests, *Citrobacter* is PYRase positive, and *Proteus* exhibits NPA activity.

A detailed discussion of the tests available is not appropriate for this book; however, where the laboratory staff encounter such tests, care should be taken to ensure that the manufacturer's instructions are understood and implemented correctly. Almost all of these tests are designed to identify an isolate in the shortest possible time, using simple techniques; therefore, they are often designed to detect a supposedly unique characteristic of a particular organism to give a speedy and simple result. However, because different strains of almost any species of microorganism can exhibit diverse reactions, even the most commercially successful tests can often only identify the majority of, but not all, strains encountered. The result from a single test type therefore needs to be interpreted with caution and should be combined with other information about a microbial isolate before a definitive conclusion is reached. Baylis (2000) lists in *The Catalogue of Rapid Microbiological Methods* many of the currently available proprietary test kits and systems for the identification of microorganisms and this catalogue is regularly updated.

11.8   Control cultures and other performance checks

To ensure that colony morphologies and confirmation test reactions obtained are those expected for the organisms sought, cultures of reference organisms (see Chapter 7) should be used as positive and, where appropriate, negative control cultures for the tests employed alongside the test isolates. In the case of diagnostic reactions and confirmation tests, the range of control organisms that might be considered appropriate could be extensive and might even exceed the number of test samples examined. Each laboratory, therefore, needs to consider how much resource it should devote to such procedures. This does not mean that controls may be abandoned entirely and, for most tests, it is likely that at least one relevant positive control organism will be used.

11.8.1   *Control cultures for diagnostic reactions on isolation media*

In order to interpret accurately the diagnostic reactions on isolation media, it is essential to examine a positive control culture that is expected to show typical reactions. This control organism should have passed through the entire test procedure and not simply have been streaked on to the isolation medium (although this may be done in addition as part of the media quality control operation). This is because the overall ability of a detection method to isolate a target organism from the mixed culture derived from a test sample needs to be demonstrated (see Chapter 7); additionally, the stress of growing in a selective enrichment broth followed by isolation on a selective agar can adversely affect the reactions shown by some organisms, e.g. *Salmonella*.

   For species of microorganisms such as *Staph. aureus* that may show 'typical' or 'atypical' reactions, it may also be necessary to examine two or more positive control cultures to demonstrate a range of possible diagnostic reactions and, here again, the value of examining a number of positive controls needs careful consideration.

   Another decision for the food microbiologist is whether or not to use negative control cultures; this is because many different reactions might be considered negative which could result in the use of numerous negative control cultures. Thus, the number of negative control cultures perhaps needs to be limited to those that demonstrate negative results for the prime reactions on any particular medium, e.g. a lysine decarboxylase-negative reaction on XLD agar for the isolation of *Salmonella* or an aesculin-negative reaction on Oxford medium for the isolation of *Listeria* spp. Alternatively, negative control cultures might be used to demonstrate selectivity, e.g. inhibition of a Gram-positive organism on XLD agar. However, when the expected reaction is lack of growth, a separate test might then be needed to demonstrate that the negative control culture was in fact alive and would have grown had the selectivity of the medium been defective.

   It is for the senior laboratory staff to determine the appropriate application of positive and negative control cultures and these should be documented in the laboratory's methods.

11.8.2   *Control cultures for basic confirmation tests*

Positive and negative control cultures should always be used for basic confirmation tests, although this is often overlooked. For the Gram staining technique, known Gram-positive and known Gram-negative bacteria should be examined alongside the test culture smears; *Staph. aureus* and *E. coli* are useful in this respect as they demonstrate coccal and rod cell morphologies respectively as well as the two Gram reactions. If the two organisms are prepared as a mixed culture in the same smear, the results, when viewed microscopically, can be very effective.

Ideally, similar performance checks should also be made for spore stains, by examining a known spore-forming organism, normally a *Bacillus* sp. In this case, however, it may be difficult to obtain a culture that sporulates consistently and retains the spores as intracellular inclusions, so it is important to select an appropriate positive control strain carefully.

Depending upon the significance of the result, the use of positive control cultures may or may not be beneficial for motility tests. If all that is required is a demonstration that an organism is motile, then a positive control culture may not be needed, especially if the 'test' organism is displaying a clearly positive reaction. However, for organisms such as *Listeria* spp., where the **style** of motility is important, then the use of a positive control culture is important as, in the case of *Listeria* spp., it is important to clearly demonstrate a tumbling motion.

11.8.3   *Control cultures for biochemical confirmation tests*

For biochemical confirmation tests, the use of control cultures depends on whether the media were prepared by the laboratory staff or were purchased as pre-prepared growth media or as proprietary test kits. If the media were prepared in-house, the use of both positive and negative control cultures would be appropriate, as they should form part of the laboratory's media quality control system.

For pre-prepared growth media, and especially for proprietary test kits, it may be appropriate to rely upon a written assurance of quality from the supplier, although this, of course, cannot demonstrate its correct use in the laboratory. Given the large number of organisms that may be differentiated by one test kit, more than one control organism may be necessary to demonstrate the positive reactions of all the test kit components. Where supplies are sourced from known, accredited and reputable suppliers (see Chapter 7), then reliance on the manufacturer's quality assurance and quality control systems may be the most appropriate option supported by occasional in-house checks of batches received using control cultures and use of the appropriate control organisms alongside test 'suspects' for identification.

11.8.4   *Other performance / quality monitoring procedures*

For many of the traditional biochemical confirmation tests, sterility controls should be set up alongside the test cultures. This is especially important when media or reagents have been aseptically dispensed, e.g. when filter-sterilised sugar solutions have been added to broths for carbohydrate fermentation tests after the basal media have been sterilised by autoclaving.

Another performance check that can be applied to some related colony counts employing diagnostic media is to compare counts obtained for the same test sample on different media. A test sample examined for counts of coliforms and *E. coli* using different media provides a good example; if the coliform result is < 10 cfu/g then an *E. coli* count of, say, 100 cfu/g is theoretically impossible because *E. coli* should be detected by the coliform test. In practice, however, this does occur from time to time. Minor differences between the two counts, e.g. 50 cfu/g versus 100 cfu/g might be attributable to differences in the selectivity of the two growth media used or even the randomness of organism distribution in the test sample, but if such results are observed frequently, or if the differences in count are large, then an investigation should be initiated.

It is important that all food microbiology laboratory staff who are appropriately trained, participate in proficiency testing schemes (see Chapter 7) as these challenge not only the efficacy of the enumeration or detection method used but also the confirmation tests, as well as the ability of the laboratory staff to perform the tests and interpret the results correctly.

11.9   Further reading

Anon (2000) *Compendium of Microbiological Methods for the Analysis of Food and Agricultural Products.* (CD-ROM) AOAC International, Gaithersburg, USA.

Baylis, C. (2000) *Catalogue of Rapid Microbiological Methods, Review No. 1*, 4th edn. Campden & Chorleywood Food Research Association, Chipping Campden, UK.

Baylis, C.L. (ed.) (2003) *Manual of Microbiological Methods for the Food and Drinks Industry, Guideline No. 43*, 4th edn. Campden & Chorleywood Food Research Association, Chipping Campden, UK.

Blazevic, D.J. & Ederer, G.M. (1975) *Principles of Biochemical Tests in Diagnostic Microbiology.* John Wiley & Sons, New York, USA.

Board, R.G., Jones, D. & Skinner, F.A. (eds) (1992) *Isolation Methods in Applied and Environmental Microbiology.* Society for Applied Bacteriology Technical Series No. 29. Blackwell Scientific Publications, London, UK.

Corry, J.E.L., Roberts, D. & Skinner, F.A. (eds) (1982) *Isolation and Identification Methods for Food Poisoning Organisms.* Society for Applied Bacteriology Technical Series No. 17. Academic Press, London, UK.

Cowan, S.T. (ed.) (1974) Cowan & Steel's *Manual for the Identification of Medical Bacteria*, 2nd edn. Cambridge University Press, Cambridge, UK.

Environment Agency (2002) *The Microbiology of Drinking Water – Parts 4–10, Water Quality and Public Health. Methods for the Examination of Waters and Associated Materials.* Environment Agency, Nottingham, UK.

Food and Drug Administration (1995) and Revision A (1998) *Bacteriological Analytical Manual*, 8th edn. Association of Official Analytical Chemists International, Arlington, Virginia, USA.

Downes, F.P. & Ito, K. (eds) (2001) *Compendium of Methods for the Microbiological Examination of Foods.* Compiled by the APHA Technical Committee on Microbiological Methods for Foods. American Public Health Association, Washington D.C., USA.

Gibbs, B.M. & Shapton, D.A. (eds) (1968) *Identification Methods for Microbiologists: Part B.* Society for Applied Bacteriology Technical Series No. 2. Academic Press, London, UK.

Gibbs, B.M. & Skinner, F.A. (eds) (1966) *Identification Methods for Microbiologists: Part A.* Society for Applied Bacteriology Technical Series No. 1. Academic Press, London, UK.

Hocking, A.D., Arnold, G., Jenson, I., Newton, K. & and Sutherland, P. (eds) (1997) *Foodborne Microorganisms of Public Health Significance*, 5th edn. Australian Institute of Food Science and Technology, NSW Branch, Food Microbiology Group, Sydney, Australia.

Holbrook, R. (2000) Detection of Microorganisms in Foods – Principles of Culture Methods. In: *The Microbiological Safety and Quality of Food Volume II* (eds B.M. Lund, T.C. Baird-Parker, G.W. Gould), Aspen Publishers Inc., Maryland, USA. pp. 1761–1790.

Krieg, N.R. (ed.) (1984) *Bergey's Manual of Systematic Bacteriology, Volume 1.* Williams & Wilkins, Baltimore, USA.

Radcliffe, D.M. and Holbrook, R. (2000) Detection of Microorganisms in Food – Principles and Application of Immunological Techniques. In: *The Microbiological Safety and Quality of Food Volume II* (eds B.M. Lund, T.C. Baird-Parker, G.W. Gould), Aspen Publishers Inc., Maryland, USA. pp. 1791–1812.

Roberts, D., and Greenwood, M. (eds) (2003) *Practical Food Microbiology.* 3rd edn. Blackwell Publishing Ltd, Oxford, UK.

Sneath, P.H.A. (ed.) (1986) *Bergey's Manual of Systematic Bacteriology, Volume 2.* Williams & Wilkins, Baltimore, USA.

12 Introduction to 'Alternative' Microbiological Methods

12.1 Introduction

The principles upon which traditional microbiological techniques and methods are based date back to the nineteenth century and the work of Louis Pasteur, Robert Koch and others. These principles have served food and public health microbiologists extremely well over the years and, indeed, they remain fundamentally unchanged today. Traditional microbiological methods employ inexpensive consumable items, e.g. Petri dishes and agar-based media, test tubes and liquid broths, as well as relatively simple equipment and they can be very sensitive and very specific. For example, it is possible to detect very low numbers of a sub-lethally damaged, target organism, e.g. <1 *Salmonella* per gram of test sample, in the presence of high numbers of an actively growing competitive microflora.

These attributes are highly desirable, but they are achieved at the expense of time and labour. Traditional microbiological methods usually have long elapse times, i.e. the time between first preparing the test sample (weighing and diluting) and obtaining the test result, especially when the method has to be very sensitive; this is because microbial cell amplification is only achieved by growth. In order to be able to see a *Salmonella* colony on an agar plate for example, the organism must be able to grow from perhaps a single cell that was inoculated onto the agar to around 10^9–10^{11} cells to produce a visible colony and such growth takes time. Despite much research, it has proved difficult to find ways that can effectively speed up the rate of cell division, especially when the organism has been previously sub-lethally damaged by a food process.

Many methods require sub-culture of the growing organisms from one medium to another in order to select and purify the target organisms and to induce the relevant metabolic processes that produce the diagnostic reactions, e.g. distinctive colours of colonies or medium, by which they can be recognised. Sub-culturing is undertaken usually at 18–24 hour intervals or multiples thereof and much of the work of preparing and inoculating growth media is labour intensive which, despite the relatively low cost of tools, equipment and consumables used, results in the relatively high cost of operating a food microbiology laboratory.

In the latter part of the twentieth century and, indeed now, in the twenty-first century, much of the food industry operates on very short production and distribution timescales, particularly the chilled food product sector. Also, to minimise the cost of storing ingredients, many food companies operate on the 'just-in-time' (JIT) principle in which their suppliers are required to deliver the raw materials at a specific

Plates 30, 31 and 35–44 are located in the colour plate section at p. 228.

time, in some cases just a few hours before they are needed for a production process. This means that there is no opportunity to obtain microbiological test results before the ingredients are used, as it may take 2 days or more before the results become available using conventional tests. Similarly, many food companies take 'in-process' factory samples for examination, i.e. samples of mixes or part-assembled product collected at different stages throughout the food manufacturing process, in order to verify that the quality standards of the process are being maintained, and that the product being manufactured has not become contaminated at an early stage. However, conventional test results are, again, not available until some time (up to several days) after the process is finished.

Many finished perishable products have a very short shelf life for which microbiological results may only become available after the product is consumed. Other products destined for consumption by vulnerable groups, e.g. babies, are subject to a positive release programme, requiring product batches to be withheld from distribution pending the availability of satisfactory microbiological test results. In each of these cases, there is a significant benefit to be gained from increasing the speed of availability of reliable results.

It is very expensive to have to discard contaminated food at the end of a production process in response to 'out of specification' microbiological test results and it is more cost effective to monitor and confirm that satisfactory control systems and hygiene standards are being maintained if possible, in real time, throughout all stages of production (see Chapter 5).

However, there is still a need for microbiological information to support and verify that such process controls are effective, so, to help in the response to increased commercial pressures that have impacted on the food industry since the 1970s, microbiological test results are required more rapidly than is achievable by conventional methods and, at lower cost; much research time has therefore been invested in the development of so-called microbiological 'rapid methods'. Although there is now a wide variety (and growing number) of these methods available, some are actually more expensive, and / or less sensitive and less specific than the traditional methods. Also, even though results may be available more rapidly than from conventional methods, this is not always at convenient times, i.e. during the normal working time of the laboratory, so some of the potential benefit for using such methods is lost.

Despite these limitations, the quest for useful applications of these newer methods has continued and accordingly the terminology has evolved; the original term 'rapid methods' has been replaced by a variety of terms, all of which more accurately reflect the wide diversity of technologies that are now applied to the microbiological examination of foods and food-related samples as replacements of traditional microbiological methods. These include descriptions such as 'modern', 'automated', 'alternative' and 'proprietary' methods, or even 'new directions in microbiological analysis of foods'. Some modern alternative methods are in widespread use, while others are used by relatively few enthusiasts. There is no common theme to the principles behind these alternative test methods, so the following sections describe the main categories of alternative method that are likely to be encountered in today's food microbiology laboratories. Regular updates on commercially available alternative microbiological test methods are available together with useful key references and the validation status of the methods (Baylis, 2000).

12.2   The evolution of 'alternative' microbiological methods

In order to understand the potential applications of alternative microbiological methods in the food industry, it is necessary to understand not only the principles upon which a specific method is based, but also its purpose and limitations, and this is most conveniently approached by highlighting some of the 'milestones' in the development of alternative methods for food microbiology.

Although the practical development of alternative methods accelerated considerably during the 1970s, some of the principles are not new; direct microscopic counts and dye reduction tests, for example, have been used in the dairy industry since the early part of the twentieth century. The principle of obtaining a biochemical profile of an organism by observing the results of its growth in different substrate-based media (originally carried out using hundreds of test tubes containing media) is one that has been applied since very early in the development of microbiology, and a modern proprietary version of this is the well-known and widely used API 20E gallery (bioMérieux SA, France) used for the identification and differentiation of Enterobacteriaceae (see Chapter 11). The distinction between traditional and alternative methods is therefore not always especially clearcut. Figure 12.1 provides an overview of some of the alternative technologies and techniques that are available and the test purposes for which they may be applied in the modern food microbiology laboratory.

The early approaches to the development of 'rapid methods' were largely aimed at reducing the time taken to obtain a total aerobic mesophilic colony count (TCC) which, by conventional methods, takes up to 3 days to obtain a result. The technologies explored and developed to achieve this have included:

- Microscopy, where, by direct observation of microorganisms from a test sample under the microscope, the operator was expected to count the number of (usually) bacteria observed (see 12.3.1 'Direct counts').
- Automatic measurement of changes in the electrical properties of a growth medium that occur as a result of microbial growth (see 'Electrical methods' in Section 12.3.2).
- Measurement of the presence and levels of adenosine tri-phosphate (ATP) in microbial cells by means of the luciferin-luciferase enzyme reaction of the firefly (see 12.3.3 'Bioluminescence').
- The *Limulus* amoebocyte lysate test which allows a measurement of the content of lipopolysaccharide, a component of the Gram-negative bacterial cell wall, to be made. The method is most commonly used to determine contamination levels of Gram-negative spoilage bacteria, especially *Pseudomonas* spp., present in a food sample (see 12.3.4 'Limulus amoebocyte lysate test').

At the same time as these developments were occurring, methods were also being explored for automating some of the repetitive aspects of traditional plating techniques and equipment such as the spiral plater (see Plate 31. See colour plate sections for all colour plates), the gravimetric diluter (see Plate 30) and automated media pourers was developed. Also, simplified, miniaturised versions of the conventional pour plate, e.g. Petrifilm™, became available. These developments did not provide a result more quickly but could allow samples to be processed more efficiently (see Chapter 10).

By the mid-1980s, it was becoming clear that, for a variety of reasons, the traditional

Figure 12.1 Some typical applications of 'alternative' techniques for the detection and enumeration of microorganisms and their toxins.

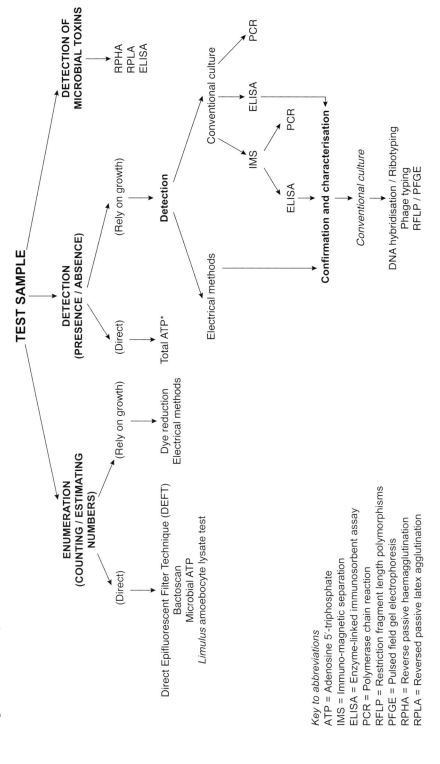

Key to abbreviations
ATP = Adenosine 5'-triphosphate
IMS = Immuno-magnetic separation
ELISA = Enzyme-linked immunosorbent assay
PCR = Polymerase chain reaction
RFLP = Restriction fragment length polymorphisms
PFGE = Pulsed field gel electrophoresis
RPHA = Reverse passive haemagglutination
RPLA = Reversed passive latex agglutination

* ATP 'Hygiene tests' measure microbial ATP **and** somatic ATP from food debris.

total colony count (TCC) method was generally preferred to alternatives by many laboratories and subsequent research and the development of alternative methods targeted the detection of specific groups of organisms, especially pathogens, as these are of greater commercial and public health importance than the TCC. Selective media were developed for use with electrical methods to facilitate the detection of specific organisms or groups of organisms. In particular, selective media were developed for the detection of *Salmonella* using the Bactometer®, Malthus System™ or RABIT conductimetry systems (see 12.3.2 'Electrical methods'), and immunological methods, notably the enzyme-linked immunosorbent assays (ELISA), became available for the specific detection of organisms such as *Salmonella* and *Listeria monocytogenes*.

Flow cytometry is an alternative microbiological technique that has evolved over several decades and incorporates developments from other sciences and branches of microbiology. It originated from a technique developed for counting particles in a very thin stream of liquid (approximately 30 μm diameter) as they passed through an electrostatically charged field but, for microbiological purposes nowadays, flow cytometry methods utilise fluorescent-labelled antibodies that are attached to the specific target organisms, fluorescence being induced by means of laser light. The technology can discriminate between particles of different size, enabling yeasts, for example, to be counted separately from bacteria.

A more recent application of the antigen–antibody reaction, however, is the use of immuno-magnetic separation in which antibody-coated, magnetic beads specifically bind and thus concentrate a 'target' genus or species of organism and then, by the use of a magnet, it is separated from any competitive microflora. This development largely took place in the 1990s and is significant because the prime objective of this 'modern' method is not necessarily to obtain a faster result or to automate a test procedure; instead it is used primarily to improve the sensitivity of a 'traditional' technique and this is, perhaps, one of the few definite modern improvements upon the fundamental approaches to the isolation of bacteria used by the early microbiologists. An immuno-magnetic separation step is an integral part of methods developed for the detection of the relatively 'new' food-borne pathogen *Escherichia coli* O157, an organism which, although it shares many metabolic and biochemical properties with other strains of *E. coli*, is not detected by the common conventional methods for isolating and enumerating 'normal' *E. coli*.

Probably the most significant developments in modern microbiological methodology are those of molecular methods, i.e. DNA-based methods, some of which can be used as 'alternatives' to traditional methods for detecting a microbial species whilst others have been newly developed for sub-species characterisation and do not have counterparts in 'conventional' microbiology. At present, such methods are largely confined to use in research laboratories and large commercial laboratories that can most effectively and economically apply them in a routine manner. However, an increasing number of companies are developing methods using this technology and it is anticipated that more cost-effective, robust and easier-to-use equipment and reagents will allow the technology to be more widely used in food industry laboratories. There is a variety of molecular methods available and only the most widely known and currently used techniques are described in this chapter (see '12.3.5 Fingerprinting methods'). For more comprehensive, but very readable, reviews of this technology and the techniques available, the reader is referred to Hocking *et al.* (1997) and Jones (2000).

Alongside these developments of alternative methods for the detection and enumeration of bacteria, there has been a parallel evolution of methods for detecting yeasts and moulds. Many of the rapid methods developed for bacteria are applicable to yeasts; however, moulds pose particular problems, largely because their mycelial growth habit creates considerable difficulty in enumerating them and assessing the extent of mould contamination of a sample. For a more detailed discussion of alternative methods for detecting moulds, the reader is referred to Williams (1989).

The potential presence of microbial toxins in food is also of concern to the food industry and these toxins can now be detected and quantified by methods that are more convenient than the traditional chemically based techniques. Although toxins occur as a result of the growth of microorganisms, the traditional methods of detection have really been more appropriate to the work of a chemistry laboratory. However, the development of modern test methods that are based largely upon immunological techniques (antigen–antibody reactions), e.g. ELISA and latex bead agglutination tests, has now placed the responsibility for their detection reasonably firmly within the sphere of activity of the food microbiologist.

12.3　The principles and applications of some 'alternative' microbiological methods

As indicated above, traditional microbiological techniques employ relatively inexpensive consumable items, e.g. Petri dishes and media, as well as relatively simple equipment; by far the most expensive items of equipment for the small laboratory are the laboratory autoclave and incubators. In contrast, chemistry laboratories regularly employ sophisticated equipment that is expensive to buy but allows many samples to be processed by few technicians and these laboratories therefore have relatively low operating costs. The development of alternative microbiological methods in some ways replicates the approach taken in chemistry i.e. to reduce the elapse time of microbiological tests as well as to make microbiological testing more cost-effective. Alternative microbiological methods have been developed to detect the microorganisms themselves, their components or the products of their metabolism.

12.3.1　*Direct counts*

In direct detection methods, the microorganisms are 'observed' directly, either by the laboratory technician using a light microscope or automatically by a modern electronically operated instrument. In some cases, the count obtained includes dead cells as well as live cells, whilst in others, so-called 'vital stains' react with the genetic material of dead and live cells in different ways so that dead and live organisms can be distinguished by different colours.

Direct examination techniques using light microscopy are limited by the fact that they have high limits of detection, i.e. the lowest concentration (number) of bacteria that can be enumerated is usually around 500 000 per ml because of the mathematical limitations imposed by the area of the specimen that can be viewed under the microscope. The simplest of the direct counts is the 'Breed smear' that has been used for assessing the hygienic quality of milk. From this most simple, and old, technique, sample preparation methods have been developed that concentrate the organisms from a sample, and these are then observed by microscopy; this approach has made such tests more sensitive.

If the organisms in a test sample can be concentrated and then suitably stained, then a direct viable organism count can be made. In the first modern development of this technique, the Direct Epifluorescent Filter Technique (DEFT) (Pettipher, 1983), the test sample (raw milk) is treated with enzymes and detergent to 'clarify' the liquid and the organisms then concentrated by filtration. The clumps of bacteria on the filter are then stained with acridine orange and viewed using an epifluorescence microscope; viable and non-viable organisms can be either counted manually or automatically using computerised, image analysis equipment.

A significant development of the DEFT is the Bactoscan instrument (Foss, Denmark) which is probably the most widely used example of a proprietary technique that employs the direct detection of microorganisms. It is used in many countries to assess the hygienic quality of ex-farm milk, farmers being paid more when lower numbers of organisms are detected. In the Bactoscan 8000S instrument, the milk sample is digested in a pre-treatment to allow extraction of the microorganisms by centrifugation. These are then stained with acridine orange; the individual stained cells are applied to the edge of a rotating disc and the dye in the cells is 'excited' by light of a specific wavelength to induce fluorescence; fluorescing particles are then counted via a photomultiplier. The number of electronic impulses recorded is then converted into the number of individual bacterial cells per millilitre of milk. It takes about 5 minutes to process a sample and the detection limit (sensitivity) is around 10^4 cells/ml. A more recent development, the Bactoscan FC, employs flow cytometry to detect the individual, viable cells.

Attempts have also been made to develop tests capable of directly detecting specific target organisms, e.g. *Salmonella* in test samples, by means of fluorescent-labelling of antibodies that attach specifically to the target organism. However, by and large these have been unsuccessful as they lack sufficient sensitivity and / or specificity, leading to the possibility of incorrect results. In the case of tests such as the TCC, incorrect results are undesirable but generally not disastrous for the food processor; however, in the case of pathogenic bacterial contaminants, the consequences of incorrect results are very much more severe and, to date, direct methods have been unable to provide the level of security required by the food industry and the consumer.

Nevertheless, specific organisms can be 'tagged' using biochemical markers that detect or measure enzyme activity and cell membrane integrity; these markers are, in turn, detected using sophisticated sensors, such as laser-based, optical-electronic analysers with computerised data handling systems. For example, the Chem*Scan*® and D-*Count*® techniques (Chemunex SA, France) employ flow cytometry in the detection of viable, stained cells. Chem*Scan*® is used for filterable samples whilst the D-*Count*® has been developed for samples that cannot be filtered. Viable microorganisms are stained with 'viability reagents', i.e. enzyme substrates that are transported through the cell membrane and are cleaved by cytoplasmic enzymes to produce a fluorescent compound. After the 'staining' step, the sample material is injected into a flow cell and the microorganisms are focused into a narrow stream where excitation by a laser beam causes any fluorescent molecules present to fluoresce. This is detected by photomultiplier tubes whose signal is then subjected to digital data analysis to discriminate between microorganisms and autofluorescing particles. The test time depends on the sensitivity required, i.e. whether the food sample is examined directly or after a conventional enrichment step; for direct detection, a test result can be obtained within 1 hour compared with 20 hours for the more sensitive enrichment protocol.

12.3.2   *Detection of metabolic activity during sample incubation*

Dye reduction tests

The simplest form of microbiological technique that detects metabolic activity is the dye reduction test, which has been used by the dairy industry for many decades. Most 'dairy' bacteria have a fermentative metabolism and, when present in high numbers, are capable of lowering the oxidation-reduction (redox) potential of their growth medium because they produce 'reducing compounds'.

Solutions of dyes such as resazurin and methylene blue change colour at low redox potential, or E_h value, and, if a sample of milk is mixed with resazurin or methylene blue and incubated (at 37°C), the dye decolourises within the test time (up to 2 hours depending on the test requirement) if a high number of viable microorganisms is present, but remains the same colour if the numbers of viable organisms are relatively low. These tests have been incorporated as part of the quality monitoring system for raw milk supplied to dairies, being used as simple and rapid hygiene / microbial quality indicators for raw milk. If a sample decolourises under the test conditions applied, the original sample is deemed to be of unacceptable quality and, conversely, a sample that remains the same colour throughout the incubation period is deemed of acceptable microbial quality. Although these tests were used for many years as an industry standard, they are not totally reliable because different bacteria cause different rates of reduction of the dyes under the same test conditions. In many modern dairy companies today, raw milk quality is controlled by a combination of quality assurance programmes supported by conventional microbiological tests and direct fluorescence detection-based tests to assess total viable microbial loads.

The dye 2,3,5-triphenyltetrazolium chloride (TTC) has been incorporated into agar media to assist total colony count estimations. As organisms metabolise and grow, the TTC is reduced to formazan which is a red-coloured compound that can allow the colonies to be seen somewhat earlier in the incubation period. This technique, however, is not widely practised in food microbiology laboratories today.

Electrical methods

Electrical methods for detecting microbial activity have one feature in common with dye reduction tests in that the number of organisms initially present in a sample determines whether or not a response is triggered during the incubation period; however, instead of triggering a colour change, organisms are detected by their ability to increase or decrease the electrical conductivity of the growth medium.

The test sample inoculum is dispensed into a well or cell containing two electrodes, as well as a growth medium especially designed to maximise any changes in electrical signal caused through microbial metabolism. Every few minutes during incubation (commonly every 6 or 12 minutes), an alternating current is passed between the electrodes and the conductivity of the medium is recorded. As incubation proceeds, conductivity changes are plotted graphically against time and the 'detection time' is calculated. The detection time is the earliest time that an accelerating change in conductivity is detected; it is inversely related to the numbers of organisms present in the original sample. The longer the detection time, the fewer organisms were present and so a quantitative microbiological result can be obtained (Figure 12.2). Electrical methods provide a result that is an approximation of the TCC of a test sample; however, since bacteria tend to increase conductivity by the production of ionic molecules whilst, in many media, fungi and yeasts tend to decrease the conductivity, possibly by absorbing ionic molecules, different samples of similar material do not always give repeatable results because they may contain mixed cultures that are

Figure 12.2   An illustration of the application of an electrical method, adapted from Easter and Gibson (1989).

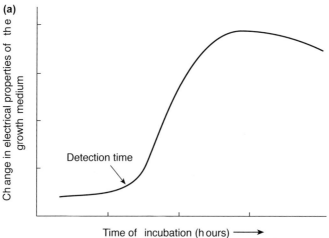

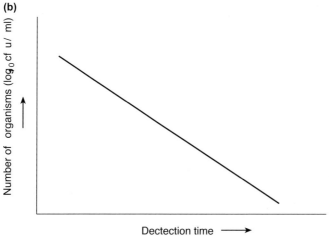

(a) A 'detection curve' indicating the detection time, i.e. the time of incubation at which there is an accelerating change in the electrical properties of the growth medium.
(b) A 'calibration curve' showing the relationship between the numbers of organisms in a test sample and the detection time in hours. Note the higher the number of organisms present, the shorter the detection time.

unlikely to be consistent between one sample and the next. The principles of electrical methods have been discussed in detail by Firstenberg-Eden and Eden (1984) and Easter and Gibson (1989).

By adding selective agents and / or specific substrates to the growth medium and by detecting the signal produced by the organisms capable of growth, it is possible to detect specific organisms, e.g. *Salmonella*. For this organism, detection is accomplished by adding sodium biselenite as a selective agent and by measuring the conductance change in the medium due to the specific conversion of trimethylamine oxide to trimethylamine (Easter and Gibson, 1985). For such tests, the organisms are not quantified; their detection or non-detection is merely determined by recording a conductivity change or lack of change during a specified incubation period. Pathogen detection tests are therefore a useful application of this technology, especially as the commercial instruments are designed with incubators that can accommodate 128 or even 256 samples, and several incubators can be linked to the computer monitoring and recording system, allowing many tests to be undertaken simultaneously with results recorded automatically.

Three main proprietary instruments employing electrical detection systems are used in food microbiology, although other systems are available. The Bactometer® (bio-Mérieux SA, France) was probably the first instrument to be developed commercially although this was quickly followed by the Malthus System™ (IDG (UK) Ltd) and, a little later, by the RABIT (Don Whitley Scientific Ltd, UK). All three instruments employ similar principles but differ mainly in the frequency of electric current that is passed through the test material and in the type of electrical signal detected, i.e. conductance, capacitance or impedance (Firstenberg-Eden and Eden, 1984).

12.3.3 *Detection of metabolic activity without sample incubation*

Bioluminescence

Dye reduction tests and electrical methods detect some aspects of microbial metabolism and so employ an incubation step, which means that the test result only becomes available after the required incubation time, which for electrical methods can be over 24 hours. However, it is possible to measure the levels of adenosine 5'-triphosphate (ATP) in a test sample within a few minutes, and this can provide an estimate of the total microbial population present in the test sample.

ATP is universally present in all living cells and is involved in a range of reactions related to the conversion or expenditure of energy. The quantity of microbial ATP within a microbial population is usually proportional to the number of living organisms present. ATP is also involved in the reaction that leads to light emission in fireflies, the 'luciferin-luciferase' enzyme reaction, known as bioluminescence (Figure 12.3). It is this reaction that is used in tests to quantify the level of microbial ATP present after its extraction from the microorganisms in a test sample, the amount of light produced being proportional to the ATP concentration and hence to the number of organisms present in the original sample. A photo-multiplier tube is used to detect the amount of light emitted; this information is converted to digital data that are recorded and can be used to estimate the number of microorganisms present in the original sample.

Figure 12.3   An illustration of the firefly luciferin-luciferase reaction used to measure microbial ATP.

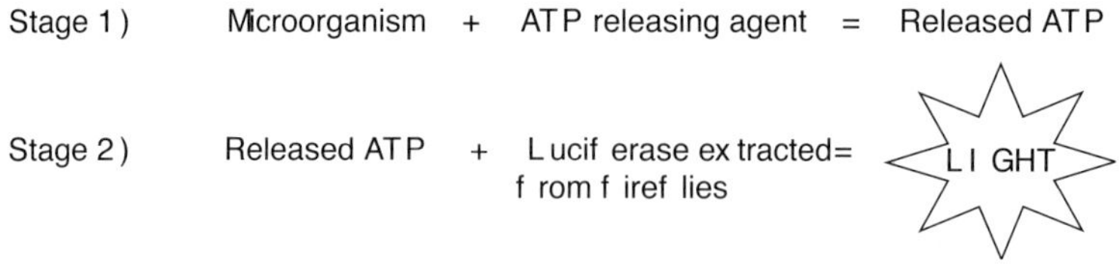

Stage 1)   Microorganism   +   ATP releasing agent   =   Released ATP

Stage 2)   Released ATP   +   Luciferase extracted=   ⟨LIGHT⟩
                                     from fireflies

Note that ATP is first released from the microorganisms in the test sample (Stage 1) and it then reacts with the enzyme luciferase (ATP reagent) to produce light (Stage 2). This is measured in a luminometer.

However, non-microbial ATP, e.g. somatic ATP, may also be present in the test sample originating from the food material or other sources and this has to be destroyed or removed from the test sample before the microbial ATP is released from the microorganisms and measured. This is done either by filtration or by enzyme pre-treatment; in the latter case, the enzyme then has to be removed or inactivated before the level of microbial ATP can be measured after its release from the cells. ATP-based methods have been reviewed in detail by Stannard (1989).

The sensitivity of ATP-based tests for estimating microbial numbers in test samples is quite low at around 10^3 to 10^4 organisms and this, combined with the practical issues of reliably separating microbial from somatic ATP, has hindered the development of a simple test protocol for use with microorganisms in foods; the use of ATP-based methods for assessing the microbial load of food samples has therefore not been widely taken up by the food industry. A simple ATP-based method, however, is very widely used in the food industry today for 'hygiene monitoring' of equipment surfaces after cleaning.

In any food processing environment, the production equipment becomes contaminated both with food debris and with microorganisms during the production operation. The purpose of a cleaning and disinfection procedure is two-fold: firstly, in the cleaning stage, to remove food residues and the majority of microorganisms both physically and chemically from the equipment surfaces and secondly, to destroy any remaining microorganisms, usually by chemical disinfection or by heat treatment. Since many food residues contain ATP as do viable microorganisms, any traces of ATP that remain on equipment surfaces after cleaning can indicate that the procedure has not been effective. Thus, a simple method that detects both microbial and somatic ATP on a swab sample taken from production equipment after cleaning can indicate to production personnel very quickly (within a few minutes) whether or not the cleaning procedure has been satisfactory.

Since the original commercial development of ATP bioluminescence assays, undertaken principally by two companies (Biotrace Ltd, UK and Celsis Ltd, UK), a number of companies have marketed a range of similar tests for use in the laboratory or in production areas. An automated, benchtop luminometer, used to measure the light produced in an assay to assess the numbers of microorganisms contained in a test sample, is shown in Plate 35. A variety of commercial hygiene monitoring tests / kits is available, and these are used to indicate whether or not a food product contact-surface has been cleaned adequately by measuring the total ATP collected on a swab taken from processing equipment, usually just before production begins. Plate 36 shows a hygiene monitoring swab used in ATP-assays, whilst a portable, hand-held luminometer is shown in Plate 37. Because bioluminescence tests for detecting and measuring ATP can provide a result within minutes, they have become widely used in food industry hygiene monitoring programmes.

12.3.4   *Detection of cell components*

Limulus *amoebocyte lysate test*

The *Limulus* amoebocyte lysate (LAL) test has been used in the pharmaceutical industry for many years to confirm the safety of solutions for injection with respect to the absence of pyrogenic (pyrogens cause fever in the patient) substances. It detects the lipopolysaccharide (LPS) contained within the cell walls of Gram-negative bacteria. LPS (known as a pyrogen) reacts with an enzyme present in the blood of the horseshoe crab (*Limulus polyphemus*) to produce a clotting factor that makes the blood gel. This forms the basis of a gel titration test, although a version with a colorimetric end-point has also been developed. A range of tests, in different formats, is produced by Associates of Cape Cod, Inc., USA.

In the gel titration test, dilutions of the test sample are prepared and added to the LAL in the wells of a micro-titre plate which is then incubated for 1 hour. The wells are observed for gel formation and the most dilute inoculum (highest dilution) that has caused a gel to form is recorded. From this dilution, the concentration of LPS, and hence the concentration of Gram-negative bacteria, in the original sample can be calculated.

Although Gram-negative bacteria are destroyed by even a mild heat treatment, LPS is highly heat-resistant and can survive the Ultra Heat Treatment (UHT) used to sterilise milk. The test therefore detects LPS from both dead and live organisms and has been used in the examination of UHT milk to estimate the levels of Gram-negative psychrotrophs present in the milk before it was sterilised, i.e. to determine the hygienic quality of the raw milk from which the UHT milk was made. This is important as some Gram-negative organisms produce heat-resistant protease enzymes that also survive UHT processing and can spoil the milk even though the milk has remained sterile. Such a test is potentially highly relevant to many other areas of the modern food industry since many foods are subject to spoilage by Gram-negative psychrotrophic bacteria, such as *Pseudomonas* spp.

Immunological assays

Immunology is the study of the immune response that a host animal produces to the presence of antigens (agents that cause an immune response), which can be either soluble, e.g. toxins, or particulate, e.g. microorganisms. When an antigen enters an animal's blood stream, the animal (or 'host') produces antibodies (proteins) that usually specifically bind to the antigen and help to prevent it causing damage or illness to the host. It is the specificity of the antibody–antigen reaction that can be used in tests to detect specific microorganisms, since all microorganisms carry antigens on their cell surfaces. In order to manufacture a proprietary test, an antibody specific to the target organism, e.g. *Salmonella*, is required and this is done by injecting animals with an inactivated culture of the organisms so that they produce antibodies in their bloodstream; these antibodies are then harvested, purified and used to manufacture the immunological test.

Serology

The simplest form of immunological reaction used in microbiological tests is the direct mixing of an antibody with the suspected antigen. If the antibody 'recognises' the antigen (an antibody interacts specifically with the antigen that caused (elicited) its production) then the two fit together like a 'lock and key' (Jones, 2000) and the binding of many antibodies to the bacterial cells causes the bacteria to agglutinate and form 'clumps' in a cell suspension that can be seen with the naked eye within a few seconds. This reaction is normally used in confirmation tests for the identification of bacteria, especially *Salmonella* (see '11.5.5 Slide agglutination tests' in Chapter 11) and is not categorised as an 'alternative' microbiological technique. The same interaction, however, is employed in several tests used as alternatives to conventional microbiological methods in modern food microbiology.

Enzyme-linked immunosorbent assays

Enzyme-linked immunosorbent assays (ELISAs) employ the antigen–antibody reaction but, instead of visualising the clumping interaction, the antibodies are linked to an enzyme (commonly horseradish peroxidase or alkaline phosphatase) that catalyses a colorimetric reaction. The colour intensity produced is proportional to the concentration of antigen present and, in theory, it is possible to make such tests quantitative. However, there are a number of difficulties in obtaining a reliable, quantitative result from these tests and most are used as qualitative screening tests for the detection of specific pathogens in foods.

There are a number of different commercially available formats of ELISAs (Jones, 2000), but one of the most common is the 'sandwich' ELISA (Plate 38). First, the specific antibodies are bound to an inert surface (also called the 'solid phase'), such as the wall of the well of a micro-titre plate; this is done during the manufacture of the test kit. When the ELISA is used in the laboratory, the test sample is prepared according to the manufacturer's instructions, an aliquot of the preparation is added to the well and (if present) the antigen binds to the specific antibodies in the well; any unbound antigen is then washed off and an antibody–enzyme conjugate (prepared as a reagent by the manufacturer) is added and allowed to bind to the bound antigen, creating a 'sandwich'. Finally, the enzyme substrate is added and incubated to allow the reaction to occur and the colour to develop. The colour intensity produced in this 'non-competitive' sandwich ELISA is proportional to the antigen concentration present. By contrast, in a 'competitive' ELISA, the antigen is bound to the solid phase and antibody is added to the well together with the test sample preparation; more antibody binds to the bound antigen if there is little or no antigen in the test sample and *vice versa*; therefore, a high colour intensity develops when antigen is absent from the test sample and the colour intensity is reduced by competitive binding if the antigen is present.

Most ELISAs take less than 30 minutes to complete and have found particular application in tests for the detection of *Salmonella* and *Listeria monocytogenes*, but, as with many alternative methods, they have a low sensitivity, requiring around 10^4–10^6 of the target organisms/ml to produce a positive result. They are therefore almost always used after an enrichment step(s) similar to those used for conventional methods (see Chapter 10), which improves both the sensitivity and the specificity of the ELISA test by raising the numbers of the target organism and reducing the numbers of or eliminating related organisms that might cross-react with the specific antibody causing a false positive reaction. Enrichment steps involve incubation, which usually takes at least 24 hours, and, for the detection of pathogens, ELISAs essentially replace the plating stage, thus saving the time required for plate incubation. It is important to note that any positive ELISA results usually require confirmation by conventional microbiological methods.

The commercial development of ELISAs in food microbiology has been undertaken principally by three international companies (Organon-Teknika, bioMérieux SA, France and Tecra Diagnostics UK), and reagent kits are available for the detection of most of the common food-borne pathogens (Plate 39). Samples of enrichment broth (heat-treated to release antigens), together with positive and negative controls, are dispensed into the wells of the micro-titre plate, already coated with the specific antibody for the target organism (Plate 40) and any antigen in the test sample attaches to the antibody in the plate. The plate is then washed to remove unbound material, either manually or using an automated plate washer (Plate 41). The antibody conjugate is then added and allowed to bind to the bound antigen during a short incubation. The plate is washed again, the enzyme substrate is added and colour develops in those wells where binding has occurred; results can be 'read' visually (Plate 42) or using an automated plate reader (Plate 43) that provides a numerical, printed result.

In addition to micro-titre plates, ELISA-based tests have now been developed in a range of different formats. Some companies have produced fully automated instruments suitable for use in large laboratories, e.g. the EiaFoss instrument (Foss, Denmark) or the VIDAS® instrument (bioMérieux SA, France) in which the prepared sample is inoculated onto a reagent strip that is then processed automatically by instrument; a positive reaction is indicated by fluorescence that the instrument reads and records automatically. Other companies have developed extremely simple procedures, e.g. immuno-chromatographic units (Path-Stik (Celsis Ltd, UK)) in which the reagents are immobilised on a 'dipstick' that is dipped into suitable sample material that is drawn over the reagents by capillary action (Plate 44) and a coloured line develops if the specific target antigen, e.g. *Salmonella*, is present (Radcliffe and Holbrook, 2000).

Immuno-magnetic separation

Immuno-magnetic separation (IMS) is an enrichment technique designed to increase the sensitivity of conventional microbiological methods rather than to automate the test or to provide a more rapid result. Small ($< 100\,\mu$m) magnetic beads are coated with antibodies to the target organism. The coated beads are used to capture specifically the cells of the target organism from enrichment broths. The magnetic property of the beads is then used to separate the cells from the broth and these are removed and plated out onto selective agar. The conventional approach is then followed to isolate and identify the organism present.

Immuno-magnetic beads (Dynabeads) have been developed by Dynal Biotech (Norway) for the separation of bacteria and parasites from food and environmental samples. The technique is probably most widely used for the detection of *E. coli* O157 which is otherwise difficult to detect using conventional enrichment and plating methods, probably because it is a poor competitor and rapidly becomes overgrown in the enrichment broth by other organisms. However, similar products are also available for application in the isolation of *Salmonella*, *Listeria* and *Cryptosporidium*. In most laboratories, the immuno-magnetic step is done manually but an automated instrument capable of processing 15 samples in 20 minutes is also available.

For the isolation and detection of *E. coli* O157, the sample is incubated in an enrichment broth for 6 hours (and usually a further 22 hours) during which time the organism multiplies. After both 6 and 28 hours incubation, an aliquot of the incubated broth is pipetted into a capped tube; commercially available immuno-magnetic particles are then added; the target organism binds to the antibodies on the magnetic particles, then a magnet is applied to the outside of the tube to immobilise the beads at the side of the tube. The supernatant liquid is then removed and the beads are washed several times to remove unbound organisms before they are finally re-suspended in sterile buffer solution and spread onto a conventional selective diagnostic agar that is incubated for 18–24 hours and then assessed conventionally. Any suspect colonies are further identified using biochemical tests and by antibody-based agglutination tests (Radcliffe and Holbrook, 2000)

12.3.5 *Fingerprinting methods*

Because of the importance of bacterial pathogens to the food industry, it is sometimes necessary to further discriminate the identity of the strain isolated to sub-species level, particularly in surveillance programmes of foods or the environment as well as in the investigation of outbreaks of illness and for tracing organisms in food processing and environmental / ecological situations. There are a number of methods used for sub-classifying pathogenic bacteria including biotyping, serotyping, phage typing, antibiotic resistance patterns (resistotyping), various molecular typing methods including pulsed-field gel electrophoresis (PFGE) and ribotyping. This latter technique has been automated and is quite commonly used to generate comparative deoxyribonucleic acid (DNA) profiles of microbial strains in problem-solving investigations and strain tracing studies.

Methods are now commercially available for classifying pathogenic bacteria at levels below the species level, i.e. sub-species classification. Because these techniques are more highly specialised than basic microbiological techniques, they are not used in most food industry / contract testing service laboratories; when this level of differentiation is required, it is recommended that the services of specialist laboratories, e.g. national public health sector or research institutes / associations, are employed for their expertise in these techniques and the interpretation of the results obtained. Further information about these techniques can be obtained from Old and Threlfall (1998), Hill and Jinneman (2000) and Jones (2000).

DNA/RNA hybridisation ('gene probes')

Deoxyribonucleic acid (DNA) is a very long, 'ladder-like' molecule composed of two complementary polynucleotide chains, normally coiled in a double helix, that carries all the information necessary to make a specific living organism. Ribonucleic acids

(RNA) are similar molecules but usually single stranded, that 'convert' information stored in DNA for building proteins. The complementary polynucleotide chains of DNA can be separated into their single-stranded components and hybridisation is the artificial formation of the complementary chain from an existing single-stranded polynucleotide chain to produce another double stranded nucleic acid molecule.

DNA and RNA hybridisation techniques are both used in the laboratory to detect specific sections of target DNA or RNA and can thus confirm the identity of specific pathogens, e.g. *Salmonella* and *Listeria monocytogenes*. The methods exploit the fact that single strands of these molecules can hybridise with their complementary strand and, because a single strand of DNA will hybridise with a complementary strand of RNA, a DNA probe can be used to detect RNA which is present in cells in much higher numbers of copies than DNA which is often present as only one copy per cell. First, single-stranded sections of DNA or RNA, complementary to a gene sequence that uniquely identifies the target organism, are prepared artificially by the test manufacturer; these are known as DNA or RNA 'probes' and these are coupled to a detection system similar to those used for immuno-assays, e.g. an enzyme-linked system that yields a coloured or fluorescent end-product or a radioisotope. In the laboratory, a test sample is treated in a specific series of steps according to the test kit manufacturer's instructions and mixed with the probe which 'seeks' its complementary strand within the sample and then binds to it. The specific target gene sequence in the test sample therefore becomes 'labelled' and a coloured end-product, fluorescence or radioactive reaction (depending on the label used) can be detected using appropriate equipment.

Because a particular DNA target sequence may be present as only a single copy per cell, relatively large numbers of bacteria are needed to provide enough genetic material for the results of any hybridisation to be detectable; however, cells usually carry multiple copies of RNA, and RNA probes are therefore potentially more sensitive. To increase the sensitivity of DNA and RNA hybridisation techniques further, they are generally applied either to an aliquot of an enrichment broth after incubation or to a colony that has grown on a selective diagnostic agar.

One of the first commercially available DNA hybridisation tests is the AccuProbe® (Gen-Probe, USA) that is especially useful for the rapid identification of *Listeria monocytogenes* when a sample has yielded suspect *Listeria* colonies on diagnostic agar. A simple test procedure results in a luminescent reaction, within 1 hour or so; the reaction is 'read' in a luminometer that clearly displays the result as 'positive' or 'negative'.

Restriction fragment length polymorphism analysis and ribotyping

Restriction Fragment Length Polymorphism (RFLP) is the basis of 'genetic finger-printing'. RFLP analysis employs 'restriction enzymes' to cut DNA at specific sites resulting in sections (or 'fragments') of particular sizes. These DNA fragments are then subjected to an electrophoretic process that separates them on the basis of size. To do this, the DNA fragments are placed onto one end of a strip of gel and an electric current is applied which draws them towards the opposite end at different velocities depending on their molecular size, the smaller fragments moving more quickly than the larger sections of DNA. After separation, the fragments are visualised, usually by means of a suitably labelled DNA probe. This reveals a pattern of bands that can be used to compare different isolates of the same species.

Ribotyping is essentially RFLP analysis of a specific set of microbial genes that carry the information for ribosomal RNA. This technique has been commercialised through the development of automated equipment, e.g. the RiboPrinter® (Qualicon, USA) that can generate DNA profiles and produce a genetic fingerprint of an organism in around 8 hours; automation ensures a high degree of reproducibility and enables banding patterns to be compared with a database or ribotype library of many hundreds of patterns held within the instrument, using computerised software.

One limitation of RFLP is that large DNA fragments cannot be separated efficiently by conventional (unidirectional) electrophoresis. Greater separation can be achieved in the presence of a pulsed, or alternating, electric field, applied before the DNA probe is used, hence the term pulsed field gel electrophoresis (PFGE). Eukaryotic organisms, e.g. yeasts and fungi, have a number of chromosomes containing DNA in their nuclei and can often be differentiated by applying PFGE to a whole cell lysate, producing banding patterns by which different isolates can be compared without the need to cut the DNA into fragments. Closely related strains, however, may yield banding patterns that are very similar and, in this case, 'rare-cutting' enzymes can be used to improve the discrimination of the technique. Rare-cutting enzymes are restriction enzymes but, unlike those used in RFLP, produce only a few, large fragments of DNA per chromosome. In the case of prokaryotes, i.e. bacteria, restriction enzymes are always used because these microorganisms have only a single or very few chromosomes of DNA and so cannot be differentiated without the preparation of small chromosomal fragments.

The polymerase chain reaction

The polymerase chain reaction (PCR) is a technique that employs an enzyme (*Taq* polymerase) and short pieces of commercially-prepared DNA (primers) to amplify by thousands of times the number of copies of a pre-selected region of DNA from the target organism to a level sufficient for testing. Like hybridisation techniques, it employs the fact that DNA is a double-stranded molecule that can be separated into two complementary strands. First, the double-stranded DNA from organisms in the test sample mixed with the PCR reagents is separated into its two complementary strands by heating the 'reaction mixture'; then, secondly, the mixture is cooled slightly, the primers bind to each DNA strand either side of the target region and act as starting blocks for the enzyme (polymerase) to make copies of the target sequence. This results in two double-stranded pieces of target DNA. The temperature is then raised and the cycle begins again; often 20 or more cycles are carried out, after which thousands of copies of the target sequence are present and a detection procedure is then applied so that the amplified material can be visualised (Jones, 2000).

There are commercially available PCR-based tests for detecting and identifying some bacterial pathogens in foods, e.g. *Salmonella*, but as yet the technique is not in widespread use in food industry laboratories. BAX® systems (Qualicon, USA) are PCR-based screening methods for detecting the presence of food-borne pathogens in test samples. The first reagents developed by this company were designed for use with manual, gel-based detection techniques but now, a temperature cycler and detector have been developed to automate the process for large numbers of samples, providing results within 4 hours. Where such techniques are used, the specific manufacturer's instructions must be followed to ensure reliability of results.

Enrichment of food and food-related samples

Molecular microbiology (genetic) techniques have provided a very significant step forward in allowing the characterisation of microbial strains to the sub-species level, and the tests described above are usually applied to pure cultures of microorganisms previously isolated from foods by traditional microbiological techniques. However, it is also possible, in theory at least, to apply molecular techniques directly to food samples or to enrichment cultures after incubation.

   In these cases, it is important to interpret any positive result with caution because in general, molecular methods do not distinguish between the DNA from dead or dying cells and 'live cell DNA'. Where such a technique is applied directly to a food sample (as opposed to a culture isolated from the food), a positive result might wrongly suggest that the food was unfit for consumption or 'out of specification', e.g. if the food had been heat-processed to kill any pathogens present in the raw material; a positive result might be due to an organism that contaminated the sample before heat treatment and had subsequently been killed / inactivated in the process, rather than being a viable post heat-process contaminant.

   One means of overcoming this complication is to apply the molecular method to an enrichment culture of the sample after incubation. In an enrichment culture, any dead cells present in the original sample are likely to be at a level too low to detect and a positive reaction may thus be presumed to be due to a live organism that has grown to a detectable number during incubation. Further tests can verify this.

12.4 Microbial toxins

Tests are available for the detection of microbial toxins in foods, one of the earliest developed being the double immuno-diffusion technique for the detection of staphylococcal enterotoxins (Holbrook and Baird-Parker, 1975). In this technique, a highly purified extract from the food sample is inoculated into wells cut into the agar in a Petri dish (known as an Ouchterlony plate) in close proximity to wells containing individual antitoxins prepared against the different types of staphylococcal enterotoxin. The plate is incubated overnight or for several days to allow both the toxin, if present, and antitoxin to diffuse towards each other. Where a toxin encounters its homologous antitoxin, a precipitation reaction occurs that can be seen as a white precipitin line between the two wells.

   A more rapid (approximately 4-hour) test, which needs less sample preparation to extract and purify the toxin, is reverse passive haemagglutination (RPHA). In this test, dilutions of the test sample extract are dispensed into the V-shape bottomed wells of a micro-titre plate. A suspension of washed, sheep red blood cells, that have been sensitised by coating with antibody prepared against a specific staphylococcal enterotoxin, is then added to each well and the plate is incubated. If the sample contains no enterotoxin, the red blood cells settle to the bottom of the well and form a tight 'button' in the V shape. If the sample contains homologous enterotoxin, an immune reaction occurs with the IgG antitoxin, and this cross-links the red blood cells, forming a diffuse matrix that prevents the cells from settling, and no 'button' is observed. The test is described as 'reverse' because in RPHA the purified antibody (to the enterotoxin) is attached (coupled) to a particle (treated red-blood cell) whilst the antigen (the enterotoxin) remains in solution i.e. it is soluble, which contrasts with conventional serological agglutination assays, e.g. to detect *Salmonella*, where the antigen is particulate i.e. the bacterial cell and the antibody is in solution; also, RPHA is described as 'passive' because the particles e.g. red-blood cells, play no direct part in the immune reaction and simply provide a means of visualising the antigen–antibody response.

Several commercial tests have been developed that employ inert, latex beads instead of sheep red blood cells but otherwise use the principles of RPHA. These are known as reversed passive latex agglutination (RPLA) tests and are available to detect staphylococcal enterotoxins A, B, C and D, *Vibrio cholerae* enterotoxin and *E. coli* heat-labile (LT) enterotoxin, *Clostridium perfringens* enterotoxin, *Bacillus cereus* diarrhoeal toxin and *E. coli* Vero cytotoxins VT1 and VT2.

12.5 Further reading

Baylis, C. (2000) *Catalogue of Rapid Microbiological Methods, Review No. 1*, 4th edn. Campden & Chorleywood Food Research Association, Chipping Campden, UK.
Grange, J.M., Fox, A. & Morgan, N.L. (eds) (1987) *Immunological Techniques in Microbiology*, Society for Applied Bacteriology Technical Series No. 24, Blackwell Scientific Publications, Oxford, UK.
Hill, W.E. & Jinneman, K.C. (2000) Principles and Applications of Genetic Techniques for Detection, Identification, and Subtyping of Food-Associated Pathogenic Microorganisms. In: *The Microbiological Safety and Quality of Food* (eds. B.M. Lund, T.C. Baird-Parker & G.W. Gould), Aspen Publishers, Inc., Maryland, USA, pp. 1813–1851.
Hocking, A.D., Arnold, G., Jenson, I., Newton, K. & Sutherland, P. (eds) (1997) *Foodborne Microorganisms of Public Health Significance*, 5th edn. Australian Institute of Food Science and Technology, NSW Branch, Food Microbiology Group, Sydney, Australia.
Jones, L. (2000) *Molecular methods in food analysis: principles and examples*. Key Topics in Food Science and Technology, No. 1. Campden & Chorleywood Food Research Association, Chipping Campden, UK.
Old, D.C. and Threlfall, E.J. (1998) Salmonella. In: *Topley and Wilson's Microbiology and Microbial Infections*, 9th edn. *Volume 2, Systematic Bacteriology* (eds. A. Balows and B.I. Duerden), Arnold, London, UK.
Pettipher, G.L. (1983) *The Direct Epifluorescent Filter Technique for the Rapid Enumeration of Microorganisms*. Research Studies Press Ltd, Letchworth, UK.
Radcliffe, D.M. & Holbrook, R. (2000) Detection of Microorganisms in Food – Principles and Application of Immunological Techniques. In: *The Microbiological Safety and Quality of Food* (eds. B.M. Lund, T.C. Baird-Parker & G.W. Gould), Aspen Publishers, Inc., Maryland, USA, pp. 1791–1812.
Stanley, P.E., McCarthey, B.J. & Smither, R. (eds.) (1989) *ATP Luminescence: Rapid Methods in Microbiology*, Society for Applied Bacteriology Technical Series No. 26. Blackwell Scientific Publications, Oxford, UK.
Stannard, C.J., Petitt, S.B. & Skinner, F.A. (eds) (1989) *Rapid Microbiological Methods, Beverages and Pharmaceuticals*, Society for Applied Bacteriology Technical Series No. 25. Blackwell Scientific Publications, Oxford, UK.
Wyatt, G.M., Lee, H.A. & Morgan, M.R.A. (1992) *Immunoassays for Food Poisoning Bacteria and Bacterial Toxins*. Chapman & Hall, London, UK.

Glossary of Terms

Accuracy The ability of a measuring device to provide a true estimate of a value (measurement), on an average of multiple readings, irrespective of precision. (Precision is the ability of a device to provide values in close agreement with each other, on an average of multiple readings, irrespective of accuracy.)
The 'target diagram' below illustrates the difference between precision and accuracy.

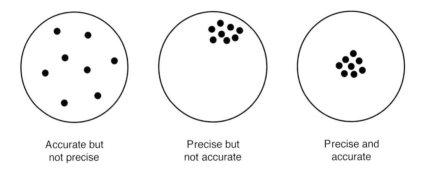

| Accurate but not precise | Precise but not accurate | Precise and accurate |

The ideal measurement is both accurate and precise.

Aerobe A microorganism that is able to grow in the presence of oxygen at levels found in air i.e. approximately 21%.

Aerobic Describes an environment in which oxygen is present at a similar partial pressure (percentage) to that of air.

Aerotolerant Describes an organism that is an anaerobe but is able to grow in the presence of small quantities of oxygen.

Agar A gel composed of carbohydrate and derived from seaweed that is used to solidify liquid, nutrient media and thus enables bacteria, yeasts and moulds to form colonies on or in a solid matrix. (See also Slope / Slant.)

Algae Unicellular and multicellular eukaryotic microorganisms that possess chlorophyll and produce oxygen during photosynthesis.

Anaerobe Describes an organism that is only able to grow in an environment from which oxygen is absent.

Anaerobic Describes an atmosphere or environment from which oxygen is absent.

Antibody An immunoglobulin formed in direct response to the introduction of an antigen into humans or animals. Antibodies combine with their specific antigens. The combination of antigen and antibody causes physical (often visible) clumping of the combined antigen–antibody complexes.

Antigen   Any agent capable of causing an immune response (the production of an antibody) in humans and animals. Antigens are usually large molecules such as proteins or lipopolysaccharides, including some toxins as well as bacterial cell walls.

Aseptic   For microbiological test purposes, this refers to the prevention of contamination; aseptic techniques are work practices and precautionary measures that prevent extraneous contamination of cultures and media and infection of workers.

Autoclave   An item of laboratory equipment, whose principle of operation is similar to that of a pressure cooker. It comprises a pressure vessel in which the air is displaced by steam, enabling the contents of the chamber to be raised to a temperature greater than the boiling point of water, as required for the thermal inactivation of bacterial spores. An autoclave is used to sterilise laboratory media and equipment or to decontaminate waste materials by means of defined time–temperature autoclave processes.

Bacillus   a)   bacillus – a rod-shaped bacterium.
  b)   *Bacillus* – a bacterial genus comprising aerobic, Gram-positive rods that produce heat-resistant spores.

Bacteriocins   Protein antibiotics produced by a variety of strains of bacteria. They are inhibitory or lethal to other bacteria (often related species).

Bacteriophage   Any virus whose host is a bacterium.

Biotyping   The conventional method for distinguishing between bacterial types using their metabolic and / or physiological characteristics.

Buffer   A substance or substances that allow the pH value of a solution to be maintained at a nearly constant value, very often a weak acid in the presence of one of its salts, e.g. citric acid and sodium citrate. Also used are disodium hydrogen phosphate and sodium dihydrogen phosphate or some proteins. When extra acid is added, the buffer reacts with the hydrogen ions of the acid and thus maintaining the pH value so that it does not decrease. In other words, there is a buffer against acidification.

Calibrate (Calibration)   In relation to laboratory equipment, calibration is the process of establishing the accuracy of a laboratory measuring device, e.g. a thermometer, a temperature probe, a balance, by comparison with a reference device of known accuracy that is usually traceable to a national standard. Any inaccuracies found are taken into account on subsequent use of the laboratory device.

CAMP test (Christie, Atkins, Munch-Petersen test)   A synergistic, haemolytic reaction, which is a characteristic enhancement of the conventional haemolysis reaction that occurs on sheep blood agar when, for example, *Listeria monocytogenes* is grown in close proximity to a **specific**, weakly beta haemolytic strain of *Staphylococcus aureus* but not when *Listeria monocytogenes* is grown in close proximity to *Rhodococcus equi*.

Cocco-bacillus   A bacterium whose cells are intermediate in shape between those of a rod and a coccus.

Coccus   A bacterium whose cells are approximately spherical (see Rod, Cocco-bacillus).

Colony-forming unit (cfu)   The unit of quantification for microbiological enumeration (counting) tests. When microorganisms, especially bacteria and yeasts, are inoculated into or onto agar, they may be deposited as either separate cells or as small clumps of cells, both of which give rise to individual discrete colonies; the term colony-forming unit (cfu) is used because it is not possible to distinguish between colonies derived from single cells and colonies derived from small clumps.

Commensal In microbiology, a microorganism (commensal) living in close association with an animal or man (host) in the intestines, on the skin, etc. The commensal derives benefit from the host (often nutritional benefit) and the host either is not affected or not harmed, and may derive some benefit, e.g. vitamins, from the commensal organism.

Commercial sterility A condition in which a food product may contain viable microorganisms yet is microbiologically stable, i.e. no microbial growth occurs under the conditions found within the product and / or the product storage conditions. This term usually applies to some canned products that are microbiologically stable when stored at ambient temperature in temperate climates but may become unstable when stored in tropical conditions.

Conductivity (or resistivity) units Measurement units used for monitoring water quality. Water containing chemical contaminants has a higher conductivity than water that has been purified. Units of conductivity are microsiemens.cm (μs.cm) and units of resistivity are megohms.cm. These measurements are temperature dependent and are usually expressed as μs.cm at 25°C or megohms.cm at 25°C. Tap water may have a conductivity of 80–800 μs.cm at 25°C depending on the nature of the water source. Reverse Osmosis (RO) water typically has a conductivity of 2–40 μs.cm at 25°C, again depending on the nature of the tap water used. Distilled water usually has a conductivity of around 4 μs.cm at 25°C. The quality of highly purified water, having a conductivity $<$1 μs.cm at 25°C, is usually measured in terms of resistivity. A higher resistivity indicates higher quality; 1 μs.cm is equivalent to 1 megohm.cm and there is a reciprocal relationship between the two units, i.e. conductivity $=$ 1/resistivity.

Cyst (of protozoa) A specialised cell, produced either in response to adverse environmental conditions or as a normal part of the life cycle of protozoa.

D-value The time required (usually expressed in minutes) at a given temperature to reduce the number of viable cells or spores of a given microorganism to 10% of the initial population (i.e. to reduce by 90%), e.g. the time required to reduce the numbers of a microorganism from 10^5 cfu/ml to 10^4 cfu/ml. Usually expressed as e.g. $D_{60} = 2$ minutes where $_{60}$ is the temperature in °C.

De-ionisation A means of purifying water by the removal of ions from a solution.

Distillation The process of converting a liquid into a vapour, e.g. by boiling water to create steam, then condensing the vapour (steam) and collecting the condensed liquid / distillate, i.e. distilled water. This process purifies water by separating it from the substances dissolved in it.

Electrolyte A substance or solution that contains free positively and negatively charged ions and is therefore able to conduct an electric current.

Enzyme A proteinaceous catalyst, produced by living organisms, that acts on one or more specific substrates.

Equilibrium Relative Humidity (ERH)   The vapour pressure of water in the atmosphere, expressed as a percentage. Air that is completely saturated with pure water has an ERH of 100%; air that is completely dry has an ERH of 0% (see Water activity).

Eukaryote   A self-replicating organism that possesses, within itself, a clearly defined nucleus containing its genetic material separated from its cytoplasm by a nuclear membrane. (See also Prokaryote.)

Exposure assessment   The qualitative and / or quantitative evaluation of the likely intake of biological, chemical, and physical agents via food as well as exposures from other sources if relevant (Codex Alimentarius Commission, 1999).

F   The equivalent, in minutes at some given reference temperature, of the total heat supplied, with respect to its capacity to destroy spores or vegetative cells of a particular organism.

F_0 The severity of a heat process with respect to a reference temperature of 121.1°C.

F_0 = **3 minutes**   A heat process of 121.1°C for 3 minutes, or a process at a different temperature of equivalent lethality determined according to the z-value for a specified microorganism. A process of F_0 = 3 minutes is used in canning to achieve a > 12 log cycle reduction in the numbers of mesophilic *Clostridium botulinum* spores, which is considered to provide an adequate margin of safety.

Facultative   'Optional life-style' associated with a mode of life not normally adopted, e.g. facultative anaerobe = usually grows aerobically but can grow anaerobically.

Family   A taxonomic group of organisms consisting of related genera. Families are grouped into orders.

Fungus   A unicellular or multicellular, eukaryotic microorganism that does not contain chlorophyll and has a rigid cell wall composed of cellulose- or chitin-based polymers. The 'body' of a fungus is normally composed of filaments known as hyphae which accumulate into a larger mass known collectively as mycelium. Macro-fungi produce visible fruiting bodies, examples of which are mushrooms and toadstools. Moulds (micro-fungi) are filamentous fungi, most species of which do not produce visible fruiting bodies. Fungi generally prefer to grow in or on solid materials. (See also Yeasts).

Genotyping   Methods used to differentiate bacteria and other microorganisms based on the composition of their nucleic acids.

Genus (plural genera)   A taxonomic group consisting of closely related species. Genera are grouped into Families.

Gram reaction   A classification system for bacteria based upon their cell wall composition. The reaction to the Gram stain is observed by light microscopy and distinguishes Gram-negative bacteria from Gram-positive bacteria; the cell walls of Gram-positive bacteria are resistant to decolourisation after staining with a specific dye whilst the cell walls of Gram-negative bacteria can be decolourised easily.

Gray (Gy)   The unit of energy absorbed from ionising radiation by the matter through which the radiation passes. A radiation dose of 1 Gy involves the absorption of 1 joule of energy by each kilogram of matter through which the radiation passes. Large multiple units are frequently used to express the radiation dose in food irradiation, e.g. 1000 Gy = 1 kGy.

HACCP   A practical food safety management system that identifies and evaluates hazards that are significant for food safety and enables the determination of appropriate control measures and monitoring procedures. The initials stand for 'Hazard Analysis Critical Control Point'. HACCP is a structured approach to identifying real and potential food safety hazards in the food chain (Hazard analysis), deciding how and where to control them (Critical Control Points) and then defining and implementing systems to manage, monitor and record their control.

Haemolysis   The breakdown or 'lysis' of red blood cells (erythrocytes) by bacteria. When blood is incorporated into agar media, the result is an opaque red-coloured agar. If bacteria cause haemolysis, then around the colony appears a zone of discolouration (greening) or clearing where the red cells are ruptured. Various types of haemolysis have been described according to the size of the zone of clearing and any colouring within the zone of haemolysis.

Halophile   An organism that can tolerate and grow optimally in the presence of high levels of salt (sodium chloride) in its environment. The term is no longer widely used for yeasts and moulds because most, so-called, halophiles are tolerant of low water activity in general and should therefore be called xerophiles. Some bacteria such as some *Vibrio* spp. and a few moulds are, however, true halophiles and grow only poorly in the absence of salt.

Hazard   A biological, chemical or physical agent in, or condition of, food with the potential to cause an adverse health effect (Codex Alimentarius Commission, 1999).

Hazard characterisation   The qualitative and / or quantitative evaluation of the nature of the adverse health effects associated with the hazard. For the purpose of microbiological risk assessment the concerns relate to microorganisms and / or their toxins (Codex Alimentarius Commission, 1999).

Hazard identification   The identification of biological, chemical or physical agents capable of causing adverse health effects and which may be present in a particular food or group of foods (Codex Alimentarius Commission, 1999).

Humectant   A soluble substance that binds water and makes the water unavailable for microbial use / growth.

Hydrophobic   Repels water. Fatty foods and the waxy surfaces of some microorganisms are hydrophobic and therefore difficult to wet. Detergents are added to diluents used for fatty test samples to aid their dispersion.

Infection   In the context of bacterial food-poisoning, infection is a condition in which the pathogen multiplies in the host's body and becomes established in or on the cells or tissues of the host.

Intoxication   In the context of food poisoning, intoxication is an adverse health effect due to the action of toxins (poisons) either produced by the organism in the food prior to consumption or produced by the organism in the consumer's intestine.

Isomers   Chemical compounds that share the same atomic structure but differ from each other in the conformation (shape) of the molecule, e.g. mirror images.

Metabolism   The biochemical changes (constructive and destructive) that occur in living organisms. Metabolism results in energy production and growth, and involves nutrient uptake into cells and the excretion of waste end-products.

Micro-aerobe   An organism that is able to grow optimally in a microaerobic environment.

Microaerobic   Describes an environment or atmosphere in which oxygen is present at a lower partial pressure (percentage) than that in air (usually 5–10% as opposed to 21% in air).

Micron   One thousandth of a millimetre; often written as 'μm'.

Morphology   The appearance (shape, size, form) of a microbial cell or colony, hence, cell morphology, colony morphology.

Mould   (See Fungus.)

Nanometre   One thousandth of a micron; often written as 'nm'.

Obligate   A required attribute, e.g. obligate aerobe = grows only under aerobic conditions.

Osmophile   An organism that can tolerate high levels of sugars in its environment. The term is no longer widely used because most osmophiles are tolerant of low water activity in general and should therefore be called xerophiles.

Pasteurisation   A heat treatment that kills vegetative pathogens and most spoilage microorganisms in milk and other foods but does not inactivate bacterial spores or some non-pathogenic vegetative organisms; a common pasteurisation process for milk is 71.7°C for 15 seconds.

'Phage (bacteriophage)   A virus that infects bacteria.

'Phage typing   A method used to distinguish between bacteria within the same species on the basis of their susceptibility to a range of bacterial viruses (bacteriophages).

Phenotype   The observable characteristics of an organism, which include biotype, serotype, phage type and bacteriocin type.

Photooxidation   Oxidation of chemicals, e.g. in laboratory media, catalysed by visible or ultra-violet light.

Polymerase Chain Reaction (PCR)   A technique used to amplify the number of copies of a pre-selected region of DNA to a sufficient level for testing.

Polymorphism   Individuals of the same species that appear in two or more morphologically distinct types (morphotypes) exhibit polymorphism.

Positive release   Refers to the process of holding batches of food product in 'quarantine' under appropriate storage conditions until the results of specified tests (usually microbiological) are available and signed by an authorised person as satisfactory, at which point the 'quarantine' is lifted and the products made available for despatch.

Prokaryote   A self-replicating organism that does not possess any specialised membrane separating its chromosomes from its cytoplasm.

Protozoa   A diverse group of eukaryotic, mostly unicellular, microorganisms, with a defined life-cycle, ranging in size from about 1 μm to just visible to the naked eye.

Pulsed Field Gel Electrophoresis (PFGE)   A technique that allows chromosomal restriction fragment patterns to be produced.

Quality assurance   All those planned and systematic actions necessary to provide adequate confidence that a product or service will satisfy given requirements for quality.

Quality control   The operational techniques and activities that are used / carried out to fulfil requirements for quality.

Quantitative Risk Assessment   A Risk Assessment that provides numerical expressions of risk and indication of the attendant uncertainties (Codex Alimentarius Commission, 1999).

Reference cultures   Cultures of microorganisms usually obtained from a recognised national collection.

Repeatability   An assessment of the variation between microbiological test results obtained by a single operator in one laboratory examining replicate test samples from the same test material at the same time using the same methods and equipment.

Reproducibility   An assessment of the variation between test results obtained by different operators in the same or in different laboratories, each examining replicate test samples from the same test material using the same methods and the same or different equipment.

Restriction Enzyme Analysis (REA)   A method for discriminating between isolates of the same species on the basis of patterns obtained from the separation of DNA fragments in agarose gel after digestion with one or more restriction enzymes. Differences in the banding profiles of two isolates is referred to as a restriction fragment length polymorphism (RFLP).

Restriction enzymes (Restriction endonucleases)   Enzymes that attack DNA. Each enzyme recognises a particular and different nucleotide sequence and cuts the DNA at a specific site.

Reverse osmosis   A means of purifying water in which water molecules are forced through a semi-permeable membrane under pressure from a high solute concentration to a low solute concentration, i.e. in the opposite direction from conventional (natural) osmosis.

Ribotyping   A method for characterising bacterial isolates according to their ribosomal RNA pattern (ribotype) and identifying the isolate by comparing the pattern obtained with a database of patterns.

Risk   A function of the probability of an adverse health effect and the severity of that effect, consequential to a hazard(s) in food (Codex Alimentarius Commission, 1999).

Risk analysis   A process consisting of three components: *risk assessment, risk management* and *risk communication.*

Risk assessment   The scientific evaluation of known or potential adverse health effects by means of *hazard identification* (what is the hazard?), *hazard characterisation* (what type of adverse effects are caused?), *exposure assessment* (what is the likelihood of it being consumed and how much will be consumed?) and *risk characterisation* (what is the effect on a given population of exposure to the hazard?).

Risk characterisation   The process of determining the qualitative and / or quantitative estimation, including attendant uncertainties, of the probability of occurrence and severity of known or potential adverse health effects in a given population based on hazard identification, hazard characterisation and exposure assessment (Codex Alimentarius Commission, 1999).

Risk communication   An interactive process of exchange of information and opinion on risks.

Risk management   The process of accepting, minimising or reducing assessed risks.

Rod   A bacterium whose cells are significantly longer in one axis than the other two giving it an approximately cylindrical or 'sausage' shape (see Bacillus, Coccus, Cocco-bacillus).

Selectivity   The ability of a growth medium to restrict the growth of organisms that would otherwise compete with the target organism for nutrients etc. but allow the target organism to grow well.

Sensitivity   The ability of a method to detect slight variations in the number of microorganisms within a given matrix e.g. a food.

Sequelae   A morbid affliction (illness) occurring as the result of a previous disease / infection.

Serotype (serovar)   A variety of microorganism, within a species, that is serologically distinct from all other members of the species. Most commonly used for distinguishing or comparing different isolates of *Salmonella* and *E. coli*.

Serotyping   A method of distinguishing bacteria on the basis of their antigenic properties, i.e. their surface proteins or other components.

Serovar   (See Serotype.)

Sine wave (simple harmonic motion)   The amplitude (height) and the frequency (distance of separation of waves) determine the characteristics of a wave. Sine wave describes the type of wave that occurs in the motion of light (in its wave-like form). These characteristics are exploited in phase contrast microscopy.

Slope (Slant)   A solid agar medium (usually 7–10 ml) that has been allowed to set in a diagonally oriented tube, e.g. capped test tube, Universal bottle (see Plate 10, see colour plate sections for all plates), or similar. There are two components to a slope – the slope itself with a large surface in contact with the air, and the butt which is the deep agar below the slope. Inoculation may involve streaking the slope, to grow and maintain cultures, as well as stabbing to the bottom of the butt with an inoculated wire, to enable growth in reduced oxygen conditions (see Chapter 9).

Solute   A substance that has been dissolved in a solvent to form a solution, e.g. salt (solute) is dissolved in water (solvent) to form brine (solution). Similarly sugar is dissolved in water to form a syrup.

Solution   The result of dissolving a solute in a solvent (see Solute).

Solvent   A substance, such as water, in which solutes may be dissolved (see Solute).

Species   A taxonomic unit within a genus. A species may contain particular varieties within it, based on serotypes, phage types, etc. All related species are grouped into a genus.

Specificity   The degree to which a method will provide confidence that the target organism will be found i.e. a high specificity provides a high level of confidence in the result and the percentage of false 'positives' will be very low.

Stab culture   Inoculation of a slope or deep agar by stabbing with an inoculated straight wire or needle to the bottom of the butt of the slope or deep agar. Growth may occur at various depths, depending on the oxygen requirements of the organism (see Chapter 9).

Strain   An isolate or group of isolates that can be distinguished from other isolates of the same genus and species by either phenotypic and / or genotypic character-istics.

Substrate   A substance that is utilised in chemical or biochemical reactions.

Symbols $<$ = less than
$>$ = greater than
$\geq$ = greater than or equal to
$\leq$ = less than or equal to
$\pm$ = plus or minus
$\sim$ = approximately

Taxonomy   The arrangement and classification of living organisms.

Thermoduric (thermotolerant) organism   A vegetative organism able to survive heat processes that are usually lethal to vegetative bacteria but that is not as heat resistant as a bacterial spore. In dairy microbiology the term refers to pasteurisation survivors such as the enterococci.

Trend analysis   The examination of data collected over a period of time, for a single activity or operation, to determine whether they remain consistent or show any changes (for better or worse). Examples of such data might be laboratory results, process control charts or complaints records. Analysis might be done by using simple graphs or more complex statistical techniques.

Ultra Heat Treatment (UHT)   A high temperature heat treatment (usually 138–142°C for 2–5 seconds) applied to liquid foods, usually followed by aseptic packaging for the production of long life, ambient-stable, low acid products.

Validation (of microbiological methods)   The confirmation, by detailed examination and the provision of objective evidence, that the particular requirements for a specific intended use are fulfilled (European Standard BS EN ISO / IEC 17025 (2000)). In simple terms, 'Can it work?' / 'Does it work in the laboratory under the required conditions?'

Vegetative cell   The 'normal' growing form of a bacterial, yeast or mould cell, usually lacking properties such as high heat-resistance.

Vero cytotoxigenic   Organisms that produce a toxin capable of killing Vero cells, which is an established cell line derived from African Green Monkey kidney. Vero cytotoxigenic strains of *Escherichia coli* cause a severe form of gastrointestinal disease in humans. The difficulties that have evolved with the nomenclature of enterohaemorrhagic *E. coli* have been clarified by the International Life Sciences Institute (ILSI) (2001).

Viruses   Non-cellular microorganisms consisting simply of protein and nucleic acid (either DNA or RNA) that lack the ability to self-replicate. Viruses replicate only within a specific host animal, plant or microbial cell by using the host's genetic mechanisms to make multiple copies of their own genome. Viruses are very small, around 20–500 nanometres (nm) in size and can only be viewed using the very high magnification of an electron microscope, e.g. $\times$ 50 000.

Water activity (a_w)   A measure of the availability of water for the growth and metabolism of microorganisms. Water activity is expressed as the ratio of the vapour pressure of water in a food or solution to the vapour pressure of pure water at the same temperature, on a scale of 0–1. The presence of solutes, e.g. salt, sugar, in a food depresses the water activity by 'binding' water and thus restricts microbial growth without reducing the moisture content of the food. Water activity is a theoretical concept that cannot be measured; to determine the water activity of a food, a sample is enclosed in a small chamber at a constant temperature and the vapour pressure of water in the headspace is allowed to equilibrate with the vapour pressure of water in the food, i.e. the equilibrium relative humidity (ERH) is established; the ERH of the headspace is then measured. $a_w = \text{ERH} / 100$. (See Equilibrium Relative Humidity and Figure 2.4.)

Xerophile   An organism that can grow at low water activity (0.85 or below). Water activity is reduced by high levels of a solute such as salt, sugar or glycerol in the aqueous phase of the food. (See Halophile, Osmophile, Water activity.)

Yeasts   Fungi that have evolved a mostly single-celled life-style and reproduce asexually by producing buds or occasionally by fission (in the same way that bacteria divide). Yeasts grow well in liquid environments especially if nutritionally rich, and some can grow anaerobically using a fermentative metabolism. Some yeasts can adopt a filamentous, mould-like form, just as some moulds can grow in a yeast-like form (see Fungus).

z-value   A term used in heat process calculations that expresses e.g. the number of centigrade degrees (C°) increase required to achieve a 10-fold decrease in the *D*-value of an organism (see *D*-value, $F_0 = 3$ minutes).

References

Adams, M.R. & Moss, M. O. (2000) *Food Microbiology*, 2nd edn. The Royal Society of Chemistry, London, UK.

Advisory Committee on Dangerous Pathogens (ACDP) (1995) *Categorisation of Biological Agents according to Hazard and Categories of Containment*, 4th edn. HSE Books, Sudbury, UK.

Advisory Committee on Dangerous Pathogens (ACDP) (2001) *The Management, Design and Operation of Microbiological Containment Laboratories*. HSE Books, Sudbury, UK.

Advisory Committee on the Microbiological Safety of Food (1992) *Report on Vacuum Packaging and Associated Processes*. HMSO, London, UK.

Anon (1974) *The Health and Safety at Work Act*. HMSO, London, UK.

Anon (1993) Council Directive 93/43/EEC of 14th June 1993 on the hygiene of foodstuffs. *Official Journal of the European Communities* 19.7.93, No. L175, 1–11.

Anon (1994) *Health and Safety. The Control of Substances Hazardous to Health (COSHH) Regulations 1994*. Statutory Instrument No. 3246. HMSO, London, UK.

Anon (1995) *The Food Safety (General Food Hygiene) Regulations 1995*, Statutory Instrument No. 1763. HMSO, London, UK.

Anon (1999) Guidelines:
  - (1)   *Recommended checks for mechanically driven spiral platers*
  - (1A) *Recommended checks for automatic spiral platers*
  - (2)   *Recommended checks for autoclaves*
  - (3)   *Recommended checks for balances*
  - (4)   *Quality assurance of laboratory consumables*
  - (5)   *Media storage and preparation*
  - (6)   *The maintenance and handling of reference organisms*
  - (7)   *Temperature monitoring*
  - (8)   *Equipment calibration*
  - (9)   *Recommended checks for pH meters*
  - (10) *Recommended checks for water activity measuring instruments*
  - (11) *Recommended checks for ELISA washer and reader equipment*
  - (12) *Recommended checks for pipettes*
  - (13) *'Uncertainty of Measurement in Food Microbiology' by analysis of variance.*

Professional Food Microbiology Group of the Institute of Food Science and Technology, London, UK.

Anon (2000) *A report of the study of infectious intestinal disease in England*. The Stationery Office, London, UK.

Baylis, C. (2000) *Catalogue of Rapid Microbiological Methods, Review No. 1*, 4th edn. Campden & Chorleywood Food Research Association, Chipping Campden, UK.

Baylis, C.L., Jewell, K., Oscroft, C.A. and Brookes, F.L. (eds) (2001) *Guidelines for Establishing the Suitability of Food Microbiology Methods*. Guideline No. 29 Campden & Chorleywood Food Research Association Group, Chipping Campden, UK.

Bell, C. & Kyriakides, A. (1998) *E. coli: a practical approach to the organism and its control in foods*. Blackwell Science Ltd, Oxford, UK.

Bell, C. & Kyriakides, A. (2000) Clostridium botulinum: *A practical approach to the organism and its control in foods*. Blackwell Science Ltd, Oxford, UK.

Bell, C. & Kyriakides, A. (2002) Salmonella: *A practical approach to the organism and its control in foods*. Blackwell Science Ltd, Oxford, UK.

Blazevic, D.J. & Ederer, G.M. (1975) *Principles of Biochemical Tests in Diagnostic Microbiology*. John Wiley & Sons, New York, USA.

Board, R.G., Jones, D. & Skinner, F.A. (eds) (1992) *Isolation Methods in Applied and Environmental Microbiology*. Society for Applied Bacteriology Technical Series No. 29. Blackwell Scientific Publications, London, UK.

Bridson, E.Y. (1998) *The Oxoid Manual*, 8th edn. Oxoid Ltd, Basingstoke, UK.

British Standard BS 3145 (1978, confirmed 1999) *Specification for laboratory pH meters*. British Standards Institution, London, UK.

British Standard BS 700-1 (1982, confirmed 1993) / ISO 835/1 (1981) *Graduated pipettes. Specification for general requirements*. British Standards Institution, London, UK.

British Standard BS 700-3 (1982, confirmed 1993) / ISO 835/4 (1981) *Graduated pipettes. Specification for blow-out pipettes*. British Standards Institution, London, UK.

British Standard BS 5732 (1985, confirmed 1997) *Specification for glass disposable Pasteur pipettes*. British Standards Institution, London, UK.

British Standard BS 6706 (1986, confirmed 1992) *Specification for disposable glass serological pipettes*. British Standards Institution, London, UK.

British Standard BS 1132 (1987, confirmed 1993) *Specification for automatic pipettes*. British Standards Institution, London, UK.

British Standard BS 593 (1989, confirmed 1994) *Specification for laboratory thermometers*. British Standards Institution, London, UK.

British Standard BS 2646 (1991, confirmed 1998) Part 4. *Autoclaves for sterilization in laboratories. Guide to maintenance*. British Standards Institution, London, UK.

British Standard BS 5726 (1992) Part 4. *Microbiological safety cabinets. Recommendations for selection, use and maintenance*. British Standards Institution, London, UK.

British Standard BS 2646 (1993, confirmed 2000) Part 1. *Autoclaves for sterilization in laboratories. Specification for design, construction, safety and performance*. British Standards Institution, London, UK.

British Standard BS 2646 (1993, confirmed 2000) Part 3. *Autoclaves for sterilization in laboratories. Guide to safe use and operation*. British Standards Institution, London, UK.

British Standard BS 2646 (1993, confirmed 2000) Part 5. *Autoclaves for sterilization in laboratories. Methods of test for function and performance*. British Standards Institution, London, UK.

British Standard BS 7653-3 (1993) *Piston and/or plunger operated volumetric apparatus. Methods of test*. British Standards Institution, London, UK.

British Standard BS 7012-1 (1998) / ISO 8039 (1997) *Light microscopes. Specification for the magnifying power of microscope imaging components*. British Standards Institution, London, UK.

British Standard BS EN 12547 (1999) *Centrifuges. Common safety requirements*. British Standards Institution, London, UK.

British Standard BS EN 12469 (2000) *Biotechnology. Performance criteria for microbiological safety cabinets*. British Standards Institution, London, UK.

British Standard BS EN ISO 8655 (2002) *Piston Operated Volumetric Apparatus*. British Standards Institution, London, UK.

Brown, K.L. (ed) (1994) *Guidelines for the Design and Safety of Food Microbiology Laboratories. Technical Manual No. 42*. Campden & Chorleywood Food Research Association, Chipping Campden, UK.

Campden Food & Drink Research Association (1990) *Evaluation of Shelf Life for Chilled Foods*. Technical Manual No. 28. CFDRA, Chipping Campden, UK.

Chilled Food Association (1997) *Guidelines for Good Hygienic Practice in the Manufacture of Chilled Foods*, 3rd edn. Chilled Food Association, London, UK.

Codex Alimentarius Commission (1999) Alinorm 99/13 Appendix IV, Proposed Draft Principles and Guidelines for the Conduct of Microbiological Risk Assessment (at step 5 of the procedure). Report of the thirtieth session of the Codex Committee on Food Hygiene, Washington, D.C., 20–24 October 1997. Joint FAO/WHO Food Standards Programme, Codex Alimentarius Commission, twenty-third session, Rome, Italy.

Corry, J.E.L., Roberts, D. & Skinner, F.A. (eds) (1982) *Isolation and Identification Methods for Food Poisoning Organisms*. Society for Applied Bacteriology Technical Series No. 17. Academic Press, London, UK.

Cowan, S.T. (ed.) (1974) Cowan & Steel's *Manual for the Identification of Medical Bacteria*, 2nd edn. Cambridge University Press, Cambridge, UK.

D'Aoust, J-Y. (1997) *Salmonella* species. In: *Food Microbiology – Fundamentals and Frontiers* (eds. M.P. Doyle, L.R. Beuchat, and T.J. Montville). ASM Press, Washington, USA, pp. 129–158.

Downes, F.P. & Ito, K. (eds) (2001) *Compendium of methods for the microbiological examination of foods* (4th edn.). Compiled by the APHA Technical Committee on Microbiological Methods for Foods, American Public Health Association, Washington, D.C., USA.

Durham, H.E. (1898) A simple method for demonstrating the production of gas by bacteria. *British Medical Journal* i, 1387.

Easter, M.C. & Gibson, D.M. (1985) Rapid and automated detection of *Salmonella* by electrical measurements. *Journal of Hygiene, Cambridge*, **94**, 245–262.

Easter, M.C. & Gibson, D.M. (1989) Detection of microorganisms by electrical measurements. In: *Rapid Methods in Food Microbiology. Progress in Industrial Microbiology, Volume 26* (eds. M.R. Adams and C.F.A. Hope). Elsevier Science Publishers BV, Amsterdam, The Netherlands, pp. 57–100.

European Standard BS EN ISO / IEC 17025 (2000) *General Requirements for the Competence of Testing and Calibration Laboratories*. European Committee for Standardization, Brussels, Belgium.

Firstenberg-Eden, R. & Eden, G. (1984) *Impedance Microbiology*. Research Studies Press Ltd, Letchworth, UK.

Friedrichsen, G.W.S. (1973) *The Shorter Oxford English Dictionary on Historical Principles*. (Originally prepared by W. Little, H.W. Fowler & J. Coulson, revised and edited by C.T. Onions) Clarendon Press, Oxford, UK.

Gibbs, B.M. & Shapton, D.A. (eds) (1968) *Identification Methods for Microbiologists: Part B*. Society for Applied Bacteriology Technical Series No. 2. Academic Press, London, UK.

Gibbs, B.M. & Skinner, F.A. (eds) (1966) *Identification Methods for Microbiologists: Part A*. Society for Applied Bacteriology Technical Series No. 1. Academic Press, London, UK.

Hawker, L.E., Linton, A.H., Folkes, B.F. *et al.* (1967) *An Introduction to the Biology of Microorganisms*. Edward Arnold, London, UK.

Hill, W.E. & Jinneman, K. C. (2000) Principles and Applications of Genetic Techniques for Detection, Identification, and Subtyping of Food-Associated Pathogenic Microorganisms. In: *The Microbiological Safety and Quality of Food* (eds. B.M. Lund, T.C. Baird-Parker & G.W. Gould). Aspen Publishers, Inc., Maryland, USA, pp. 1813–1851.

Hocking, A.D., Arnold, G., Jenson, I., Newton, K. & Sutherland, P. (eds) (1997) *Foodborne Microorganisms of Public Health Significance*, 5th edn. Australian Institute of Food Science and Technology, NSW Branch, Food Microbiology Group, Sydney, Australia.

Holbrook, R. & Anderson, J.M. (1980) An improved selective and diagnostic medium for the isolation and enumeration of *Bacillus cereus* in foods. *Canadian Journal of Microbiology*, **26** (7), 753–759.

Holbrook, R. & Baird-Parker, A.C. (1975) Serological methods for the assay of staphylococcal enterotoxins. In: *Society for Applied Bacteriology Technical Series No. 8, Some Methods for Microbiological Assay* (eds. R.G. Board & D.W. Lovelock). Academic Press, London, UK, pp. 108–128.

Holmes, B. & Costas, M. (1992) Identification and typing of Enterobacteriaceae by computerised methods. In: *Society for Applied Bacteriology Technical Series No. 29, Identification Methods in Applied and Environmental Microbiology* (eds. R.G. Board, D. Jones & F.A. Skinner). Blackwell Scientific Publications, London, UK, pp. 127–149.

Institute of Food Science and Technology (UK) (1993) *Shelf Life of Foods – Guidelines for its Determination and Prediction*. IFST (UK), London, UK.

Institute of Food Science and Technology (UK) (1998) *Good Manufacturing Practice – A Guide to its Responsible Management*, 4th edn. IFST (UK), London, UK.

Institute of Food Science and Technology (UK) (1999) *Development and Use of Microbiological Criteria for Foods*. Monograph of the IFST (UK), London, UK.

International Commission on Microbiological Specifications for Foods (ICMSF) (1978). *Microorganisms in Foods 1: Their Significance and Methods of Enumeration*, 2nd edn. University of Toronto Press, Toronto, Canada.

International Commission on Microbiological Specifications for Foods (ICMSF) (1980) *Microbial Ecology of Foods Volume 1: Factors Affecting Life and Death of Micro-organisms*. Academic Press, London, UK.

International Commission on Microbiological Specifications for Foods (ICMSF) (1986) *Microorganisms in Foods 2: Sampling for Microbiological Analysis: Principles and Specific Applications* (2nd edn). University of Toronto Press, Toronto, Canada.

International Commission on Microbiological Specifications for Foods (ICMSF) (1988) *Microorganisms in Foods 4: Application of the Hazard Analysis Critical Control Point (HACCP) System to Ensure Microbiological Safety and Quality*. Blackwell Scientific Publications, Oxford, UK.

International Commission on Microbiological Specifications for Foods (ICMSF) (1996) *Microorganisms in Foods 5: Microbiological Specifications of Food Pathogens*. Blackie Academic & Professional, London, UK.

International Commission on Microbiological Specifications for Foods (ICMSF) (1998) *Microorganisms in Foods 6: Microbial Ecology of Food Commodities*. Blackie Academic & Professional, London, UK.

International Life Sciences Institute (1993) A scientific basis for regulations on pathogenic microorganisms in foods. Summary of a workshop held in May 1993 organised by ILSI, Europe, Brussels, Belgium.

International Life Sciences Institute (2001) Report on the *Approach to the control of Entero-haemorrhagic Escherichia coli (EHEC)*, ILSI Europe, Brussels, Belgium.

International Standard ISO 7218 (1996) *Methods for Microbiological Examination of Food and Animal Feeding Stuffs. Part 0. General laboratory practices*. International Organization for Standardization, Geneva, Switzerland.

International Standard BS EN ISO 9001 : 2000 (2000) *Quality Management Systems*, International Organization for Standardization, Geneva, Switzerland.

Jones, L. (2000) *Molecular methods in food analysis: principles and examples. Key Topics in Food Science and Technology, No. 1*. Campden & Chorleywood Food Research Association, Chipping Campden, UK.

Kennedy, D.A. & Collins, C.H. (2000) Microbiological safety cabinets selection, installation, testing and use. *British Journal of Biomedical Science*, **57**, 330–337.

Kyriakides, A., Bell, C. & Jones, K. (eds) (1996) *A Code of Practice for Microbiology Laboratories Handling Food Samples – (Incorporating Guidelines for the Preparation, Storage and Handling of Microbiological Media), Guideline No. 9*. Campden & Chorleywood Food Research Association, Chipping Campden, UK.

Mossel, D.A.A., Corry, J.E.L., Struijk, C.B. *et al.* (1995) *Essentials of the Microbiology of Foods. A Textbook for Advanced Studies*. John Wiley & Sons, Ltd, Chichester, UK.

Old, D.C. & Threlfall, E.J. (1998) Salmonella. In: *Topley and Wilson's Microbiology and Microbial Infections, Volume 2, Systematic Bacteriology*, 9th edn. (eds. A. Balows & B.I. Duerden). Arnold, London, UK.

Pettipher, G.L. (1983) *The Direct Epifluorescent Filter Technique for the Rapid Enumeration of Microorganisms*. Research Studies Press Ltd, Letchworth, UK.

Prentice, G.A. & Neaves, P. (1992) The identification of *Listeria* species. In: *Society for Applied Bacteriology Technical Series no. 29, Identification Methods in Applied and Environmental Microbiology* (eds. R.G. Board, D. Jones & F.A. Skinner). Blackwell Scientific Publications, London, UK, pp. 283–296.

Radcliffe, D.M. & Holbrook, R. (2000) Detection of Microorganisms in Food – Principles and Application of Immunological Techniques. In: *The Microbiological Safety and Quality of Food* (eds. B.M. Lund, T.C. Baird-Parker & G.W. Gould). Aspen Publishers, Inc., Maryland, USA, pp. 1791–1812.

Roberts, D., and Greenwood, M. (eds) (2003) *Practical Food Microbiology*. 3rd edn. Blackwell Publishing Ltd, Oxford, UK.

Snell, J.J.S., Brown, D.F.J. and Roberts, C. (1999) *Quality Assurance – Principles and Practices in the Microbiology Laboratory*. Public Health Laboratory Service, London, UK.

Snyder, O.P. (1996) Redox potential in deli foods: botulism risk? *Dairy, Food and Environmental Sanitation*, **16** (9), 546–548.

Stannard, C.J. (1989) ATP estimation. In: *Rapid Methods in Food Microbiology. Progress in Industrial Microbiology, Volume 26* (eds. M.R. Adams & C.F.A. Hope). Elsevier Science Publishers BV, Amsterdam, The Netherlands, pp. 1–18.

Williams, A.P. (1989) Fungi in foods – rapid detection methods. In: *Rapid Methods in Food Microbiology. Progress in Industrial Microbiology, Volume 26* (eds. M.R. Adams & C.F.A. Hope). Elsevier Science Publishers BV, Amsterdam, The Netherlands, pp. 255–272.

Wilson, S. & Weir, G. (1995) *Food and Drink Laboratory Accreditation – a Practical Approach*. Chapman & Hall, London, UK.

Index